LOSS PREVENTION AND SAFETY CONTROL

Terms and Definitions

Occupational Safety and Health Guide Series

Series Editor

Thomas D. Schneid
Eastern Kentucky University
Richmond, Kentucky

Published Titles

Corporate Safety Compliance: OSHA, Ethics, and the Law
by Thomas D. Schneid

Creative Safety Solutions
by Thomas D. Schneid

Disaster Management and Preparedness
by Thomas D. Schneid and Larry R. Collins

Loss Prevention and Safety Control: Terms and Definitions
by Dennis P. Nolan

**Managing Workers' Compensation: A Guide to Injury Reduction
and Effective Claim Management**
by Keith R. Wertz and James J. Bryant

Motor Carrier Safety: A Guide to Regulatory Compliance
by E. Scott Dunlap

Occupational Health Guide to Violence in the Workplace
by Thomas D. Schneid

Physical Hazards of the Workplace
by Larry R. Collins and Thomas D. Schneid

LOSS PREVENTION AND SAFETY CONTROL
Terms and Definitions

Dennis P. Nolan

CRC Press
Taylor & Francis Group
Boca Raton   London   New York

CRC Press is an imprint of the
Taylor & Francis Group, an **informa** business

CRC Press
Taylor & Francis Group
6000 Broken Sound Parkway NW, Suite 300
Boca Raton, FL 33487-2742

First issued in paperback 2017

© 2011 by Taylor and Francis Group, LLC
CRC Press is an imprint of Taylor & Francis Group, an Informa business

No claim to original U.S. Government works

ISBN 13: 978-1-138-11800-3 (pbk)
ISBN 13: 978-1-4398-3363-6 (hbk)

Library of Congress Cataloging-in-Publication Data

Nolan, Dennis P.
    Loss prevention and safety control : terms and definitions / author, Dennis P. Nolan.
      p. cm. -- (Occupational safety and health guide series)
    "A CRC title."
    Includes bibliographical references.
    ISBN 978-1-4398-3363-6 (hard back : alk. paper)
    1. Industrial safety. 2. Loss control. I. Title. II. Series.

T55.N594 2011
620.8'6--dc22
                                      2010018407

Visit the Taylor & Francis Web site at
http://www.taylorandfrancis.com

and the CRC Press Web site at
http://www.crcpress.com

Dedicated to
Kushal, Nicholas, and Zebulon

Words are, of course, the most powerful drug used by mankind.

—Rudyard Kipling (1923)

Notice

Reasonable care has been taken to ensure the book's content is authentic, timely, and relevant to the industry today; however, no representation or warranty is made for its accuracy, completeness, or reliability. Consequentially, the author and publisher shall have no responsibility or liability to any person or organization for loss or damage caused, or believed to be caused, directly or indirectly by this information. In publishing this book, the publisher is not engaged in rendering legal advice or other professional services. It is up to the reader to investigate and assess his own situations. If such a study discloses a need for legal or other professional assistance or services, then the reader should seek and engage the services of qualified professionals.

Contents

Preface

Safety and loss prevention terminology is available from various sources, but most provide a simple definition of their meaning and related topics are not completely included. Having worked in the safety and loss prevention profession for over three decades, I still find terminology that is unfamiliar. Additionally, with the advance of technology, scientific investigations, and globalization of standards, new terms and methods are constantly being developed, and obsolete or archaic terms are being retired. Older dictionaries and references have also unwittingly omitted numerous terms. The scope of this book is to provide a reference source for the background, meaning, and description of safety and loss prevention terms being used in government, industry, research, and education today. Additionally, pictures, diagrams, and tables or graphs are provided to further enhance the reader's knowledge of the subject.

The field of safety and loss prevention encompasses various unrelated industries and organizations, such as the insurance field, research entities, process industries, and educational organizations. Many of these organizations may not realize that their individual terminology may not be understood by individuals or even compatible with the nomenclature used outside their own sphere of influence. It is therefore prudent to have a basic understanding of these individual terms in order to resolve these concerns.

The fire protection and environmental fields also use identical and similar terminology, which may be slightly different in selected applications. This book focuses on terminology that is applied and used in loss prevention and control. Therefore, Occupational Safety and Health Administration (OSHA) standards and interpretations are utilized as guidelines for the definitions and explanations.

This book is based mainly on the terminology used in U.S. codes, standards, and regulations. It should be noted that some countries may use similar terminology, but the terminology may be interpreted differently.

The term "accident" often implies that the event was not preventable. From a loss prevention perspective, use of this term is discouraged, since occupational injuries and illnesses should always be considered preventable, and the use of "incident" has been recommended instead. Therefore, the term accident has generally been replaced by incident; however, for the definitions where accident terminology is still utilized, these terms are identified for explanation.

Acknowledgments

The following individuals or organizations have graciously provided information, advice, or assistance in the preparation of this book and are acknowledged below.

Jennifer Ahringer, Project Coordinator, Editorial Project Development, Taylor & Francis Group; Cindy Carelli, Senior Acquisitions Editor, CRC Press; Rebecca Rieansnider, Marketing & Communications, ABS Group of Companies, Inc.; Stephen Dunn, Vice President, International Sales and Alliances, Dyadem; Rita L. Williams, Communications Manager, ACGIH®; Laura Cooper, Manager, Member Relations, National Floor Safety Institute; Lynn Strother, Executive Director, Human Factors and Ergonomics Society (HFES); Ted Lopatkiewicz, National Transportation Safety Board (NTSB); Josiane Domenici, Project Coordinator, National Fire Protection Association (NFPA); Jack Watts, Director, Fire Safety Institute (FSI); Clayton Doak, National Institute of Occupational Safety and Health (NIOSH); Lynn C. O'Donnell, Executive Director, American Board of Industrial Hygiene (ABIH).

Also and most importantly, I would like to thank my wife, Kushal, for diligently assisting me in the concept and development of this book.

About the Author

Dennis P. Nolan has had a long career devoted to fire protection engineering, risk engineering, loss prevention engineering, and systems safety engineering. He holds a Doctor of Philosophy degree in Business Administration from Berne University, and a Master of Science degree in Systems Management from the Florida Institute of Technology. His Bachelor of Science degree is in Fire Protection Engineering from the University of Maryland. He is also a registered Professional Engineer (Fire Protection Engineering) in the State of California.

Dr. Nolan is currently associated with the Loss Prevention executive management staff of the Saudi Arabian Oil Company (Saudi Aramco). He is located in Abqaiq, Saudi Arabia, the site of some of the largest oil and gas operations in the world. The magnitude of the risks, worldwide sensitivity, and foreign location make this one of the most highly critical operations in the world. He has also been associated with Boeing, Lockheed, Marathon Oil Company, and Occidental Petroleum Corporation in various fire protection engineering, risk analysis, and safety roles in several locations in the United States and overseas. As part of his career, he has examined oil production, refining, and marketing facilities under severe conditions and in various unique worldwide locations, including Africa, Asia, Europe, the Middle East, Russia, and North and South America. His activity in the aerospace field has included engineering support for the NASA Space Shuttle launch facilities at Kennedy Space Center (and for those undertaken at Vandenburg Air Force Base, California) and "Star Wars" defense systems.

He has received numerous safety awards and is a member of the American Society of Safety Engineers, National Fire Protection Association, Society of Petroleum Engineers, and the Society of Fire Protection Engineers. He was a member of the Fire Protection Working Group of the U.K. Offshore Operators Association (UKOOA). He is the author of many technical papers and professional articles in various international fire safety publications. He has also written several other books, which include *Application of HAZOP and What-If Safety Reviews to the Petroleum, Petrochemical, and Chemical Industries* (1st and 2nd editions); *Handbook of Fire and Explosion Protection Engineering Principles for Oil, Gas, Chemical, and Related Facilities* (1st and 2nd editions); *Fire Fighting Pumping Systems at Industrial Facilities*, and *Encyclopedia of Fire Protection* (1st and 2nd editions). Dr. Nolan has also been listed for many years in *Who's Who in California,* was included in the sixteenth edition of *Who's Who in the World*, and was listed in "Living Legends" (2004) published by the International Biographical Center, Cambridge, England.

Abate

To eliminate a hazard to comply with a regulatory standard that is being violated, e.g., to correct deficiencies identified during an inspection conducted under the authority of the Occupational Safety and Health Act.

Abatement

The Occupational Safety and Health Administration (OSHA) has defined Abatement as action by an employer to comply with a cited standard or regulation or to eliminate a recognized safety or health hazard identified by OSHA during an inspection. Examples of methods commonly used to abate cited hazards include the use of engineering controls, correction of a deficiency in a program, or the use of permissible equipment to avoid a hazard.

Abnormal Use

Using tools and equipment for a purpose other than that for which they were intended to be used, and one that may not reasonably have been foreseen.

Abrasion

A personal injury consisting of the loss of partial thickness of skin from rubbing or scraping on a hard or rough surface. Personal protective equipment is usually worn to protect against abrasive injuries.

Absolute Liability

Liability without fault. It is imposed in various states when the actions of an individual or business are deemed contrary to public policy, even though an action may not have been intentional or negligent. For example, in product liability, manufacturers and retailers have been held strictly liable for products and materials that have caused injuries and have been demonstrated to be defective, even though the manufacturer or retailer was not proven to be at fault or negligent. It may also be known as Liability Without Regard to Fault or Strict Liability.

Absorbent

A substance that allows another material to penetrate into its interior structure. Absorbent materials are utilized in the cleanup of hazardous material spills.

Absorption

Penetration of a chemical substance, a pathogen, or radiant energy through the skin or mucous membrane.

Acceptable Level of Risk

The consequences and frequency of a hazard or loss that is considered to be as low as reasonably practical or tolerable by an individual, organization, society, or authorities in view of the social and political implications, scientific and technical review, and economic cost-benefit analysis. It may also be called Acceptable Risk.

Accident

An occurrence in a sequence of events that may produce unintended injury, illness, death, and/or property damage. The term often implies that the event was not preventable. From a loss prevention perspective, use of this term is discouraged since occupational injuries and illnesses should be considered preventable, and the use of "incident" is recommended instead. See also **Incident**.

Accident and Sickness Benefits (Non-occupational)
Periodic payments to workers who are absent from work due to off-the-job disabilities through accident or sickness.

Accident Cause (Causal Factor)
One or more factors associated with an incident or a potential incident. Causal factors may be identified as time-sequenced events or may be categorized as being related to human or environmental (e.g., equipment, machinery, atmospheric contaminant, temperature, etc.) influences and their interactions. See also **Causal Factor**; **Root Cause**.

Accident Chain
A term referring to the concept that many contributing factors typically lead to an incident, rather than one single event. Every incident as a result of a hazard is preventable by breaking the accident chain before the last link. Breaking the chain is known as intervention, which is reactive, whereas reducing the potential for an accident chain occurring is mitigation, which is proactive. An example of a reactive step (i.e., post-incident) is the creation of a barrier at an excavation perimeter after someone has fallen into it. An example of a proactive step is a program requiring wearing of personal protective equipment such as a hard hat at a construction site. See also **Domino Theory**.

Accident Costs
Monetary losses associated with an incident. These costs include direct and indirect costs.

Accident Experience
One or more indices describing incident performance according to various units of measurement (e.g., disabling injury frequency rate, number of lost-time accidents, disabling injury severity rate, number of first-aid cases, or dollar loss). It may also refer to a summary statement describing incident performance.

Accident-Free
A record of no incidents, sometimes of specified types, relating to an operation, activity, or worker performance during a specified time period.

Accident Hazard
A situation present in an environment or connected with a job procedure or process which has the potential for producing an incident.

Accident Insurance
Normally refers to insurance coverage for bodily injury and death resulting from accidental means (other than natural causes). For example, an insured person is critically injured in an incident. Accident insurance can provide income and/or a death benefit if death ensues.

Accident Investigation
A detailed, defined, and recorded review of an incident undertaken to identify and record the causes and contributing factors and their relationships, which led up to and caused the incident. Accident Investigation is a technique that allows an organization to learn from its experience. The intent of an incident investigation is for employers to learn from past experiences and thus avoid repeating past mistakes. See also **Incident Investigation Team**.

Accident Location
The exact position of the key event that produced an incident.

Accident Potential
A behavior(s) or condition(s) having a likelihood of producing an incident, which therefore require a review and improvement of hazard control measures.

Accident Prevention
The application of programs and countermeasures, e.g., behaviors and conditions, designed to reduce incidents or incident potential within a system or organization.

Accident Prevention Tag

A

As defined by Occupational Safety and Health Administration (OSHA) regulation 1910.145, tags used to identify hazardous conditions and provide a message to employees with respect to hazardous conditions. They contain a signal word, i.e., Danger, Caution, Warning, and indicate the specific hazardous condition or the instruction to be communicated to the employee such as "High Voltage," "Close Clearance," "Do Not Start," or "Do Not Use," or a corresponding pictograph used with written text or alone. They are used as a means to prevent injury or illness to employees who are exposed to hazardous or potentially hazardous conditions, equipment, or operations that are out of the ordinary, unexpected, or not readily apparent. The tags are to be used until the identified hazard is eliminated or the hazardous operation is completed. OSHA does not required tags to be used where signs, guarding, or other positive means of protection are being provided.

Accident Probability

The likelihood of a worker, operation, or item of equipment becoming involved in an incident. The probability of a set of unsafe conditions and/or unsafe acts producing an incident.

Accident Prone Theory

A theory of accident (incident) causation proposed in the early twentieth century, by various researchers, which has been discredited. Originally it used to attribute the cause of accidents to personality traits of individuals or groups of workers. In 1971, L. Shaw and H. S. Sichel published a comprehensive analysis of accident proneness. It concluded that many characteristics of behavior change with age, and that it is not true that most accidents are sustained by a small number of people. It examined in detail earlier studies and demonstrated the importance of certain factors such as attention (defined as the ability to choose quickly and perform a correct response to a sudden stimulus), the stability of behavior, and the involuntary control of motor behavior. It supported a study of car drivers—"a man drives as he lives"—and the conclusion that a bad civil record tends to indicate a bad incident risk. Some recent researchers view accident proneness as being associated with the propensity of individuals to take risks or to take chances. See also **Habitual Violator**.

Accident Rate

Incident experience in relation to a base unit of measure (e.g., number of disabling injuries per 1,000,000 employee-hours exposure, number of incidents per 1,000,000 miles traveled, total number of incidents per 100,000 employee-days worked, number of incidents per 100 employees, etc.).

Accident Ratios

A tiered relationship of the severity of incidents to their frequency of occurrence. Various studies (Heinrich: Heinrich, H. W., *Industrial Accident Prevention: A Scientific Approach*, McGraw-Hill, New York, NY, 1931; Bird et al.: Bird, Frank E., Germain, George L., *Loss Control Management: Practical Loss Control Leadership,* Det Norske Veritas (USA), Inc, 1996; the British Safety Council; and the UK Health and Safety Executive (HSE): Stranks, Jeremy, *Health and Safety at Work, Key Terms,* Butterworth-Heinemann, Oxford, UK, 2002) have concluded that for major injuries (i.e., fatal or serious), a multiple number of minor injuries typically occur and numerous non-injury events occur. There has not been a definitive numerical relationship established (e.g., 1 to 30 to 300 or 1 to 10 to 600); however, the principle and the general magnitudes have been accepted by the industry and the safety profession. They are studied so that by taking action to prevent near misses, the more serious, moderate, and minor types of incidents can be avoided. See also **Accident Triangle**; **Domino Theory**.

Accident Records
Reports and other recorded information concerning employee incident experience.

Accident Report
A document containing the information and facts about an individual incident put in chronological order to provide a complete picture as to what happened. The report is useful for the investigation to help establish the root cause of the incident.

Accident Reporting
Collecting information for, and/or preparing and submitting to a designated individual or agency, an official report of an incident.

Accident Severity
The extent of a loss caused by an incident. It is used by insurers in predicting the number of losses upon which the insurance premium is based.

Accident Site
The location of an unexpected occurrence, failure, or loss, either at a plant or along a transportation route, resulting in a release of hazardous materials.

Accident Statistics
Descriptive or inferential data that provide information about incident occurrences.

Accident Susceptibility
See **Accident Prone Theory**.

Accident Triangle
A statistical ratio of incidents, which highlights the frequency and severity. The relationship is depicted in a graphical form of a triangle (see Figure A.1). Sometimes depicted as a loss pyramid, with frequencies on one side and severities on the adjacent side. See also **Accident Ratios**.

Accident Type
A description of the occurrences directly related to the source of injury classification and explaining how that source produced the injury. Accident type answers the question: How did the injured person come in contact with the object, substance, or exposure named as the source of injury, or during what personal movement did the bodily injury occur?

Accidental Death and Dismemberment
A term used to describe an insurance policy that pays additional benefits to the beneficiary if the cause of death is due to a non-work-related incident. Fractional amounts of the policy will be paid out if the covered employee loses a bodily appendage or sight because of an incident.

Acid
A compound consisting of hydrogen plus one or more other elements, which readily releases hydrogen when mixed with water or some solvents. It has a pH of less than 7. Materials with a pH range of 0 to 2 are considered corrosive and will cause severe damage to the eyes and skin. See also **Acute Health Effects**; **Alkali**.

Acoustic Trauma
Hearing loss caused by a sudden loud noise in one ear, or by a sudden blow to the head. In most cases, hearing loss is temporary, although there may be some permanent loss. Welding sparks (to the eardrum), blows to the head, and blast noise are examples of events capable of producing acoustic trauma.

Acoustics
The science of sound, including the generation, transmission, and effects of sound waves, both audible and inaudible. It may also refer to the physical qualities of a room or other enclosure (such as size, shape, amount of noise) that determine the audibility and perception of speech and music within the room.

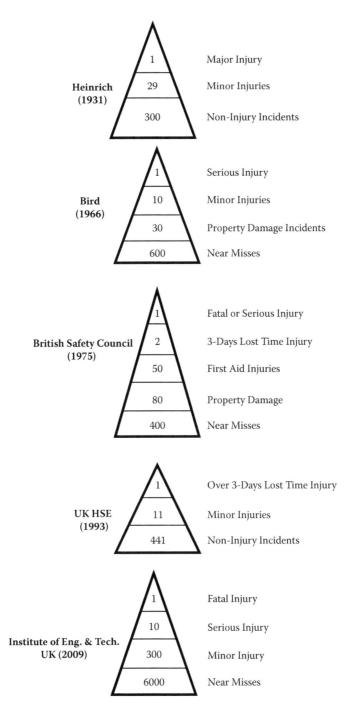

FIGURE A.1   Various accident ratio/triangle studies.

Acrid

A harsh, irritating odor or taste.

Act of God

Terminology used to describe events that are considered beyond human control and generally of a natural occurrence (i.e., earthquake, floods, lightning, hurricane, tornado, etc.). Historical and predictive frequencies of occurrence are used to determine the risk from these events. Although the natural phenomenon itself may be beyond control, its effect can be controlled in many instances. Thus, the act of God classification should not be accepted as an excuse for not taking proper precautions or properly designing structures to reduce the severity of a so-called act of God occurrence. Increasingly this terminology is considered outdated, as almost all incidents are considered preventable. This term is commonly used in the insurance industry.

Actinic Keratoconjunctivitis

See **Welder's Flashburn**.

Action Error Analysis (AEA)

A type of safety identification review that methodically analyzes the interactions between individuals and machines. It reviews the operation phase to operational phase, while considering the consequences of operator-system faults at each operating step within each phase. This analysis allows for the recognition of threats from equipment faults that may coexist with operator errors. It is considered similar to a Failure Mode and Effects Analysis (FMEA), but with increased emphasis on the steps in human procedures rather than viewing hardware exclusively. See also **Failure Mode and Effects Analysis (FMEA); Job Safety Analysis (JSA)**.

Action Level

Terminology used by the Occupational Safety and Health Administration (OSHA) and the National Institute for Occupational Safety and Health (NIOSH) to state the level of toxicant that requires medical surveillance and training to further protect employees. It is usually one-half the level of the permissible exposure limit. Action levels exist for only a few air contaminants, such as lead, cadmium, and benzene.

Activated Charcoal

An amorphous form of carbon made by the burning of wood, nutshells, animal bones, or other carbonaceous materials. Charcoal becomes activated by heating with steam to 800°C–900°C. During the heating process, a porous, submicroscopic internal structure is formed that provides an extensive internal surface area. Activated charcoal is commonly used as a gas or vapor adsorbent in air purifying respirators and as a solid sorbent in air sampling.

Activity Sampling

A measurement technique for evaluating potential incident-producing behavior. It involves the observation of worker and organizational behavior at random intervals and the instantaneous classification of these behaviors according to whether they are safe of unsafe.

Actual Exposure Hours

Employee-hours of exposure to workplace hazards taken from payroll or time clock records, wherever possible, and including only actual straight time worked (in hours) and actual overtime hours worked.

Acuity

The sharpness of a sense. Pertaining to the sensitivity of hearing, vision, smell, or touch.

Acute

Acute means sudden or brief. Acute can be used to describe either an exposure or a health effect. An acute exposure is a short-term exposure. Short-term means lasting for minutes, hours, or days. An acute health effect is an effect that develops either immediately or a short time after an exposure. Acute health effects may appear minutes, hours, or even days after an exposure. See also **Chronic**.

Acute Effect

An adverse reaction on the human body as a result of an exposure, with several symptoms developing rapidly and coming quickly to a crisis. Acute effects usually manifest within 72 hours of exposure.

Acute Exposure

Severe, usually critical, often dangerous exposure in which relatively rapid changes occur. An acute exposure normally runs a comparatively short course, and its effects can be easier to reverse in contrast with a chronic exposure. Exceptions include exposure to electrical energy and radiation.

Acute Exposure Guideline Level (AEGL) Value

An acute exposure level used to describe the risk to humans resulting from once-in-a-lifetime, or rare, exposure to airborne chemicals developed and published by the Environmental Protection Agency (EPA). The National Advisory Committee and National Research Council Committee on AEGLs are developing these guidelines to help both national and local authorities, as well as private companies, deal with emergencies involving spills or other catastrophic exposures. See also **Emergency Response Planning Guideline (ERPG) Value**; **Temporary Emergency Exposure Limit (TEEL) Value**.

Acute Health Effects

Health effects that usually occur quickly due to short-term exposures. Examples include irritation, corrosivity (tissue destruction), narcosis, and death.

Acute Mountain Sickness

A disorder occurring in individuals exposed to high altitudes (about 3048 m [10,000 ft]) for relatively long periods (24 hours or more). It presents as malaise, headache, and vomiting attributed to cerebral and pulmonary edema. It is probably related to both hypoxia and decreased atmospheric pressure. It may be prevented by proper acclimatization.

Acute Radiation Syndrome

The organic illness that follows exposure to relatively large, acute doses of ionizing radiation. It presents as one of three separately identified dose-dependent syndromes. Central nervous system syndrome follows doses of 2000 rem (20 sievert [Sv]) or above, producing relatively prompt shock and coma, uniformly fatal within hours to a day. Gastrointestinal syndrome follows doses of 500 to 2000 rem (5–20 Sv) producing nausea, vomiting, and diarrhea within hours to days with death usually occurring within a week. Hematopoietic syndrome follows doses of 100 to 500 rem (1–5 Sv), producing early nausea and vomiting followed by a lag period of several weeks after which infection and hemorrhage occur due to hematopoietic suppression. At doses of 400 to 500 rem (4–5 Sv), death will occur in about 50 percent of cases within 30 days. With adequate therapy the majority of affected individuals in this category can recover.

Acute Toxicity

The ability of a substance to cause severe biological harm or death soon after a single exposure or dose. Also, any poisonous effect resulting from a single short-term exposure to a toxic substance.

Ad Hoc Investigation

An incident investigation that is undertaken from the immediate information and concerns. It is typically performed when there is no prior investigative procedure and is usually considered an unsystematic investigation.

Adjustable Guard

A machinery protective device (e.g., barrier, shield, etc.) that incorporates an adjustable element, and after being adjusted, remains in place during a particular operation. For example, the guard on a circular table saw, which is adjusted for the wood being cut. See also **Machine Guarding**.

Adjuster, Insurance

An individual who investigates and determines the amount of loss suffered in an insurance claim.

Adjustment Stress Theory

A theory of incident causation developed circa 1957, which postulates that individuals who cannot achieve some form of adjustment with their work and work environment will tend to have more incidents than others. The failure to adjust is caused by a range of physical and psychological stressors.

Administrative Controls

Methods of controlling employee exposures. These controls include administrative procedures, location and proximity, time periods away from the hazard, training in work practices to avoid the hazard, work assignments, etc. See also **Control Methods or Technology; Engineering Controls**.

Adsorption

The adhesion of molecules of a gas, liquid, or dissolved material to a surface.

Advanced Life Support (ALS)

Advanced life support stabilizes a patient while therapy continues to be provided. Providers assess and treat the underlying causes of the patient's condition so as to prevent regression.

Aerosols

Liquid droplets or solid particles dispersed in air that are of fine enough particle size to remain so dispersed for a period of time. Fog and smoke are examples of natural aerosols, while fine sprays such as perfumes, insecticides, and inhalants are examples of aerosols that are manufactured.

Agent or Agency

The principle object, such as a tool, machine, or equipment, involved in an incident.

Aggregate Limit

Normally refers to liability insurance and indicates the amount of coverage that the insured has under contract for a specific time period, no matter how many separate incidents may occur.

Agricultural Safety

Agriculture ranks among the most hazardous industries. Farmers are at very high risk for fatal and nonfatal injuries. Farming is one of the few industries in which the families (who often share the work and live on the premises) are also at risk for fatal and nonfatal injuries. National Institute for Occupational Safety and Health (NIOSH) records indicate that approximately 1,750,000 full-time workers were employed in production agriculture in the United States in 2007. During this same year, 411 farmers and farm workers died from a work-related injury for a fatality rate of 23.5 deaths per 100,000 workers. Between 1992 and 2007, 8088 farmers and farm workers died from work-related injuries in the United States. The leading cause of death for these workers was tractor overturns, accounting for an average of 96 deaths annually. The most effective way to prevent tractor overturn deaths is the use of a Rollover Protective Structure (ROPS; Figure A.2). In 2006, only 59 percent of tractors used on farms in the United States were equipped with ROPS. If ROPS were placed on all tractors manufactured since the mid-1960s used on U.S. farms, the prevalence

FIGURE A.2 Tractor with rollover protection structure.

of ROPS-equipped tractors could be increased to over 80 percent. On average, 113 youths less than 20 years of age die annually from farm-related injuries (1995–2002), with most of these deaths occurring to youths 16–19 years of age (34 percent). Of the leading sources of fatal injuries to youths on U.S. farms, 23 percent involved machinery (including tractors), 19 percent involved motor vehicles (including all-terrain vehicles [ATVs]), and 16 percent were due to drowning. Every day, about 243 agricultural workers suffer lost-work-time injuries. Five percent of these injuries result in permanent impairment. A new set of national occupational safety and health goals for the agricultural production industry have been developed as part of the National Occupational Research Agenda process, and are available for public review and comment. See also **Rollover Protective Structure (ROPS)**.

Air Ambulance

A helicopter outfitted to transport injured individuals to medical treatment in an expeditious manner due to the patient's medical condition or because the injury site is inaccessible by motor transport. See also **Ambulance**.

Air Bag

A type of vehicle restraint system that consists of inflatable bags mounted in the interior of a vehicle that automatically inflate upon a collision to protect the occupants from injury. They are designed for frontal impact crashes. This type of crash accounts for more than half of all passenger vehicle occupant deaths. Air bags are designed to limit head and chest injuries. They are considered only supplemental to safety belts; they do not replace them.

Air Cleaner

A device designed to remove atmospheric airborne impurities. These include dusts, gases, vapors, fumes, and smoke. See also **Air Filter**.

Air Conditioning

The process of treating air for temperature, humidity, cleanliness, and its distribution to meet requirements of the conditioned space it is supplied to.

Air Contaminants

Airborne particulates, gases, and vapors that are capable of causing illness or injury to workers upon ingestion, inhalation, or skin or mucous membrane contact.

Air Filter

An air cleaning device to remove light particulate matter from normal atmospheric air.

Air Line Breathing Apparatus

A form of respiratory protective equipment comprising a full or half face mask connected by a hose to an independent source of air supply via a filter and demand valve.

Air Monitoring

The sampling and measuring of the atmosphere to determine if pollutants are present.

Air-Purifying Respirator

A type of respirator that uses chemical cartridges or canisters (sorbents) and/or filters to remove harmful substances or contaminants from the supplied air.

Air Sampling

A method of determining quantities and types of atmospheric contaminants by measuring and evaluating contaminants in a known volume of air. It may be undertaken on a short- or long-term basis. Short-term sampling involves taking an immediate sample of air and typically passing it through a particular chemical reagent that responds to the contaminant being monitored. Long-term sampling is undertaken using a personal sampling instrument or dosimeter that is attached to the individual, and by the use of static sampling equipment located in the work area.

Air-Supplied Respirator

A respirator that provides breathable air from a clean source outside of the work area, usually provided by a compressor, compressed gas cylinders, or an air mover.

Alarm

A signal indicating an emergency requiring an immediate action, such as an alarm for fire from a manual fire alarm box, a fire sprinkler system waterflow alarm, toxic or combustible gas detection, or an alarm from an automatic detection system. Alarms can be visual (flashing or strobe light/beacon), audible (bells, horns, buzzers, etc.), or both.

Alarm Condition

A predefined change in the state or condition of equipment, or the failure of the equipment to respond correctly. Such an indication may be audible or visual or both.

Alkali

A material that has a pH value of greater than 7. Materials with a pH range of 12 to 14 are considered corrosive and will result in severe damage if exposed to the eyes or skin. Sometimes they are also referred to as Bases. See also **Acid**; **Acute Health Effects**.

All Risks Insurance

Insurance provided against loss of or damage to property arising from any fortuitous cause except those that are specifically excluded in the insurance contract. An All Risk Insurance is termed to be an All Risks policy.

Alliance Program, OSHA

A program for alliances with all organizations (e.g., businesses, labor organizations, educational institutions, government agencies, professional and trade organizations) committed to workplace safety and health to collaborate with the Occupational Safety and Health Administration (OSHA) to prevent injuries and illnesses in the workplace. OSHA and its allies work together to reach out to, educate, and lead the nation's employers and their employees in improving and advancing workplace safety and health. OSHA and the participating organizations must define, implement, and meet a set of short- and long-term goals that fall into three categories: training and education, outreach and communication, and promotion of the national dialogue on

A

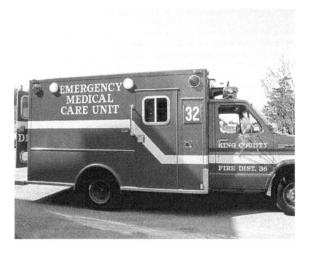

FIGURE A.3   Ambulance.

workplace safety and health. See also **Safety and Health Achievement Recognition Program (SHARP)**; **Voluntary Protection Program (VPP), OSHA**.

Ambient Noise

The total noise present in an environment. It usually is a composite of sounds from many sources, including characteristic and background noise. See also **Noise**.

Ambulance

A specialized motor vehicle equipped to transport sick or injured individuals to medical facilities while providing initial care. They are typically fitted with alerting devices (i.e., coloring, flashing lights, sirens) to facilitate expeditious movement in traffic (see Figure A.3). The term "ambulance" is derived from the Latin word ambulare, meaning to walk or move about. The word originally meant a moving hospital that follows an army in its movements. During the American Civil War vehicles for conveying the wounded off the battlefield were called ambulance wagons. See also **Air Ambulance**.

American Board of Industrial Hygiene (ABIH)

A nonprofit corporation, organized in 1960, it has been the world's largest, premier organization for certifying professionals in the practice of industrial hygiene by awarding the Certified Industrial Hygienist (CIH) designation. Eligibility requirements include education, experience, and successful completion of a comprehensive examination process. The ABIH is responsible for ensuring certification application and examination processes, certification maintenance, and ethics governance and enforcement. It works closely with the American Industrial Hygiene Association to achieve its goals. See also **American Industrial Hygiene Association (AIHA)**; **Certified Industrial Hygienist (CIH)**.

American Conference of Governmental Industrial Hygienists (ACGIH®)

Defining the Science of Occupational and Environmental Health, the ACGIH® is a member-based organization that advances occupational and environmental health. The ACGIH® is one of the industry's leading publications resources, with approximately 400 titles relative to occupational and environmental health and safety, including the renowned Threshold Limit Values (TLVs®) and Biological Exposure Indices

(BEIs®). The ACGIH® is registered with the U.S. Patent and Trademark Office. See also **Industrial Hygiene (IH)**.

American Industrial Hygiene Association (AIHA)

A professional society of industrial hygienists founded in 1939. Its goal is to promote the study and control of environmental factors affecting the health and well-being of workers. The AIHA is an international nonprofit organization that works in conjunction with the American Board of Industrial Hygiene to promote certification of industrial hygienists. It administers comprehensive education programs that keep occupational and environmental health and safety professionals current in the field of industrial hygiene and operates several well-recognized laboratory accreditation programs, based on the highest international standards. These programs help ensure the quality of the data used in making critical worker protection decisions. It works closely with the American Board of Industrial Hygiene to achieve its goals. See also **Industrial Hygiene (IH)**.

American Institute of Stress (AIS)

A nonprofit organization established in 1978 to serve as a clearinghouse for information on all stress related subjects. It maintains a library of information and reprints on all stress-related topics from scientific and lay publications, which can be ordered. A monthly newsletter is issued that reports on the latest advances in stress research and relevant health issues, and it also sponsors an international forum for the exchange of research information. See also **Workplace Stress**.

American Insurance Association (AIA)

A leading property and casualty insurance trade association representing over 350 companies that underwrite more than $123 billion in premiums annually. AIA member companies offer all types of property and casualty insurance, including personal and commercial auto insurance, commercial property and liability coverage for small businesses, workers' compensation, homeowner's insurance, medical malpractice coverage, and product liability insurance. It was previously known as the National Board of Fire Underwriters (NBFU), which was established in 1866 and published the National Building Code in 1905. The code was used as a model for adoption by cities, as well as a basis to evaluate the building regulations of cities for insurance grading purposes. The AIA took its current form in 1964 when the old American Insurance Association merged with the NBFU and the Association of Casualty and Surety Companies. The AIA represents its members at all levels of government concerning legislative, regulatory, and legal issues. It promotes the economic, legislative, and public understanding of its members through its attention to accounting procedures, catastrophe and pollution problems, automobile insurance reform, and similar issues. The insurance industry is primarily regulated at the state level, and the AIA is represented in every state. It is the only insurance trade organization that has provided this representation.

American Ladder Institute (ALI)

A nonprofit association founded in 1947, dedicated to promoting safe ladder use. It is comprised of members from the United States and Canada who are ladder manufacturers and manufacturers of ladder components. The ALI is the American National Standards Institute (ANSI)-approved developer of ladder safety standards, which govern the construction, design, testing, care, and use of the various types of ladders. See also **Ladder Safety**.

American National Standards Institute (ANSI)

A consensus body consisting of volunteer safety professionals who create or revise documents on test requirements, procedural methods, and product specifications. It was founded in 1918 and is composed of over 1300 members who set guidelines and standards for a wide range of technical areas. It is the umbrella body for standardization in the United States.

American Society of Safety Engineers (ASSE)

Professional society of safety engineers, safety directors, and others concerned with incident prevention and safety programs. It sponsors training, including safety education seminars; bestows awards; compiles statistics; and maintains job placement services. The society was founded on October 14, 1911, in New York City as the United Society of Casualty Inspectors (USCI) with 62 members. This was just after the Triangle Shirtwaist Factory Fire in New York City that occurred on March 25, 1911, where a large loss of life occurred. In 1914 the USCI name was changed to the present American Society of Safety Engineers (ASSE) and headquarters were established in New York City. In 1919 the ASSE published *Safety Engineering*, its first official publication. Presently, the ASSE has 151 chapters, 56 sections, and 65 student sections. There are also members in 64 countries including Mexico, Ecuador, Saudi Arabia, the United Kingdom, Australia, Kuwait, and Egypt. ASSE members serve on over 40 safety and health standards committees including the International Organization for Standardization (ISO). See also **ASSE Foundation**.

AMES Test

A short-term test commonly used for preliminary screening of chemicals to see if they cause mutations in a special type of bacterial cell. It is named for one of the researchers who developed it.

Analgesia

Insensibility to pain with loss of consciousness.

Anaphylaxis

A severe, life-threatening allergic-type reaction to a substance resulting from sensitization to the substance. Prior contact with the substance is necessary.

Anchor Point

A component of a personal fall arrest system. The Occupational Safety and Health Administration (OSHA) requires that anchor points used for fall protection be capable of supporting at least 2273 kg (5000 lbs) per employee attached. See also **Personal Fall Arrest System**.

Anesthesia

Loss of sensation with or without the loss of consciousness.

Annual Survey of Occupational Injuries and Illnesses (ASOII)

A statistical report on data of workplace injuries and illness which are analyzed by the rate and number of work-related injuries, illnesses, and fatal injuries, and how these statistics vary by incident, industry, geography, occupation, and other characteristics. Conducted by the Bureau of Labor Statistics (BLS) on a national level; some employers are required to participate and others do so voluntarily.

Anomaly

An unusual, abnormal, or irregular set of circumstances that, left unrecognized or uncorrected, may result in an incident.

ANSI B11, General Safety Requirements Common to ANSI B11 Machines

Contains general safety requirements for machines used to shape and/or form metal or other materials by cutting, impact, pressure, electrical, or other processing techniques, or a combination of these processes, which is applicable to the specific ANSI B11 series of codes (i.e., B11.1 to B11.18) for metal fabrication machinery (e.g., cutting, bending, presses, grinding, drilling, etc.).

ANSI B101.0, Walkway Surface Auditing Guideline for the Measurement of Walkway Slip Resistance
A guideline providing a technical review of the science of measuring surface friction (tribometry) including slip-and-fall dynamics, its causes and contributing factors, and the testing devices and methods used to measure the slip resistance of walkway surfaces. Its purpose is to prevent or mitigate the effects of injuries and fatalities from slips, trips, and falls. See also **National Floor Safety Institute (NFSI); Slips, Trips, and Falls**.

ANSI BSR Z16.5, Standard for Occupational Safety and Health Incident Surveillance
Record keeping standard for occupational injuries, illness, and sentinel incidents. It replaces the American National Standard for Uniform Record Keeping for Occupational Injuries and Illnesses, ANSI Z16.4-1977, and USA Standard Method of Recordkeeping and Measuring Work Injury Experience, ANSI Z16.1-1967. This standard is a development of the old Z16.1, which had been in use since 1937, before it was replaced by the (for injury and illness statistical research) Occupational Safety and Health Administration (OSHA) system. The standard is useful in determining what kinds of events to evaluate. It includes statistical tools, including control charts, for data analysis.

ANSI Standard Z41, Standard for Personal Protection Protective Footwear
An obsolete American National Standards Institute (ANSI) standard on safety shoes since March of 2005. American Society for Testing Materials (ASTM) international standards F 2412, Test Methods for Foot Protection, and F 2413, Specification for Performance Requirements for Protective Footwear, have replaced the former ANSI Z41 standard, Standard for Personal Protection Protective Footwear, which has been withdrawn. Both of the new ASTM standards are under the jurisdiction of ASTM Committee F13 on Pedestrian/Walkway Safety and Footwear. See also **ASTM F 2412, Test Methods for Foot Protection, and ASTM F 2413, Standard Specification for Performance Requirements for Foot Protection**.

ANSI Standard Z49.1, Safety in Welding, Cutting, and Allied Processes
A safety standard developed with the American Welding Society highlighting the safe practices while performing oxyfuel gas welding and cutting, arc welding and cutting, resistance welding, electron beam processes, and laser beam cutting and welding processes to prevent illnesses, injury, fires, and explosions. It was originally issued in 1944 and has undergone numerous revisions. It focuses on those safety practices that are in control of the welder or weld shop management. Published by the American Welding Society (AWS).

ANSI Standard Z89.1, Protective Headware for Industrial Workers
A safety standard that provides performance and testing requirements for industrial helmets, commonly known as hard hats. It established the types and classes of protective helmets, depending on the type of hazard encountered. Industrial head protective helmets meeting the requirements of the standard are classified as Type I for top protection or Type II for lateral impact protection. Both types are tested for impact attenuation and penetration resistance. Type II helmet performance requirements include criteria for impact energy attenuation from impacts from the front, back, and sides as well as the top; off-center penetration resistance; and chin strap retention. The three classes indicate the helmet's electrical insulation rating: Class E (electrical) helmets are tested to withstand 20,000 volts; Class G (general) helmets are tested at 2200 volts; and Class C (conductive) helmets provide no electrical protection. See also **Hard Hat**.

ANSI Z16.1, Method of Recordkeeping and Measuring Work Injury Experience
An obsolete ANSI standard that defined how companies could record and track injuries and illnesses prior to the Occupational Safety and Health Administration (OSHA) regulations. This document defined terms like Permanent Total Disability, Permanent Partial Disability, and Temporary Total Disability that are used throughout safety and workers' compensation literature.

ANSI Z16.2, Method of Recording Basic Facts Relating to the Nature and Occurrence of Work Injuries
See **Occupational Injury and Illness Classification System (OIICS)**.

ANSI Z87.1, Standard for Occupational and Educational Eye and Face Protection Devices
Standard that establishes performance criteria and testing requirements for devices used to protect the eyes and face from injuries from impact, non-ionizing radiation, and chemical exposure in workplaces and schools. It covers all types of protective devices, including spectacles (plano and prescription), goggles, face shields, welding helmets and handshields, and full facepiece respirators. The standard includes descriptions and general requirements, as well as criteria for testing, marking, selection, use, and care. The sections on selection, use, and maintenance define all types of eye and face protectors, and provide guidance on hazard assessment and selection. A pull-out selection chart is included, showing recommended protector categories for various types of work activities that can produce impact, heat, chemical, dust, or optical radiation hazards. The standard requires that all frames and lenses carry a manufacturer's mark or logo, plus other markings to indicate the level of protection. Face shields are marked Z87 to indicate Basic impact, or Z87+ to indicate High impact. Other markings are "S" for special purpose lenses, and Light, Medium, Dark to indicate luminous transmittance of a face shield window. This standard does not apply to hazardous exposure to blood-borne pathogens, X-rays, high-energy particulate radiation, microwaves, high-frequency radiation, lasers, masers, or sports. See also **Eye Protection**; **Eye Safety**.

ANSI Z358.1, Standard for Emergency Eye Wash and Shower Equipment
Provisions for the design, performance, installation, use, and maintenance of various types of personal emergency water decontamination or flushing systems and equipment (showers, eye washes, drench hoses, etc.). In addition to these provisions, there are some general considerations that apply to all emergency equipment.

ANSI Z535.1, Safety Colors
Standard that provides the technical definitions, color standards, and color tolerances for safety colors. Color codes used on safety signs, labels, and tags have been developed in the past by a large number of firms and organizations. Although these color codes give satisfaction to those using them, they lack uniformity. This standard establishes a "standard for safety colors" that will alert and inform persons to take precautionary action or other appropriate action in the presence of hazards. See also **Color Coding**.

ANSI Z535.2, Environmental and Facility Safety Signs
Standard that provides guidance for industries, commercial establishments, property owners, employers, and others on environmental and facility safety signs. This standard establishes requirements for a uniform visual system of identification related to potential hazards in the environment. The design, application, and use of signs and placards employing this visual alerting system are covered as well as the safety signs used at fixed locations in the environment such as industrial facilities, large movable signs that may be used on a temporary basis, and the steps to be taken to avoid the hazard.

A

ANSI Z535.3, Criteria for Safety Symbols

Standard that provides general principles for the design, evaluation, and use of safety symbols to identify and warn against specific hazards and personal injury. This standard addresses the fact that the U.S. population is multi-ethnic, highly mobile, and derived from a multiplicity of social and educational backgrounds. Word-only signs may not be effective in promoting safety to the general population. Effective safety symbols have demonstrated their ability to provide critical information for incident prevention and for personal protection. Signs with safety symbols can promote greater and more rapid communication of the safety message, and therefore greater safety for the general population in the immediate environment and workplaces.

ANSI Z535.4, Product Safety Signs and Labels

Standard that provides a hazard communication system developed specifically for product safety signs and labels. This standard sets forth a consistent visual layout and performance requirements for the design, application, use, and placement of safety signs and labels intended to identify hazards for persons using, operating, and servicing or otherwise in the proximity of a wide variety of products, as well as a national uniform system for the recognition of potential personal injury hazards. See also **Signal Word**.

ANSI Z535.5, Safety Tags and Barricade Tapes (for Temporary Hazards)

Standard that provides the requirements for the design and use of safety tags and barricade tapes and reflects harmonization with the international standards by permitting different colors to be used with the safety alert symbols that are a means of alerting people to temporary hazards often associated with construction equipment installation, maintenance, repair, lockout, or other transient conditions. Safety tags and barricade tapes are used to identify a temporary hazard. They should be used only until such time as the identified hazard is eliminated or the hazardous operation is completed.

ANSI Z535.6, Product Safety Information in Product Manuals, Instructions, and Other Collateral Materials

Standard that provides a uniform and consistent visual layout for safety information in collateral materials for a wide variety of products. This standard minimizes the proliferation of designs for safety information in collateral materials, establishes a system for the recognition of potential personal injury hazards for those persons using products, assists manufacturers in providing safety information in collateral materials, and promotes the efficient development of safety messages in collateral materials.

Anthropometry

Applied in ergonomics, it is the study of physical dimensions in people, including the measurement of human body characteristics such as size, breadth, girth, and distance between anatomical points. Anthropometry also includes segment masses, the centers of gravity of body segments, and the ranges of joint motion, which are used in biomechanical analyses of work postures. See also **Ergonomics**.

Antidote

A specific therapeutic measure that may or may not require the utilization of a physician. It is usually a remedy to relieve, prevent, or counter the effects of a poison. See also **Poison**.

Anti-fatigue Mats

Mats or padding placed on the floor designed to reduce stresses on the feet and legs when standing for long periods. Cushioned insoles for shoes can be viewed as "portable anti-fatigue mats." They may be considered during ergonomic evaluations.

A

Antistatic Wrist Strap
An antistatic device worn by an individual on the wrist to prevent the accumulation and discharge of static electricity to static-sensitive electronic equipment by safely grounding the individual. See also **Static Sensitive Device**.

AOE/COE (Arising Out of Employment/Course of Employment)
A term used for processing workers' compensation. Claims are only accepted for compensation or disability if the incident occurred arising out of employment or the course of employment (AOE/COE).

Apparent Violation
An item observed by a Compliance Officer of the Occupational Safety and Health Administration (OSHA), and mentioned in the closing conference as a violation of federal OSHA standards.

Apportionment
A way of figuring out how much permanent disability is due to a worker.

Approval Agency
An agency designated or authorized to sanction, consent to, confirm, certify, or accept as good or satisfactory for a particular purpose or use, a method, procedure, practice, tool, or item of equipment or machinery.

Approved
Term used by national or local codes to denote equipment, devices, tools, or systems meeting acceptable standards of safety performance for a particular purpose as verified by a recognized testing or professional agency. See also **Classified**; **Factory Mutual (FM)**; **Labeled**; **Listed**; **Underwriters Laboratories (UL)**.

Approved (Method, Equipment, etc.)
A method, equipment, procedure, practice, tool, etc., which is sanctioned, consented to, confirmed, or accepted as good or satisfactory for a particular purpose or use by a person or organization authorized to make such a judgment.

Arc
A luminous, high-intensity electrical discharge in a gas or a vapor. It occurs when the voltage in a conductor is great enough to create a path between itself and another conductor that is at a lower voltage. The arc created is capable of crossing or jumping the air gap in insulation that separates the two conductors. Arcing may be a source of ignition for a fire if it is of sufficient energy. It also generates ultraviolet radiation, which can burn the retina of the eye along with the skin. It may also be called spark discharge. See also **Arc Blast**; **Arc Flash**; **Spark**.

Arc Blast
Arc Blast is a result of the arc flash, and its force is dependent on the amount of short-circuit current available and the distance from the arc source. When copper super heats, as is the case when an arc flash occurs, the copper can expand up to 67,000 times. This expansion causes molten copper to be spewed away from the source in a very forceful manner. This force or pressure can cause injury to body parts, or the body can be injured from a resultant fall or from being thrown into a nearby object.

Arc Fault Circuit Interrupter (AFCI)
AFCIs are electrical safety devices designed to automatically stop the flow of electricity in an electrical circuit when the arc fault is detected within the circuit, to prevent fires from developing. Arc faults are considered one of the major causes of residential electrical fires that occur each year in the United States. Since 2008, new homes built in the United States are required to be provided with AFCIs. See also **Ground Fault Circuit Interrupter (GFCI)**; **Grounding**.

Arc Flash

A fire flash, or arc, that can be produced during an electrical failure or fault. It can reach temperatures of 19,982°C (36,000°F). The temperature of the arc is a function of short-circuit current availability and distance from the arc source. See also **Arc**; **Arc Blast**.

Arc Flash Hazard Analysis

A review of the potential exposure of an individual to arc flash energy to prevent its occurrence or injury to personnel and determine appropriate safe work practices, arc flash protection boundary, and the necessary levels of personal protective equipment. See also **Arc Flash Protection Boundary**.

Arc Flash Protection Boundary

The approach limit at a distance from a prospective arc source or arc flash hazard within which an individual could receive a second-degree burn if an electrical arc flash incident occurs.

Arc Flash Suit

Flame-resistant clothing and equipment that encapsulates the entire body, except for the hands and feet. It includes pants, jacket, and hood with a face shield.

Arc Resistant Switchgear

Equipment that is designed to withstand an internal arcing fault and directs the internally released energy away from an individual.

Area Hazard Analysis (AHA)

A process for analyzing hazards, focused on the hazards an individual faces in their work area as opposed to hazards of individual work activities.

Arson

The burning of buildings or property with malicious or criminal design, generally utilizing highly flammable materials or explosives to spread the fire quickly, or deliberately placed obstructions to impede fire fighting. This action may be performed by the owner or others and is by law a crime subject to federal jurisdiction.

Artificial Respiration

Movement in and out of the lungs by artificial means. It is also known as Rescue Breathing, Artificial Resuscitation, and Pulmonary Resuscitation. See also **CPR (Cardiopulmonary Resuscitation)**.

As Low as Reasonably Achievable (ALARA)

The process of determining what level of protection and safety makes exposures, and the probability and magnitude of potential exposures, as low as reasonably achievable, taking into account economic (e.g., cost, effect on production) and social factors. It is typically utilized in the nuclear industry. See also **Acceptable Level of Risk**; **As Low as Reasonably Practical (ALARP)**.

As Low as Reasonably Practical (ALARP)

The principle that no activity is entirely free from risk, and that it is never possible to be sure that every eventuality has been covered by safety precautions, but that there would be a gross disproportion between the cost in (money, time, or trouble) of additional preventive or protective measures, and the reduction in risk in order to achieve such low risks. See Figure A.4. Sometimes also referred to as So Far As Is Reasonably Practical (SFAIRP). See also **Acceptable Level of Risk**; **As Low as Reasonably Achievable (ALARA)**.

Asbestos

A mineral fiber that can pollute air or water and cause cancer or asbestosis when inhaled. The Environmental Protection Agency (EPA) has banned or severely restricted its use in manufacturing and construction.

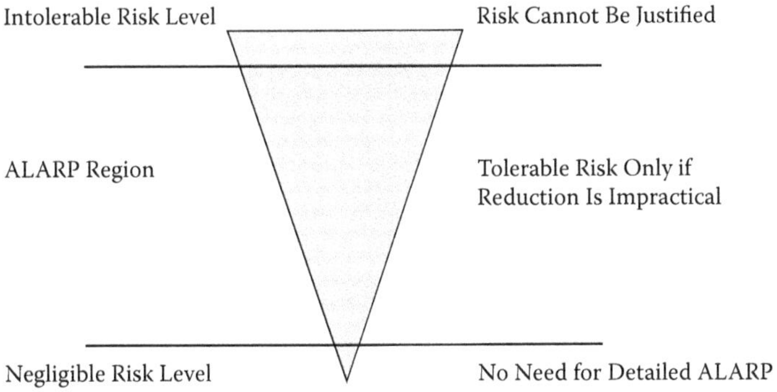

FIGURE A.4 ALARP principle.

Asbestos Abatement
Procedures to control fiber release from asbestos-containing materials in a building or to remove them entirely, including removal, encapsulation, repair, enclosure, encasement, and operations and maintenance programs.

Asbestosis
A form of lung disease (pneumoconiosis) caused by inhaling fibers of asbestos and marked by interstitial fibrosis of the lung varying in extent from minor involvement of the basal areas to extensive scarring. The disease makes breathing progressively more difficult and can be fatal.

A-Scale Sound Level
A measurement of sound approximating the sensitivity of the human ear, used to note the intensity or annoyance level of sounds.

Asphyxiant
A substance that causes chemical suffocation, i.e., prevents oxygen in sufficient quantities from combining with the blood or being used by body tissues.

Asphyxiation
Suffocation resulting from being deprived of oxygen. Simple asphyxiants act mechanically by excluding oxygen from the lungs when breathed in high concentrations (examples: nitrogen, hydrogen, carbon dioxide). Chemical asphyxiants act through chemical action preventing oxygen from reaching the tissue, or else prevent the tissue from using it even though the blood is well oxygenated (e.g., carbon monoxide, hydrogen cyanide, aniline).

ASSE Foundation
An organization to advance occupational safety, health, and environmental development, research, and education by funding scholarships, fellowships, research grants, and internships. The foundation was established in 1990 as a part of the American Society of Safety Engineers (ASSE). See also **American Society of Safety Engineers (ASSE)**.

ASSE Z359.1, Safety Requirements for Personal Fall Arrest Systems, Subsystems, and Components
This standard establishes requirements for the performance, design, marking, qualification, instruction, training, inspection, use, maintenance, and removal from service of connectors, full body harnesses, lanyards, energy absorbers, anchorage connectors, fall arresters, vertical lifelines, and self-retracting lanyards comprising personal fall arrest systems for users within the capacity range of 59 to 140 kg (130 to 310 lbs).

FIGURE A.5 Example of assembly area identification sign.

Assembly Area

A designated gathering point defined in facility evacuation procedures. Normally located so that individuals would not be affected by incidents occurring at the evacuation location and also so that emergency responders have immediate access to the incident location without impact from evacuees. Figure A.5 provides and an example of an Assembly Area identification sign. See also **Evacuation**.

Assignable Cause

A cause factor designated as having a relationship to an incident. A causal factor or several causal factors can be assigned to an incident based on a predetermined causal classification system. See also **Causal Factor (CF)**.

Assigned Protection Factor (APF)

This is a numerical value assigned to a specific class of respirator, and represents by increasing value the relative protection that the type of device affords when its fit is verified by fit testing and is worn by an employee trained in its use. APFs can be used to assist in estimating the maximum concentrations of contaminant in which a particular respirator can be used.

Assigned Risk

Many states have unsatisfied judgment—or financial responsibility laws—that make the purchase of insurance mandatory. Some motorists cannot buy insurance for some reason, such as poor accident experience. To make it possible for them to be insured, there are assigned risk plans in which such risks are insured. These risks are rotated among the subscribing companies in proportion to the amount of automobile liability insurance each writes in the state. All companies writing this class of insurance are required to participate in this activity. A comparable system operates in some states with respect to workers' compensation.

Associate in Risk Management (ARM)

A professional designation offered by the Insurance Institute of America based on experience and education in risk management. Three main courses are required: Risk Assessment, Risk Control, and Risk Financing. Risk Assessment includes Risk Management Programs; The Risk Management Process; Legal Foundations of Liability Loss Exposures; Assessing Property, Liability, Personnel, and Net Income Loss Exposures; Management Liability and Corporate Governance; Forecasting; and Cash Flow Analysis. Risk Control includes Controlling Property, Personnel, Liability, and Net Income Loss Exposures; Intellectual Property Loss Exposures; Criminal

A

Loss Exposures; Disaster Recovery for Property Loss Exposures; Understanding Claim Administration; Fleet Operations Loss Exposures; Environmental Loss Exposures; Understanding System Safety; and Motivating and Monitoring Risk Control Activities. Risk Financing includes Insurance as a Risk Financing Technique; Reinsurance and Self-Insurance; Retrospective Rating Plans and Captive Insurance Companies; Finite and Integrated Risk Insurance Plans; Capital Market Products; Forecasting Accidental Losses; Accounting and Income Tax Aspects; Claim Administration; and Allocating Risk Management Costs. See also **Insurance Institute of America (IIA)**.

Associate Safety Professional (ASP)

An interim designation to show progress toward the Certified Safety Professional (CSP®) recognition offered by the Board of Certified Safety Professionals (BCSP). See also **Certified Safety Professional® (CSP®)**.

Assumption of Risk

The legal theory that a person who is aware of a danger and its extent and knowingly exposes him- or herself to it assumes all risks and cannot recover damages, even though the person is injured through no fault of his or her own.

Asthma

A disease characterized by increased responsiveness of the trachea and bronchi to various stimuli, manifested by difficulty in breathing caused by generalized narrowing of the airways.

ASTM D 120 Standard Specification for Rubber Insulating Gloves

The primary standard covering minimum requirements for the manufacturing and testing of rubber insulating gloves for the protection of workers from electrical shock.

ASTM F 1818, Specification for Foot Protection for Chainsaw Users

Standard for chain saw cut–resistant footwear is designed to provide protection to the wearer's feet when operating a chain saw. It is intended to protect the foot area between the toe and lower leg.

ASTM F 2412, Test Methods for Foot Protection, and ASTM F 2413, Standard Specification for Performance Requirements for Foot Protection

The primary standards covering minimum requirements for the design, performance, testing, and classification of protective footwear against a variety of hazards that can potentially result in injury (see Figure A.6). These tests include impact resistance for the toe area of footwear; compression resistance for the toe area of footwear; metatarsal impact protection that reduces the chance of injury to the metatarsal bones at the top of the foot; conductive properties that reduce hazards

FIGURE A.6 Safety shoe.

that may result from static electricity buildup and reduce the possibility of ignition of explosives and volatile chemicals; electric shock–resistant non-conductive properties; static dissipative properties to reduce hazards due to excessively low footwear resistance that may exist where static dissipative footwear is required; puncture resistance of foot bottoms; chain saw cut resistance hazards; and dielectric hazard protection. All footwear manufactured to the American Society for Testing Materials (ASTM) specification must be marked with the specific portion of the standard with which it complies. See also **ANSI Standard Z41, Standard for Personal Protection Protective Footwear**.

ATHEANA (A Technique for Human Error Analysis)

A technique used in the field of Human Reliability Assessment (HRA) for the purposes of evaluating the probability of a human error occurring throughout the completion of a specific task. From such analyses measures can then be taken to reduce the likelihood of errors occurring within a system, therefore leading to an improvement in the overall levels of safety. There are three main reasons for conducting an HRA: error identification, error quantification, and error reduction. A number of techniques are used for such purposes, which can be divided into two classifications: first- and second-generation techniques. The first-generation techniques work on the basis of the simple dichotomy of "fits/doesn't fit" in the matching of the error situation in context with related error identification and quantification. The second-generation techniques are more theory based in their assessment and quantification of errors. The method is able to categorize the human factors contributing to an incident, but it fails to prioritize or establish details of the causal relationships between these factors. Thus, further work is required in order to identify the root causes of an incident from an HRA perspective. The basic technique was developed by the U.S. Nuclear Regulatory Commission in late 1970s to the mid-1980s. See also **Human Error Analysis**.

Atmosphere-Supplying Respirator

A respirator that provides breathing air from a source independent of the surrounding environment. It is typically either an air line or self-contained breathing apparatus.

Attendant, Permit Space

The attendant is the individual stationed outside a (work) permit space to perform attendant duties. The attendant's major function is to monitor and protect the authorized entrants.

Audible Range, Hearing

The frequency range over which normal ears hear. Approximately 20 Hz, through 20,000 Hz. Above the range of 20,000 Hz, the term ultrasonic is used. Below 20 Hz the term subsonic is used.

Audiogram

A chart, graph, or table record resulting from an audiometric test showing an individual's hearing threshold levels as a function of frequency (usually 500 to 6000 Hz). The audiogram may be represented numerically or graphically. The hearing level is shown on a chart as a function of frequency. Of high interest are the frequencies at which normal speech occurs, i.e., 0.5, 1, and 1.5 kHz.

Audiometer

A signal generator or instrument for objectively measuring the sensitivity of hearing in decibels.

Audiometry

The measurement of an individual's hearing acuity or ability over a range of frequencies.

Audit, Safety
A management tool used to measure the effectiveness and efficiency of the implementation of safety policy, programs, and procedures by subjecting each area of an activity to a systematic critical examination with the purposes of minimizing loss, and providing a quantified assessment of performance and actions needed to render identified hazards harmless. See also **Safety Inspection**.

Authority Having Jurisdiction (AHJ)
The legal entity or responsible organization (private or governmental) that requires, accepts, and approves safety and fire requirements for the locations within its domain based on the regulations or codes it has adopted.

Attenuation
Reduction of intensity; to render less virulent or harmful; to reduce in strength.

Attenuation (Sound)
The reduction, expressed in decibels, of sound intensity to an observer either due to the distance from the source of noise or due to a barrier or acoustically treated material. See also **Ear Muffs**; **Hearing Protection**.

Autoignition Temperature
The minimum temperature of a vapor and air mixture at which it is marginally self-igniting in the absence of an independent ignition source, such as a spark or flame.

Automated External Defibrillators (AEDs)
An AED is an electronic device designed to deliver an electric shock to a victim of sudden cardiac arrest. Ventricular fibrillation may be restored to normal rhythm up to 60 percent of the time if treated promptly with an AED, a procedure called defibrillation. The American Heart Association (AHA) estimates that approximately 890 deaths from coronary heart disease occur outside of the hospital or emergency room every day. According to the Occupational Safety and Health Administration (OSHA) in 2001 and 2002, there were reported 6628 workplace fatalities: 1216 from heart attack, 354 from electric shock, and 267 from asphyxia. The AHA and OSHA have estimated that up to 60 percent of these victims might have been saved if automated external defibrillators (AEDs) were immediately available. Chances of survival from sudden cardiac death diminish by 7 to 10 percent for each minute without immediate CPR or defibrillation. After 10 minutes, resuscitation rarely succeeds. See Figure A.7 for an AED provided in a wall-mounted case.

Automatic Darkening Welding Filter
An optical filter that automatically switches from a light state to a dark state (shade) in response to change in incident light intensity.

Automatic Fire Alarm System
A system of controls, initiating devices, and alarm signals in which all or some of the initiating circuits are activated by automatic devices, such as smoke detectors.

Automatic Guard
A protective device (e.g., barrier) that is associated with and dependent upon the mechanism of the machinery. It operates to physically remove from the danger area any part of an individual exposed to harm.

Automatic Safe Load Indicator
A device fitted to a crane to provide the operator with automatic warning of approach to an overload situation.

Automatic Seat Belt System
A protective system fitted to a vehicle that automatically moves the seat belt restraint system from its storage location to its protection location, i.e., over the shoulder of an individual, to ensure that the seat beat is worn and assist in protecting the individual

A

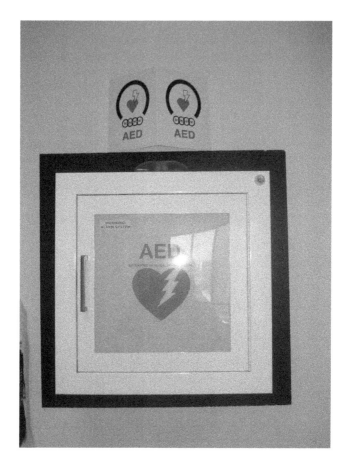

FIGURE A.7   Automated External Defibrillator (AED).

from injury if a crash occurs. Upon leaving the vehicle the belt automatically moves to its storage position.

Automatic Sprinkler System

See **Sprinkler System**.

Average Days Charged per Disabling Injury

As defined by the American National Standards Institute (ANSI) Z16.2 Standard, a measure to express the relationship between the total days charged to a disabling injury and the total number of disabling injuries. The average may be calculated by use of the following formula: Average days charged per disabling injury = Total days charged/Total number of disabling injuries. The following alternate formula may also be used to compute this measure: Average days charged per disabling injury = Standard injury severity rate/Standard injury frequency rate.

Avoidable Accident

An incident that could have been prevented by proper behavior, or by environmental modifications or controls. All incidents are considered preventable.

Awkward Posture

Posture is the position of the body while performing work activities. Awkward posture is associated with an increased risk for injury. It is generally considered that the more a joint deviates from the neutral (natural) position, the greater the risk of injury. Awkward postures typically include reaching behind, twisting forward, backward bending, pinching, and squatting. Specific postures that have been associated with injury include the following examples:

Wrist

    Flexion or extension (bending up and down)

    Ulnar or radial deviation (side bending)

Shoulder

    Abduction or flexion (upper arm positioned to the side or above shoulder level)

    Hands at or above shoulder height

Neck (cervical spine)

    Flexion or extension or bending the neck forward and to the back

    Side bending as when holding a telephone receiver on the shoulder

Low back

    Bending at the waist, twisting

B

Back Safety

It is estimated that nearly 80 percent of people will experience some type of back injury during their lifetime. This is a common injury caused by activities both on and off the job. Common causes of back injuries include lifting, twisting, and reaching. According to the Bureau of Labor Statistics (BLS), more than one million workers suffer back injuries each year, and back injuries account for one of every five workplace injuries or illnesses. One-fourth of all compensation indemnity claims involve back injuries, resulting in billions of dollars of lost time expense and medical treatments on top of the pain and suffering borne by employees. Statistics show that such strains are the most common among workers. Medical experts indicate that back injuries have little to do with the weight of the lifted objects and are the result of lifting improperly. Although lifting, placing, carrying, holding, and lowering are involved in manual materials handling (the principal cause of compensable work injuries), the BLS survey shows that four out of five of these injuries were to the lower back, and that three out of four occurred while the employee was lifting.

Background Noise

Noise occurring from sources other than the particular sound noise being monitored or investigated.

Background Radiation

Natural radiation in the environment including cosmic rays and radiation from naturally radioactive elements; it may also mean radiation from sources other than the one directly under consideration.

Backup Alarm

An automatic audible sounding alarm fitted to designated vehicles to warn whenever the vehicle is engaged to move in the reverse direction. It is commonly used on construction sites for equipment or other similar locations, where workers may not be cognizant of the various operations of vehicles in their environment, other equipment masks the normal sound of the vehicle approaching in reverse, or their hearing level may not be the norm. They are required by Occupational Safety and Health Administration (OSHA) regulation 29 CFR 1926.602 (a) (9) (ii).

Barrier

See **Guard**.

Barrier Analysis

An investigative safety review based on the premise that an energy flow is associated with all incidents. Barriers are developed and integrated into a system or work process to protect personnel and equipment from unwanted energy flows. For an incident to occur there must be a hazard that comes into contact with a target, because barriers or controls were unused or failed. It may be called a barrier and control analysis or energy trace and barrier analysis. Barriers can be Physical (guarding, shields, protective clothing), Administrative (engineering design, work procedures, work processes), or Supervisory/Management (training, knowledge, supervision, management oversight).

B

Barrier Cream

A cream for use on the skin to protect against injury (e.g., dermatitis, inflammation, etc.) from contact with specific types of harmful substances (e.g., adhesives, oils, solvents, resins, fiberglass, dyes, etc.).

Barrier Guard

A protection device for operators and other individuals from hazard points on machinery and equipment. There are several types of barrier guards: adjustable, fixed, movable, and interlocking. Adjustable barrier guards have an enclosure attached to the frame of the machinery or equipment, with front and side sections that can be adjusted. Fixed barrier guards are a point of operation enclosure attached to the machine or equipment. The gate or movable barrier guard is a device designed to enclose the point of operation completely before the clutch can be engaged. An interlocking guard is connected to a switch so that the operation of the machine or equipment cannot be started unless the guard is in place. It may also stop the machine or equipment if the guard is moved while it is in operation. See also **Interlocking Guard**.

Barrier Tape

A temporary delineation measure used to quickly and effectively highlight restricted areas, particularly for hazard avoidance or emergencies. It is composed of a high-visibility polyethylene plastic tape that is used to surround or section-off hazard areas, danger zones, and construction sites. Barrier tapes typically have a clear, bold warning message that is continuously repeated the length of the roll. They provide a visual warning message to keep out of restricted areas. It is also ideal for crowd and traffic control purposes. Commercially available tape rolls are 7.6 cm (3 inches) wide and are 2540 cm (1000 inches) in length. It also can be called Perimeter Tape or Safety Tape. See **ANSI Z535.5 Safety Tags and Barricade Tapes (for Temporary Hazards)**.

Basic Impact Lenses

Lenses are tested to withstand the impact of a 2.5-cm (1-inch) ball dropped from 127 cm (50 inches), plus a penetration test for plastic lenses as required by American National Standards Institute (ANSI) Z87 for all eye and face protectors, two-level classification for impact protection. See **ANSI Z87.1-2003, Standard for Occupational and Educational Eye and Face Protection Devices**.

Beat Disorders

A group of prescribed occupational diseases associated with external pressure and friction at or about a particular joint.

Beat Elbow

Bursitis of the elbow commonly occurs from the use of heavy vibrating tools. Sometimes it is also referred to as tennis elbow.

Beat Hand

Subcutaneous cellulitis of the hand.

Beat Knee

Bursitis or subcutaneous cellulitus of the knee joints due to friction of vibration, commonly found in the mining industry. It is sometimes called housemaid's knee.

Behavior-Based Safety (BBS)

A management safety process that focuses on what people do, analyzes why they do it, and then applies a research-supported intervention strategy to improve what people do. It begins with the understanding of the human behavior and the ability to influence it proactively, then identifying and implementing managing systems that proactively affect the safety performance. By focusing on safety-related behaviors where incidents may happen, an organization can effect changes that lead to an overall improvement in safety performance. Behavior-based safety management is a concept where workgroups can carry out safety performance management themselves. To manage their own performance, workgroups make measurements and track the rate at which they have to perform critical tasks, identified as high-risk tasks. These are task-related observable acts that may expose the workforce to injury. Utilizing the accumulated performance data, the workgroups then perform problem solving and develop action plans to reduce their exposure levels. Depending on the root cause of the behavior, the intervention methods may vary. For behaviors that are under the direct control of employees, feedback is a powerful tool for reinforcing safe behaviors. Where system factors and working conditions are found to prevent employees from working safely, actions are required to have these barriers immediately removed to enable the safety environment to be maintained. See also **Domino Theory**.

Behavior Objective

A statement of behavior that includes a measurement feature to demonstrate that learning has occurred.

Benching

A method of protecting personnel during an excavation from sidewall cave-in. It is accomplished by excavating the sides of an excavation to form one or more of a series of horizontal steps, with a vertical rise between each step. The angle used for benching is based on a ratio of horizontal to vertical cuts. It should be noted that benching is reserved only for cohesive soils as other soils do not have the ability to support a vertical wall. See also **Excavation Protective System**; **Shoring**.

Benchmark Safety Comparisons

A measurement that serves as a standard against others may be measured. Industry safety statistics (injuries, fatalities, fires, etc.) are commonly used as a safety benchmark measurement for comparison as a lagging safety key performance indicator.

Better Safe than Sorry

A cliché meaning to take it easy, steer clear of any risks. It apparently dates from about the early nineteenth century, when it was "it's better to be sure than sorry." This version, which uses safe, has been used since about 1933.

Bicycle Safety

The Consumer Product Safety Commission (CPSC) estimates that there has been about 500,000 nonfatal bicycle-related injuries treated in emergency rooms every year since the early 1970s. About 30 percent of these incidents involved head or face injuries. Additionally, it is estimated that there may be as many as 1000 bicycle-related fatalities annually, and about 90 percent of these incidents involve motor vehicles. The CPSC bicycle standard (16 CFR Part 1512), effective in 1976, set safety requirements for reflectors, wheels, tires, chains, pedals, braking and steering systems, and for the frame and fork structural components. The CPSC Bicycle Helmet Standard (16 CFR 1203) sets forth the requirements for bicycle helmets (Figure B.1).

Biohazard

A combination of the words biological and hazard. A risk to humans presented by organisms or products of organisms.

Biohazard Area

The area, operating complex, facility, room in a facility, etc., in which work has been or is being performed with biohazard agents or materials.

Biohazard Control

A set of equipment and procedures used to prevent or minimize the exposure of humans and their environment to biohazardous agents or materials.

B

FIGURE B.1   Bicycle safety helmet.

Biological Agents
   Agents that act on or within the body to produce disease or acute and chronic infec-
   tion. Infections may be caused by bacteria, viruses, *Rickettsia*, *Chlamydia*, or fungi.
Biological Effects
   Reactions of the body due to exposure to physical, chemical, or biological agents.
Biological Effects of Radiation
   Any alteration in a biological system resulting from exposure to radiation. Effects
   may or may not be hazardous or detrimental to the system.
Biological Exposure Indices (BEIs®)
   Reference values intended as guidelines for the evaluation of potential health hazards
   in the practice of industrial hygiene. BEIs® represent the levels of determinants that
   are most likely to be observed in specimens obtained from a healthy individual who
   has been exposed to chemicals to the same extent as an individual with inhalation
   exposure to the Threshold Limit Value (TLV®). TLVs® and BEIs® are registered terms
   of the American Conference of Governmental Industrial Hygienists (ACGIH®). See
   also **Threshold Limit Values (TLVs®)**.
Biological Hazard Tag
   As defined by the Occupational Safety and Health Administration (OSHA) regulation
   1910.145 (f) (8) (i), biological hazard tags are used to identify the actual or potential presence of a biological hazard and to identify equipment, containers, rooms,
   experimental animals, or combinations thereof, that contain or are contaminated with
   hazardous biological agents.

Biological Monitoring
The direct quantitative analysis of expired air, body fluids, or tissue for the presence of the hazardous agent or its metabolites and/or evidence of biologic impairment quantified by the use of physiologic, psychometric, or biochemical tests.

B

Biological Safety
A specialized area with the goal of protecting workers from agents of disease, such as bacteria and viruses, by using containment, decontamination, and Personal Protective Equipment (PPE) procedures. It is also known as Biosafety.

Biomechanics
The application of mechanical laws to living structures, specifically to the locomotor systems of the human body.

Biomechanics, Occupational
Concerned with the mechanical properties of human tissue, particularly the response of tissue to mechanical stress. A major focus is the prevention of overexertion disorders of the lower back and upper extremities.

Blast
A transient change in gas density, pressure (both positive and negative), and velocity of the air surrounding an explosive point. The most common sources of blasts are from the ignition of semi- or unconfined vapor cloud explosions, detonation of high explosives, or the rupture of high-pressure vessels. The initial change can be either gradual or discontinuous. A discontinuous change is commonly referred to as a shock wave, and a gradual change as a pressure wave. See also **Explosion**.

BLEVE (Boiling Liquid Expanding Vapor Explosion)
A specific type of fireball that can occur as the result of the situation where a vessel containing a pressurized liquid comes in direct contact with external flame. As the liquid inside the vessel absorbs the heat of the external fire, the liquid begins to boil, increasing the pressure inside the vessel to the set pressure of the relief valve(s). The heat of the external fire will also be directed to portions of the vessel where the interior wall is not "wet" with the process liquid. Since the process liquid is not present to carry heat away from the vessel wall, the temperature in this region (usually near the interface of the boiling liquid) will rise dramatically, causing the vessel wall to overheat and become weak. A short time after the vessel wall begins to overheat, the vessel can lose its structural integrity and a rupture will occur. After vessel rupture, a fireball will usually result, with the external fire available as the ignition source.

Blind Spots
Are commonly referred to as areas in vehicles that cannot be directly observed under existing circumstances, due to limitations in conventional vehicle mirror arrangements. Blind spots exist in a wide range of vehicles, including cars, trucks, motorboats, and aircraft. Vehicle driving training instruction informs individuals how to compensate for the blind spot phenomenon. The locations most commonly referred to as blind spots in automobiles are the rear quarter areas on both sides of the vehicle. Vehicles in the adjacent lanes of the road may fall into these blind spots, and a driver may be unable to see them using only the car's mirrors. Other areas that are sometimes called blind spots are those that are too low to see behind and in front of a vehicle. Also, in cases where side vision is hindered, areas to the left or right can become blind spots. Blind spots can be eliminated in automobiles by overlapping side and rear-view mirrors, or checked by turning one's head briefly, or by adding another mirror with a larger field of view. Detection of vehicles or other objects in blind spots may also be aided by systems such as video cameras or distance sensors, though these are not common in most vehicles marketed to the general public.

B

Blinding

Method of process isolation to prevent unexpected release of materials from a pipe, line, or duct by inserting a solid plate or cap, which completely covers the bore and is capable of withstanding the maximum expected pressure from the upstream side. Blinding is considered a positive means of isolation. It may also be called blanking.

Blood-Borne Pathogens

Pathogenic microorganisms that are present in human blood and can cause disease in humans. These include, but are not limited to, hepatitis B virus (HBV) and human immunodeficiency virus (HIV).

Blowout

An uncontrolled flow of gas or oil, and combination of these or other well fluids from a wellbore at the well head or into the petroleum formation during a drilling operation. It is caused by formation pressure exceeding the drilling fluid pressure.

Blowout Preventer (BOP)

A mechanism to rapidly close and seal off a well borehole, which is used to prevent a well blowout from occurring. It consists of rams and shear rams, usually hydraulically operated, which are fitted at the top of the well being drilled. It is activated if well pressures are encountered that cannot be controlled by the drilling process systems (i.e., drilling mud injection) and could lead to a blowout of the well.

Blue Cross

Typically a symbol or icon used to represent health care or health insurance coverage plans. Sometimes also shown as or with a Blue Shield. The color blue is used since it has traditionally been associated as a symbol of healing or nurturing, and the cross is a religious symbol for protection. The Blue Cross is an independent, nonprofit, membership hospital plan. Benefits provided include coverage for hospitalization expenses subject to certain restrictions, outpatient services, and supplementary care such as nursing home care. See also **Blue Cross and Blue Shield Association (BCBSA); Blue Shield; Green Cross; Red Cross**.

Blue Cross and Blue Shield Association (BCBSA)

A federation of 39 health insurance companies. The subject insurers offer some form of health insurance coverage in every part of the United States. They also act as administrators of Medicare in many states or regions of the United States, and also provide group coverage to state government employees, as well as the U.S. federal government under a nationwide option of the Federal Employees Health Benefit Plan. Historically "Blue Cross" was used for hospital coverage insurance while "Blue Shield" was used for medical coverage insurance. But this distinction is only slight nowadays. The color blue has been used to symbolize healing and truth. A shield is a protective icon and the cross symbol for belief in faith. See also **Blue Shield; Green Cross; Red Cross**.

Blue Shield

Typically a symbol or icon used to represent health care or health insurance coverage plans. The color blue is used since it has traditionally been associated as a symbol of healing or nurturing and the shield symbolizes protection. Blue Cross is an independent, nonprofit, membership medical surgery plan. Benefits cover expenses associated with medical and surgical activities. See also **Blue Cross; Blue Cross and Blue Shield Association (BCBSA); Green Cross; Red Cross**.

Board of Certified Safety Professionals (BCSP)

A nonprofit corporation with the purpose of certifying practitioners in the safety profession. It was established in 1969 and is based in Savoy, Illinois, USA. Its

B

main duties consist of setting standards related to professional safety practices, evaluating the academic and professional experience qualifications of applicants, administering examinations relating to professional safety knowledge and skills, establishing recertification standards, and authorizing individuals meeting BCSP standards to use BCSP certifications and designations. It operates solely as a peer certification board. The BCSP provides certification as a Certified Safety Professional® (CSP®), Associate Safety Professional (ASP), and as a Graduate Safety Practitioner (GSP). In 2008, BCSP acquired the Council on Certification of Health, Environmental, and Safety Technologists (CCHEST). Begun in 1985 as a joint venture of BCSP and the American Board of Industrial Hygiene (ABIH), CCHEST is now a wholly owned division of BCSP offering three accredited certifications at the para-professional level. See also **American Board of Industrial Hygiene (ABIH)**; **Certified Safety Professional® (CSP®)**.

Bodily Injury (BI)
Injury to a human being, as opposed to injury to property.

Body Belt
A strap provided with a means for securing it about the waist and for attaching it with a lanyard to a lifeline or anchorage point. It is used to provide personnel positioning limits against a fall. They are considered a fall prevention device rather than a fall protection device. It may also be called a Safety Belt.

Body Burden
The total amount of toxic material that has entered a body through inhalation, ingestion, absorption through the skin, or injection over time. Body burdens are often expressed as milligram per kilogram of body weight.

Body Harness, Full
A design of straps that may be secured about an individual in a manner to distribute the fall arrest forces over the thighs, pelvis, waist, chest, and shoulders, with a means for attaching it to other components of a personal fall arrest system. See also **Fall Restraint System**; **Personal Fall Arrest System**.

Boiler and Machinery Insurance
Now referred to by the insurance industry as Equipment Breakdown Insurance. See **Equipment Breakdown Insurance**.

Boiler Codes
Regulations and standards prescribing requirements for the design, construction, testing, and installation of boilers and unfired pressure vessels (e.g., American Society of Mechanical Engineers [ASME], etc.).

Boilover
A boiling liquid eruption in a hydrocarbon or chemical storage tank. Usually described as an event in the burning of the contents of the tank, where after a long period of time, there is a sudden increase in fire intensity with the expulsion of the contents of the tank, due to water at the bottom of the tank being heated to vaporization and causing a boiling eruption.

Bollard
A thick, low, short, post, often of iron or steel and usually used in series, provided for the purpose of excluding or diverting motor vehicles from a road or lawn, or placed around critical equipment near vehicular traffic, and therefore somewhat considered a safety barrier against these vehicles.

Bonding System, Electrical
The interconnection of objects metallically, typically by a clamp or bare wire, to equalize the electrical potential between the objects to prevent an electrical

B

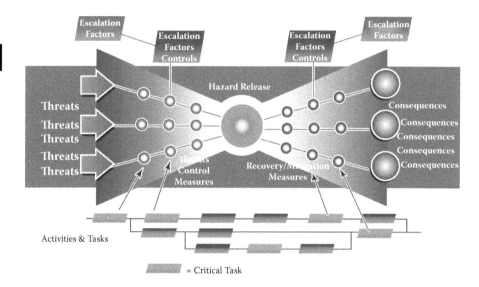

FIGURE B.2　Graphical depiction of Bow-Tie Analysis. (Image Courtesy of ABS Consulting.)

imbalance which may lead to spark, e.g., transferring of a flammable liquid from one container to another can release electrical energy if it is not bonded. See also **Grounding**.

Bow-Tie Analysis

A type of qualitative process hazard analysis (PHA). The Bow-Tie PHA methodology is an adaptation of three conventional system safety techniques: Fault Tree Analysis, Causal Factors Charting, and Event Tree Analysis. Existing safeguards (barriers) are identified and evaluated for adequacy. Additional protections are then determined and recommended where appropriate. Typical cause scenarios are identified and depicted on the pre-event side (left side) of the bow-tie diagram (see Figures B.2 and B.3). Credible consequences and scenario outcomes are depicted on the post-event side (right side) of the diagram, and associated barrier safeguards are included. One attribute of the Bow-Tie method is that in its visual form, it depicts the risks in ways that are readily understandable to all levels of operations and management. Bow-Tie reviews are most commonly used where there is a requirement to demonstrate that hazards are being controlled, and particularly where there is a need to illustrate the direct link between the controls and elements of the management system.

Brainstorming

An evaluation technique for gathering ideas, views, and suggestions from a group on a particular issue and problem solving. It endeavors to allow free creative thinking and builds on the ideas of others. It is typically employed in risk analysis such as a What If Analysis. Generally two phases are involved: a generation phase and a clarification phase. In the generation phase the team creates as many ideas as possible without hindrance. In the clarification phase the team ensures the ideas are understood and evaluated. See also **Hazard and Operability Study (HAZOP)**, **Safety Flowchart**; **What-If Analysis (WIA)**.

Breakthrough Time

Time from initial chemical contact to detection.

B

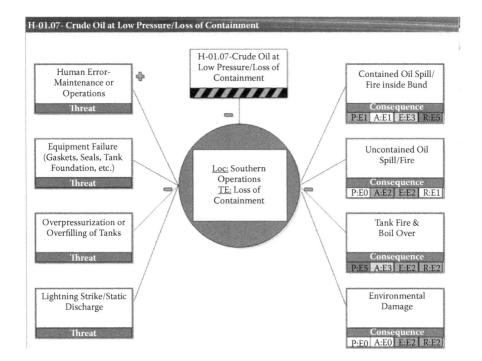

H-01.07- Crude Oil at Low Pressure/Loss of Containment

Human Error-Maintenance or Operations		H-01.07-Crude Oil at Low Pressure/Loss of Containment		Contained Oil Spill/Fire inside Bund
Threat				Consequence
				P:E1 A:E1 E:E3 R:E5

Equipment Failure (Gaskets, Seals, Tank Foundation, etc.)		Loc: Southern Operations TE: Loss of Containment		Uncontained Oil Spill/Fire
Threat				Consequence
				P:E0 A:E2 E:E2 R:E1

Overpressurization or Overfilling of Tanks				Tank Fire & Boil Over
Threat				Consequence
				P:E5 A:E3 E:E2 R:E2

Lightning Strike/Static Discharge				Environmental Damage
Threat				Consequence
				P:E0 A:E0 E:E2 R:E2

FIGURE B.3 Bow-Tie software. (Image Courtesy of ABS Consulting.)

Breathing Apparatus
   A device to replace the surrounding environment air supply and provide the user with sufficient independent air or oxygen supply to breathe normally. They can be of either the open circuit or closed circuit type. Open circuit supplies air via a lung governed demand valve or pressure reducer connected to a full face piece via a supply hose. The hose is connected to a compressed air supply carried by the user in a harness. In a closed circuit system, a purifier is used to absorb exhaled carbon dioxide, which having been purified is fed back to the respirator after mixing with pure oxygen. Air line breathing apparatus consists of a full or half mask connected to a compressed air line via a filter and a demand valve. See also **Self-Contained Breathing Apparatus (SCBA)**.

Breathing Zone
   An imaginary 61-cm (2-ft) radius globe that surrounds the head of an individual. The composition of air in this area is thought to be representative of the air that is actually breathed in by an individual.

Breathing Zone Sample
   An air sample collected in the breathing zone of employees to assess their exposure to airborne contaminants.

Brightness
   The light intensity of a surface in a given direction per unit of emulsive area as projected on a plane of the same direction. See also **Illumination**.

Broad Range Sensors
   Broad range sensors are sensors that only indicate that a hazardous threshold of a class of chemicals has been reached.

Buddy System

A system of organizing employees into work groups so that each employee in the work group is designated to be observed by at least one other employee in the work group. The purpose of the buddy system is to ensure each individual's safety while working in a hazardous environment and to provide rapid assistance to a responder in the event of an emergency.

Building Code

An assembly of regulations that sets forth the standards to which buildings must be constructed.

Bump Cap

A protective cap that provides bump and scrape protection for a variety of industries including food or beverage processing, auto repair, utility meter reading, pest control, and fruit harvesting. A bump cap is not intended for use where American National Standards Institute (ANSI) approved head protection (hard hat) is required. See also **ANSI Standard Z89.1-2003, Protective Headware for Industrial Workers; Hard Hat**.

Bump Test

See **Function Check**.

Bureau of Labor Statistics (BLS)

An agency of the federal government that collects information on labor economics and statistics. It is considered the preeminent source of injury and illness-related statistics in the United States. It publishes an annual report of workplace injuries and illnesses taken from employer reporting records. The report includes the rate and number of work-related injuries, illnesses, and fatal injuries, and how these statistics vary by incident, industry, geography, occupation, and other characteristics.

Burn Injury

An injury to the skin and deeper tissues caused by hot liquids, flames, radiant heat, and direct contact with hot solids, caustic chemicals, electricity, or electromagnetic (nuclear) radiation. A first-degree burn injury occurs with a skin temperature of about 48°C (118°F) and a second-degree burn injury occurs with a skin temperature of 55°C (131°F). Instantaneous skin destruction occurs at 72°C (162°F). Inhaling hot air or gases can also burn the upper respiratory tract. Approximately 2 million persons suffer serious burns in the United States each year; of these approximately 115,000 are hospitalized and 12,000 are fatal. The severity of a burn depends on its depth, its extent, and the age of the victim. Burns are classified by depth as first, second, and third degree. First-degree burns cause redness and pain (e.g., sunburn) and affect only the outer skin layer. Second-degree burns penetrate beneath the superficial skin layer and are marked by edema and blisters (e.g., scald by hot liquid). In third-degree burns, both the epidermis and dermis are destroyed, and underlying tissue may also be damaged. It has a charred or white leathery appearance and initially there may be a loss of sensation to the area. The extent of a burn is expressed as the percent of total skin surface that is injured. Individuals less than 1 year old and over 40 years old have a higher mortality rate than those between age 2 and 39 for burns of similar depth and extent. Inhalation of smoke from a fire also significantly increases mortality. Thermal destruction of the skin permits infection, which is the most common cause of death for extensively burned individuals. Body fluids and minerals are lost through the wound. The lungs, heart, liver, and kidneys will be affected by the infection and fluid loss. First aid for most burns is application of cool water as soon as possible after the burn. Burns of 15 percent of the body surface or less are usually treated in hospital emergency rooms by removing dead tissue (debridement), dressing with antibiotic cream (often silver sulfadiazine), and administering oral pain medication. Burns of 15 to 25 percent often require hospitalization to provide intravenous fluids and avoid complications. Burns of more than 25 percent are usually treated in specialized burn centers. Aggressive surgical management is directed toward early skin grafting and avoidance of such complications as dehydration, pneumonia, kidney failure, and infection. Pain control with intravenous narcotics is frequently required. The markedly increased metabolic rate of severely burned patients requires high-protein nutritional supplements given by mouth and intravenously. Extensive scarring of deep burns may cause disfigurement and limitation of joint motion. Plastic surgery is often required to reduce the effects of the scars. Psychological problems often result from scarring. Since over 50 percent of all burns are preventable (separation, barriers, protective clothing, etc.), safety programs can significantly reduce the incidence of burn injuries.

Burner Management System (BMS)

A control system dedicated to boiler-furnace safety, operator assistance in the starting and stopping of fuel preparation and burning equipment, and preventing mis-operation of and damage to fuel preparation and burning equipment.

Business Interruption (BI)

Anticipated or unanticipated disruption of the normal operations of an organization due to the occurrence of a peril. Insurance coverage (indemnification for the loss of profits and continued fixed costs) against business interruption by various perils can be obtained. See also **Business Interruption Insurance**; **Contingent Business Interruption Insurance**.

Business Interruption Insurance

A time element insurance coverage that pays for loss of earnings (i.e., net profit that would have been earned and the necessary expenses that continue during the "period of restoration"), when operations are suspended or curtailed because of property loss due to the insured peril. In determining the amount of loss, the insurer considers the insured's experience before the loss and probable future experience if no loss had occurred. See also **Contingent Business Interruption Insurance**.

C

Cage, Ladder Safety

A guard that is referred to as a cage or basket guard, which is an enclosure that is fastened to the side rails of a fixed ladder or to the structure to encircle the climbing space of the ladder for the safety of the person who must climb the ladder (see Figure C.1). Occupational Safety and Health Administration (OSHA), general industry regulations, 29 CFR 1910.27 (d) (1) (ii) and 29 CFR 1917118 (e) (1), state that a cage or well is required if the total length of the climb on a fixed ladder equals or exceeds 6.1 m (20 ft) and for OHSA construction regulations, 29 CFR 1926.1053 (a) (19), if the total length of ladder climb is 7.3 m (24 ft) or more. Otherwise the ladder must be equipped with ladder safety devices or self-retracting lifelines and rest platforms at intervals not to exceed 45.7 m (150 ft). A ladder with a cage or well must also be provided with multiple ladder sections, with each ladder section not to exceed 15.2 m (50 ft) in length. These ladder sections are to be offset from adjacent sections, and landing platforms must be provided at maximum intervals of 15.2 m (50 ft). See also **Ladder Safety Device**.

FIGURE C.1 Ladder cage.

Canister (Air-Purifying)

A container filled with sorbents and catalysts that remove gases and vapors from air drawn through the unit. The canister may also contain an aerosol (particulate) filter to remove solid or liquid particles. See also **Respirator**.

Canister Respirator

A form of gas respirator, which generally incorporates goggles and a visor, with an exhalation valve connected to a chemical canister filter. It normally provides protection against low concentrations of designated toxic gases and vapors. See also **Cartridge Respirator**; **Respirator**.

Carabiner

A self-closing, self-locking steel connector used to attach to an anchorage point. Used with a personal fall protection system.

Carcinogen, Carcinogenic, Carcinogenicity

A substance that can cause cancer. Carcinogenic means able to cause cancer. Carcinogenicity is the ability of a substance to cause cancer. There is no single definition of the evidence necessary to classify a substance as a carcinogen. Four sources of evidence are used: epidemiology, long-term animal testing, short-term tests (such as the AMES Test), and Structure-Activity Relationships. Separate guidelines and policies defining carcinogens have been used by the U.S. Environmental Protection Agency, the Occupational Safety and Health Administration (OSHA), the Consumer Product Safety Commission (CPSC), and the Food and Drug Administration (FDA). These agencies, and the International Agency for Research on Cancer (IARC), classify chemicals by the degree of evidence available as to their carcinogenicity. Carcinogens vary in their estimated ability to induce cancer. Under the U.S. OSHA Hazard Communication (Hazcom) Standard, materials are identified as carcinogens on Material Safety Data Sheets (MSDSs) if they are listed as either carcinogens or potential carcinogens by the IARC or the U.S. National Toxicology Program (NTP), if they are regulated as carcinogens by OSHA, or if there is valid scientific evidence in man or animals demonstrating a cancer causing potential. The lists of carcinogens published by the IARC, American Conference of Governmental Industrial Hygienists (ACGIH), and NTP include known human carcinogens and some materials that cause cancer in animal experiments. Certain chemicals may be listed as suspect or possible carcinogens if the evidence is limited or so variable that a definite conclusion cannot be made. See also **AMES Test**.

Carpal Tunnel Syndrome

A repetitive strain injury. It is caused by compression of the median nerve in the carpal tunnel. Often associated with tingling, pain, or numbness in the thumb and first three fingers. Carpal Tunnel Syndrome is the number one reported medical problem, accounting for about 50 percent of all work-related injuries. Carpal tunnel syndrome results in the highest number of days lost among all work-related injuries in the United States. Almost half of the carpal tunnel cases result in 31 days or more of work loss per the National Center for Health Statistics.

Cartridge, Air Purifying

A container with a filter, sorbent, or catalyst, or combination of these items, that removes specific contaminants from the air passed through the container.

Cartridge Respirator

A respirator that has a chemical cartridge filter and is effective for low concentrations of relatively non-toxic gases or vapors.

CAS Registry

The most authoritative collection of disclosed chemical substance information, containing more than 48 million organic and inorganic substances and 61 million sequences, that is maintained by the Chemical Abstracts Service (CAS) of the American Chemical Society (ACS). It covers substances identified from the scientific literature from 1957 to the present, with additional substances going back to the early 1900s. New substances identified by CAS scientists are continuously included in the database. The CAS maintains a 24-hour turnaround lookup service to identify CAS Registry Numbers. The CAS is the source and final authority for CAS Registry Numbers. It is recognized as the most authoritative collection of disclosed chemical substance information, including chemical structures, names, predicted and experimental properties, tags, and spectra.

CAS Registry Number

The CAS Registry Number is a number assigned to a material by the Chemical Abstracts Service (CAS) of the American Chemical Society (ACS). The CAS number provides a single unique identifier. A unique identifier is necessary because the same material can have many different names. For example, the name given to a specific chemical may vary from one language or country to another. The CAS Registry Number is similar to a telephone number and has no significance in terms of the chemical nature or hazards of the material. The CAS Registry Number can be used to locate additional information on the material, for example, when searching in books or chemical databases. It is used on Material Safety Data Sheets (MSDSs). The CAS registry number can contain up to 10 digits, divided by hyphens into three parts, the first consisting of up to seven digits, the second consisting of two digits, and the third consisting of a single digit serving as a check digit. The check digit is used to verify the validity and uniqueness of the number through a specific calculation applied to the entire number. The numbers are assigned in increasing order and do not have any inherent meaning. There are more than 48 million organic and inorganic substances and more than 61 million sequences in the CAS registry. The CAS also maintains and sells a database of these chemicals, known as the CAS registry. It may also be called CAS Number, CAS RNs, or CAS #. See also **UN Number**.

Cascade Air Breathing System

An arrangement of interconnected air cylinders that provide an increased supply of breathing air for individuals using a supplied air respiratory protection system.

Casual Connection

An act, agency, or force occurring without design or without being foreseen or expected that is a concurrent or contributing factor to the injury, but is not usually the proximate cause thereof.

Casualty Insurance

Insurance written by companies licensed under the casualty sections of state insurance laws as distinguished from that written under the fire or marine or life insurance sections. This type of coverage is principally concerned with insurance against loss due to an incident or other mishap.

Catastrophe

A loss of extraordinarily large dimensions in terms of injury, illness, death, damage, and destruction.

Catastrophic

A loss of extraordinary magnitude in terms of physical harm or distress to individuals, damage and destruction of property, and impact to the environment or continued business operations.

Causal Association

A statistical association between the occurrence of a factor and a disease or injury, in which available evidence indicates that the statistically associated factor increases the probability of occurrence of the disease or injury and that its removal decreases the probability of occurrence.

Causal Factor (CF)

The immediate major contributor to the occurrence of an incident. A combination of simultaneous or sequential circumstances directly or indirectly contributing to an incident that if they were removed would have either prevented the incident or reduced its consequences. For a typical incident event there are multiple causal factors. Causal factors were formerly called direct causes, key causes, observable causes, and assignable causes. Can be modified to identify several kinds of causes such as direct, early, mediate, proximate, distal, etc. See also **Tier Diagramming**; **Root Cause Analysis (RCA)**.

Causal Factors Chart (CFC)

The Causal Factors Chart is a formal, and systematic, incident investigation and root cause analysis technique. The technique depicts the events and conditions leading up to an incident. It combines critical thinking, logical analysis, and graphic representations to analyze and depict an incident event scenario. It helps structure the analysis and data gathering processes to ensure necessary and sufficient information is collected. The CFC also has been applied to Root Cause Analysis. The CFC is sometimes referred to as the Events and Causal Factors (ECF) chart. The ECF chart depicts the necessary and sufficient events and causal factors associated with a specific incident scenario.

Cause

An event, situation, or condition that could or does result directly or indirectly in an incident.

Cause-and-Effect Diagram

A method used to illustrate and pinpoint the likely causes of a concern. It is sometimes referred to as a Root Cause Analysis. It typically uses an Ishikawa (wish bone) or tree diagram (see Figure C.2), where the main concern is depicted at the head of the fish spine and the most likely causes are shown as the attached bones. When identified the root cause can be modified to eliminate or reduce the concern. See also **Root Cause Analysis (RCA)**.

Cause-Consequence Analysis (CCA)

A quantitative risk review technique. Cause-consequence analysis is a blend of fault tree and event tree analysis. This technique combines cause analysis (described by fault trees) and consequence analysis (described by event trees), and hence deductive and inductive analysis is used. The purpose of CCA is to identify chains of events that can result in undesirable consequences. With the probabilities of the various events in the CCA diagram, the probabilities of the various consequences can be calculated, thus establishing the risk level of the system. See also **Event Tree Analysis (ETA)**; **Fault Tree Analysis (FTA)**.

Caution

Forethought, alertness, or restraint in a hazardous situation to minimize risk. It may also be referred to as a warning against danger.

Caution Label or Sign

As defined by American National Standards Institute (ANSI) Z535.4, a sign or label to indicate a hazardous situation that if not avoided, could result in a minor or moderate injury. Minor burns, cuts, scratches, and pinch points that result in bruises, and minor chemical irritation are examples where caution identification is used. Caution is to be indicated in black letters on a yellow background. See also **Danger Label or Sign**; **Notice Label or Sign**; **Safety Alert Symbol**; **Safety Sign**; **Warning Label or Sign**.

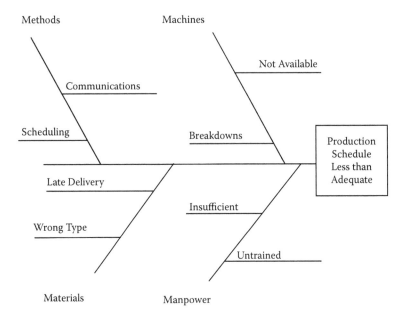

FIGURE C.2 Cause-and-effect diagram.

Caution Tag
  As defined by Occupational Safety and Health Administration (OSHA) regulation
  29 CFR 1910 (f) (6), a tag used in minor hazard situations where a non-immediate or
  potential hazard or unsafe practice presents a lesser threat of employee injury. See
  also **Danger Tag; Warning Tag**.

Cave-In
  The separation of a mass of soil or rock material from the side of an excavation, or
  the loss of soil from a support system, and its sudden movement into the excavation,
  either by falling or sliding, in sufficient quantity so that it could entrap, bury, or otherwise injure and immobilize an individual. See also **Distress**.

Ceiling Level or Limit
  Ceiling limit is an airborne concentration of a toxic substance in the work environment that should not be exceeded. In threshold limit value (TLV) and permissible
  exposure limit (PEL), or recommended exposure limit (REL), the maximum allowable concentration to which an employee may be exposed. If instantaneous monitoring is not feasible, then the ceiling is a 15-minute time-weighted average exposure not
  to be exceeded at any time during the working day.

Ceiling Value, C
  The concentration that should not be exceeded during any part of the working exposure. "An employee's exposure (to a hazardous material) shall at no time exceed the
  ceiling value" (OSHA).

Census of Fatal Occupational Injuries (CFOI)
  An annual report issued by the Bureau of Labor Statistics on the occupational fatalities that occur in industry. The CFOI Program has compiled a count of all fatal work
  injuries occurring in the United States since 1992, by using diverse data sources to
  identify, verify, and profile fatal work injuries.

C

Center for Chemical Process Safety (CCPS)

An organization formed after the Bhopal, India, incident in 1984 by the AIChE with member companies to lead a collaborative effort to eliminate catastrophic process incidents by advancing state of the art technology and management practices, serving as the premier resource for information on process safety, supporting process safety in engineering, and promoting process safety as a key industry value. Over 100 members now participate in CCPS. It now markets over 100 books and products, sponsors international conferences, and cultivates the Safety in Chemical Engineering Education university curriculum program.

Centers for Disease Control and Prevention

The federal agency charged with tracking and investigating public health trends. The stated mission of the Centers for Disease Control and Prevention, commonly called the CDC, is "[t]o promote health and quality of life by preventing and controlling disease, injury, and disability." The CDC is a part of the U.S. Public Health Services (PHS) under the Department of Health and Human Services (HHS).

CERCLA (Comprehensive Environmental Response, Compensation, and Liability Act)

A federal law enacted in 1980 to address the past disposal and clean up of inactive and abandoned hazardous waste sites. Formerly it was Public Law PL 96-510, which is stated in 40 CFR 300. It is administered by the Environmental Protection Agency. Also referred to as the Superfund Law.

Certified Functional Safety Expert/Professional (CFSE/CFSP)

Qualifications for safety engineers in process applications, machine applications, hardware, or software that demonstrates competence in safety lifecycle activities. These qualifications are administered by the non-profit CFSE Governance managed by a global consortium of vendor, user, integrator, and consultant companies.

Certified Industrial Hygienist (CIH)

An individual who has met the minimum requirements for education and experience and, through examination, has demonstrated a minimum level of knowledge in subject matters that include air sampling and instrumentation, analytical chemistry, basic science, biohazards, biostatistics and epidemiology, community exposure, engineering controls and ventilation, ergonomics, health risk analysis and hazard communication, management, noise, non-engineering controls, radiation (ionizing and non-ionizing), thermal stressors, toxicology, and work environments and industrial processes. The certification is administered and maintained through the American Board of Industrial Hygiene.

Certified Safety and Health Manager (CSHM)

An individual who has met the minimum requirements for education and experience and, through examination, has demonstrated a minimum level of knowledge of safety management through the application of management principles and the integration of safety into all levels and activities of an organization. The certification is administered and maintained through the Institute for Safety and Health Management. See also **Institute for Safety and Health Management (ISHM)**.

Certified Safety Professional® (CSP®)

A recognition title of an individual awarded by the Board of Certified Safety Professionals (BCSP) based on the individual's knowledge and experience as demonstrated to the board by the individual's application documentation and an examination based on safety fundamentals and/or comprehensive practice. Knowledge is demonstrated by educational degree accepted by the board and experience by demonstrated acceptable professional safety practice. Certified Safety Professional® and CSP® are registered certification marks awarded to BCSP by the U.S. Patent and Trademark Office. See also **Board of Certified Safety Professionals**.

CFR (Code of Federal Regulations)
Regulations that are produced under U.S. law. The CFRs are segregated into subject areas, e.g., Title 29 contains Occupational Safety and Health Administration (OSHA) regulations, Title 40 contains Environmental Protection Agency (EPA) regulations, Title 49 Department of Transport (DOT) contains regulations, etc.

Chain of Causation
The original force is responsible for every subsequent force which it puts in motion, and for the final result.

Chain Saw Protection
See **ASTM F 1818, Specification for Foot Protection for Chainsaw Users**.

Change Analysis
An investigative technique to identify all changes as well as the results of those changes. The distinction is important, because identifying only the results of change may not prompt investigators to identify all causal factors of an incident. The individual performing a change analysis systematically identifies specific elements or differences that caused the outcome of a certain task to deviate from the anticipated outcome. Change is one of the most important factors in the cause of incidents. Change is anything that disturbs the "balance" of a system operating as planned. Change is often the source of deviations in system operations. Change can be planned, anticipated, and desired, or it can be unintentional and unwanted. It is an integral and necessary part of daily business, for example, requirements change, procedures change, policies and directives change, the personnel performing certain tasks change (i.e., personnel turnover). Change can improve efficiency, productivity, and safety, or it can result in errors, loss of control, and incidents. See also **Management of Change (MOC)**.

Chartered Property and Casualty Underwriter (CPCU)
A professional designation in property-casualty insurance and risk management. The certification is based on an evaluation of experience (2 years), coursework, ethics requirement, and an examination. The curriculum includes eight courses, which generally include insurance law, history, contracts, ratemaking, risk management, finance, corporate structure, and ethics. Approximately 65,000 people have earned the designation since its inception in 1942. The prominent designees are insurance agents and brokers, insurance agency principals, claim representatives, line of business managers and executives, insurance litigators, risk managers, and underwriters. The CPCU designation is administered by the American Institute for Chartered Property and Casualty Underwriters.

Checklist, Safety
See **Safety Checklist**.

Chemical Asphyxiants
Chemical asphyxiants are a special category of toxin. They render the body incapable of using an adequate supply of oxygen.

Chemical Burns
Generally similar to burns caused by heat. Treatment is also similar. See also **Burn Injury**.

Chemical Cartridge
A filtering mechanism used with a respirator for the removal of low concentrations of specific vapors and gases. It is constructed mechanically as a cartridge for ease of replacement after use or expiry date.

Chemical Cartridge Respirator
A respirator that uses various chemical substances to purify the inhaled air of certain vapors and gases. See also **Cartridge, Air Purifying**.

Chemical Hazard Label

A label applied to containers of dangerous chemical compounds to indicate the specific risk, and thus the required precautions. There are several systems of labels. In the United States, National Fire Protection Association (NFPA) standard 704 is used to specify the label arrangement for emergency response. It consists of a diamond placard with four colored sections each with a number indicating severity 0–4 (0 for no hazard, 4 indicates a severe hazard). The red section denotes flammability. Blue is health risks. Yellow is reactivity (tendency to explode). The white section denotes special hazard information. See Figure C.3 for a depiction of this label. For workplace locations the Hazardous Materials Identification System (HMIS) is typically used, which consists of a sign with four different colored horizontal bars and severity number at the end of the top three bars for Health (blue), Flammability (red), and physical Hazard (orange). The bottom white bar is used to indicate the Personal Protection Equipment (PPE) needed. See also **Fire Hazard Identification**; **Hazardous Materials Identification System (HMIS®)**; **NFPA 704, Standard System for the Identification of the Hazards of Materials for Emergency Response**.

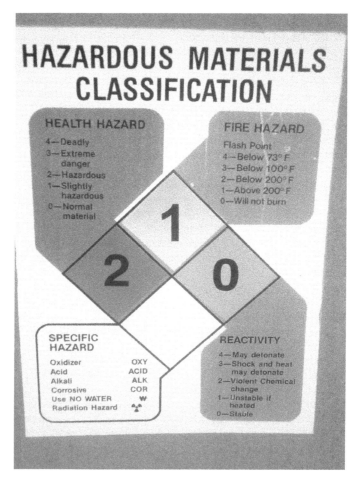

FIGURE C.3   Chemical hazard label.

C

Chemical Hygiene Plan (CHP)
Per Occupational Safety and Health Administration (OSHA) regulation 29 CFR 1910.1450, "Occupational Exposures to Hazardous Chemicals in Laboratories," is a written plan that includes specific work practices, standard operating procedures, equipment, engineering controls, and policies to ensure that employees are protected from hazardous exposure levels to all potentially hazardous chemicals in use in their work areas. The OSHA standard provides for training, employee access to information, medical consultations, examinations, hazard identification procedures, respirator use, and record-keeping practices.

Chemical Safety and Hazard Investigation Board (CSB)
An independent agency of the U.S. government chartered to investigate chemical industry incidents, determine their root cause, and publish their findings to prevent similar incidents occurring. The CSB was authorized by the Clean Air Act Amendments of 1990 and became operational in January 1998. It also collaborates with the Environmental Protection Agency (EPA), Occupational Safety and Health Administration (OSHA), and other agencies. The Board has entered into a number of memorandums of understanding (MOUs) that define the terms of collaboration. For example, in cases where several agencies are conducting investigations of a particular incident, the MOUs outline mechanisms for coordination in the field. The goal of the MOUs is to allow each agency to carry out its statutory mission efficiently and without unnecessary duplication of effort.

Chemical Splash Goggles
Goggles that provide protection to the eyes from harmful liquids such as chemicals, acids, or paint. They provide a tighter fit than safety glasses for maximum protection. They are usually designed to fit over most glasses and are typically provided with a vent feature to prevent fogging. They must meet American National Standards Institute (ANSI) Z87.1 standards. See also **ANSI Z87.1-2003, Standard for Occupational and Educational Eye and Face Protection Devices**.

CHEMTREC (Chemical Transportation Emergency Center)
A national emergency center administered by the American Chemistry Council to provide pertinent vital information concerning specific chemicals upon request by individuals. CHEMTREC provides a toll-free 24-hour telephone number (800-424-9300) to assist in responding to chemical transportation emergencies for companies that have registered with them for this service. They have responded to over one million emergencies since this service was offered in 1971. They also offer assistance to shippers of hazardous materials for compliance with government regulations, such as U.S. Department of Transportation regulation 49 CFR 172.604.

Chronic
Chronic means long-term or prolonged. It can describe either an exposure or a health effect. A chronic exposure is a long-term exposure. Long-term means lasting for months or for years.

Chronic Exposure
Repeated exposure to low-level concentrations of hazardous chemicals, whereby the symptoms are usually delayed and cumulative.

Chronic Health Effect
An adverse health effect resulting from long-term exposure or a persistent adverse health effect resulting from a short-term exposure. See also **Acute**.

Chronic Toxicity
Adverse health effects resulting from repeated does of or exposures to a substance over a relatively prolonged period of time.

Chronic Violator

The chronic or persistent violator is the individual who repeatedly violates established statutes and ordinances as in the case of Occupational Safety and Health Administration (OSHA) regulations, or in the case of company safety rules.

Circuit Breaker

A safety device for electrical circuits designed to open the circuit, i.e., from on to off position, when abnormal conditions occur (i.e., overcurrent, abnormal voltage, high temperature, grounding, etc.) to prevent damage or overheating to the system and the possible occurrence of a fire. They are usually designed to permit opening and closing of the circuits manually but will automatically open the circuit during the occurrence of an abnormal condition. The term started to be used circa 1872. Circuit breakers should be of the type and rating for the circuit or appliance it is intended to protect. See also **Fuse, Electrical**; **Overload, Electrical**.

Circumstances (of an Accident)

The set of conditions that surround an incident or led to it.

Citation

Issued by the representative of the Assistant Secretary of Labor, the Occupational Safety and Health Administration (OSHA) Area Director, which alleges conditions that violate specific maritime, construction, or general industry occupational safety and health standards (OSH Act). If OSHA issues a citation and notification of penalty, the citation should describe the specifics of the alleged violation, fix a reasonable time for abatement, and propose alternate penalties. When an employer receives a citation under the OSH Act, it must post an unedited copy of it at or near each place an alleged violation occurred or where affected employees can see it even if the employer is planning to contest it. The citation must be posted for three days or until the violation is abated, whichever is longer. See also **De Minimus Citation**; **OSHA Violation(s)**.

Claim

The amount a policyholder believes he or she has coming from an insurance company as the result of some occurrence insured against. It may also refer to a broad, comprehensive term whose meanings include, but are not limited to, cause of suit action, judgment, right, or a demand for compensation or for payment of medical expenses.

Claim Form

Paperwork used to report a work injury or illness to the employer.

Claimant

A person who makes an insurance claim.

Classified

Products or materials that are specified for use which meet standard test requirements for fire safety concerns. Standard test conditions are usually set by national or local code requirements and verified through independent testing laboratories (e.g., UL or FM). See also **Approved**; **Factory Mutual**; **Labeled**; **Listed**; **Underwriters Laboratories (UL)**.

Classified Area

An area or a zone defined as a three-dimensional space in which a flammable atmosphere is or may be expected to be present in such frequencies as to require special precautions and restrictions for the construction and use of electrical apparatus and hot surface exposures (e.g., lights) that act as an ignition source. Classified areas have specific restrictions based on the equipment involved, the gases encountered, and the probability of leakage. Several national and industry institutions have guidelines and codes for the classification of areas, e.g., defined in the United States by the National

Electrical Code (NFPA 70), Flammable and Combustible Liquids Code (NFPA 30), American Petroleum Institute (RP 500), etc. See also **Explosionproof**; **Hazardous Area, Electrical**; **Intrinsically Safe**.

Code(s)

Rules and standards that have been adopted by a governmental agency as mandatory regulations having the force and effect of law. It is also used to describe a body of standards.

Code of Safe Practices

Workplace rules on how to perform duties safely and keep the worksite safe; must be specific to the employer's operations and posted at each job site.

Cold Work

Work activities that do not produce sufficient energy to ignite flammable atmospheres or combustible materials, but could contribute to an incident where an injury may occur. Common examples of cold work activities include demolition; removal of asbestos-containing materials; work involving cryogenic materials; movement of oversized loads on roads; maintenance work at steam, sewage, utility plants not involving open flames; seal or gland replacement; and repairs on pumps or compressors. See also **Hot Work**.

Cold Work Permit

A safety process control for cold work activity to ensure proper initiation, review, and execution. See also **Hot Work**; **Work Permit**.

Collection Efficiency

The percentage of a specific substance removed and retained from the air by an air cleaning or sampling device. It is a measure of the performance of the cleaner or sampling device.

Collision Diagram

A detailed representative sketch of an intersection or section of roadway where a motor vehicle incident has occurred. The diagram shows the manner of collision, and the resting positions of vehicles and other items after the collision occurred by the use of designated symbols. It is prepared to support the incident investigation report and for traffic improvement studies by highway agencies. Its primary graphic consideration is to properly indicate the direction of original travel, coupled with a line indicating the path a vehicle would have followed, angle of collision, and location of the crash compared to the roadway. Other information important to the crash is included such as date, time of day, roadway/weather conditions, crash severity, etc. It is sometimes referred to as a traffic collision diagram.

Color Coding

A system of color identification for the marking of physical hazards, highlighting of safety equipment or areas, provision of safety signs, and similar safety functions. Specific color codes are employed in the identification of alarms, piping, compressed gas cylinders, electrical wiring, fire sprinkler temperature ratings, etc. American National Standards Institute (ANSI) standard 535.2 specifies specific safety color applications. See Table C.1. See also **ANSI Z535.2, Environmental and Facility Safety Signs**.

Combustible

In a general sense, any material capable of burning, generally in air under normal conditions of ambient temperature and pressure, unless otherwise specified. This implies a lower degree of flammability. Although there is no general industry distinction between a material that is flammable and one that is combustible (NFPA 30, Flammable and Combustible Liquids Code, defines the difference between the classification of combustible liquids and flammable liquids based on flash point temperatures

TABLE C.1
Example of Industrial Color Coding Practices

Color	Application
Red	• Danger signs
	• Stop buttons or electrical switches uses for emergency stopping of machines
	• Emergency stop handles or bars on machines
	• Hazardous operating indication lights on controls, alarm panels, or in the installation
	• Fire protection equipment and systems (e.g., fire extinguishers, hose reels, alarms, etc.)
	• Stop condition
Orange	• Warning signs
	• Marking of guards for machines
	• Wind socks
	• Personal flotation devices and lifeboats
Yellow	• Caution signs
	• Highlighting physical hazards (e.g., yellow and black striping, hazard alerting signs)
	• Cabinets for flammable liquid storage
	• Marking of containers or corrosive or unstable materials
	• Caution condition
	• Traffic or road markings
Green	• Safety instruction and safety equipment location signs
	• Marking of safety equipment (e.g., stretchers, first aid kit)
	• Marking of emergency egress and evacuation routes
	• Control panel indication of a safe status of an operating mechanism
	• Safety showers and eyewashes
	• Electrical grounding conductors
	• Safe or acceptable condition indicators
Blue	• Notice signs
	• Advisory, informational, and instructional signs
	• Mandatory action signs
White	• Road markings
	• Medical or fire suppression emergency vehicles

and vapor pressures). Combustion can also occur in cases where an oxidizer other than the oxygen in air is present, e.g., chlorine, fluorine, or chemicals containing oxygen in their structure. Combustible is a relative term as many materials will burn under one set of conditions and will not burn under others. The term combustible does not indicate the ease of ignition for a material, the burning intensity, or the rate of burning. Only when the term is modified by adjectives such as highly as in highly combustible will a distinction be made in common terminology. Therefore, additional characteristics are required to define the degree of combustibility, such as flame spread and smoke developed, are selected for building construction to reduce the risk of fire.

Combustible Dust

Normally referred to as dust with a diameter of 420 microns or smaller and can be a fire or explosion hazard when it is dispersed in the proper concentration and ignited in the air or with any other available gaseous oxidizer. Each particular dust has its own flammability limits. Materials that may form combustible dust include metals (such as aluminum and magnesium), wood, coal, plastics, biosolids, sugar, paper,

C

soap, dried blood, and certain textiles. A combustible dust explosion hazard may exist in a variety of industries, including food (e.g., candy, sugar, spice, starch, flour, feed), grain, tobacco, plastics, wood, paper, pulp, rubber, furniture, textiles, pesticides, pharmaceuticals, dyes, coal, metals (e.g., aluminum, chromium, iron, magnesium, and zinc), and fossil fuel power generation.

Combustible Gas Detector
An instrument designed to detect the presence or concentration of combustible gases or vapors in the atmosphere. It is usually calibrated to indicate the concentration of a gas as a percentage of its Lower Explosive Limit (LEL) so that a reading of 100 percent indicates that the LEL limit has been reached. They use either a solid-state circuit, infrared (IR) beam electrochemical or duel catalytic bead for the detection of gas in an area. Portable monitors are used for personnel protection, and fixed installations are provided for property protection. The Instrument Society of America (ISA) has provided a guideline for combustible gas detector utilization, ISA TR 12.13.03, Guide for Combustible Gas Detection as a Method of Protection.

Combustible Liquid
As generally defined by National Fire Protection Association (NFPA), 30, Flammable and Combustible Liquids Code, it is any liquid that has a closed-cup flash point defined as at or above 37.8°C (100°F). Combustible liquids are classified as Class II or Class III and flammable liquids are classed as IA, IB, or IC. Class II liquid: Any liquid tested with a flash point defined as at or above 37.8°C (100°F) and below 60°C (140°F). Class IIIA: Any liquid tested with a flash point defined as at or above 60°C (140°F), but below 93°C (200°F). Class IIIB: Any liquid tested with a flash point defined as at or above 93°C (200°F). See also **Flammable Liquid**.

Combustion
A rapid exothermic chemical process. It involves the evolution of radiation effects, i.e., heat and light, as a result of a reaction of an oxidizer (usually oxygen in air) with an oxidizing material. Other physical phenomena that sometimes occur during combustion reactions are explosion and detonation. Combustion is generally defined as oxidation accomplished by burning. See also **Fire Triangle**; **Fire Tetrahedron**.

Common Cause Failure
Failure of two or more structures, systems, and components due to a single specific event or cause. Typical examples include a design deficiency, a manufacturing deficiency, operation and maintenance errors, a natural phenomenon, a human-induced event, saturation of signals, or an unintended cascading effect from any other operation or failure within the plant or from a change in ambient conditions.

Common Mode Failure
Failure of two or more structures, systems, and components in the same manner or mode due to a single event or cause. Common mode failure is a type of common cause failure in which the structures, systems, and components fail in the same way. It is different from a systematic failure in that it is random and probabilistic but does not proceed in a fixed, predictable cause and effect fashion. See **Systematic Failure**.

Company Rules
An internally developed set of standards regarding company policies and requirements for safety and general conduct.

Compensable Injury
An occupational injury or illness resulting in sufficient disability to require the payment of compensation as prescribed by law. A work injury or illness for which compensation benefits are payable to the worker or beneficiary under worker compensation laws.

Compensation, Injury

Indemnity paid to an employee for disability sustained in an occupational incident.

Competent Person

An individual who is capable by experience or education or both to identify, analyze, and determine corrective measures for a hazard. An Occupational Safety and Health Administration (OSHA) "competent person" is defined in 29 CFR 1926.32 (f), as "one who is capable of identifying existing and predictable hazards in the surroundings or working conditions which are unsanitary, hazardous, or dangerous to employees, and who has authorization to take prompt corrective measures to eliminate them."

Compliance

Conforming to mandatory and voluntary regulations and standards, incident and injury histories, the courts, and custom and practices.

Comprehensive General Liability Insurance

Insurance that covers the insured against all liability for all general liability hazards, including unknown hazards, unless excluded by the policy. Examples of exposures covered are premise and operations, products and completed operations, independent contractors, and designated contractors.

Compulsory Insurance

A form of insurance required by law, e.g., workers' compensation, vehicle liability.

Condition Monitoring

A system for evaluating the physical state of machinery and equipment over time, to prevent a failure that could lead to an incident. Readings and parameters are plotted to give a characteristic trend. The objective is to identify evidence of deterioration at an early stage so that corrective maintenance can be undertaken before failure. There are a range of techniques that may be employed that include strain measurement, corrosion monitoring, vibration measurement, and acoustic emission monitoring. It is considered a type of preventive maintenance or predictive maintenance.

Conductive Hearing Loss

A type of hearing loss due to any disorder in the middle or external ear that prevents sound from reaching the inner ear. It is not caused by noise exposure.

Cone, Safety

A temporary movable warning device to highlight or alert individuals to a hazard on the ground. They are usually cone shaped and composed of a heavy plastic, and are a bright, highly visible orange color, reflective or fluorescent. They may also have a reflective strip to further increase their visibility to vehicle traffic. They are frequently utilized to alert for temporary hazards located in roadways or in pedestrian walkways. Traffic cones were originally invented in 1914 by Charles P. Rudabaker. They were first manufactured from concrete. Not all traffic cones are conical; pillar-shaped movable bollards fulfilling a similar function are often called by the same name. Sizes of 30.5 cm (12 inches) are usually utilized indoors; larger sizes are used outdoors. They are also referred to as toddlers, road cones, traffic cones, construction cones, pylons, or Witches' Hats.

Confined Space

A space that by design has limited openings for entry and exit, unfavorable natural ventilation, contains or produces dangerous air contaminants, and is not intended for continuous employee occupancy. Occupational Safety and Health Administration (OSHA) uses the term "permit-required confined space" (permit space) to describe a confined space that has one or more of the following characteristics: contains or has the potential to contain a hazardous atmosphere; contains a material that has the potential to engulf an entrant; has walls that converge inward or floors that slope downward and taper into a smaller area which could trap or asphyxiate an entrant; or contains any other recognized safety or health hazard, such as unguarded machinery, exposed live wires, or heat stress. Confined spaces include storage tanks, compartments of ships, process vessels, pits, silos, vats, degreasers, reaction vessels, boilers, ventilation and other exhaust ducts, sewers, tunnels, underground utility vaults, and pipelines.

Confined Space Hazard

See **Confined Space**.

Conflagration

A fire extending over a considerable area, and destroying numbers of buildings and/ or substantial amounts of property.

Consequence

The direct undesirable result of an incident sequence usually involving a fire, explosion, or release of toxic material. Consequence descriptions may include estimates of the effects of an incident in terms of factors such as health impacts, physical destruction, environmental damage, business interruption, and public reaction or company prestige (Table C.2). See also **Severity**.

Consequence Analysis

A type of risk analysis that considers the particular effects of a particular process failure and the damage caused by these effects. It is undertaken to evaluate potentially serious hazardous outcomes of incidents and their possible consequences for individuals and the environment.

Constraint

A restriction or a compelling force affecting freedom or action. Forcing into a holding within close bounds. An operational condition that may necessitate work performance in a less than ideal, potentially unsafe environment (e.g., building construction) and that therefore requires the provision of special safeguards.

Construction Safety

National Institute for Occupational Safety and Health (NIOSH) records indicate that between 1992 and 2005 falls to lower level, electrocution, highway incidents, and being struck by an object were the leading causes of death in construction. One-third of the fall-related deaths were falls from a roof, 18 percent were falls from scaffolding, and 16 percent were falls from ladders. Electrocutions accounted for 9 percent of the deaths in construction in 2005. The main cause of electrocution for electrical workers was direct contact with energized equipment and wiring, while over half the electrocutions of non-electrical workers were caused by contact with overhead power lines with objects including ladders, poles, and cranes.

Construction Safety Council (CSC)

A not-for-profit organization dedicated to the advancement of safety and health interests in the field of construction throughout the world. The council was founded in 1989. Its mission is to reduce the tragic and costly incidents, injuries, and illnesses to construction workers in the United States. It offers training, educational publications, and conducts research and offers consultation on construction safety matters.

Construction Safety Orders (CSO)

Occupational Safety and Health Administration (OSHA) regulations that are specific to construction operations and hazard controls.

Construction Safety Standards

Generally refers to Occupational Safety and Health Administration (OSHA) standards and requirements that are defined in 29 CFR Part 1926 Safety and Health Regulations for Construction. See also **General Industry Safety Standards**.

C

TABLE C.2
Typical Consequence Rating Descriptions

Rating	Consequence (Severity)
1	• Minor on-site injuries (first aid and non-disabling, reportable injuries) • Property damage less than base level amount • Minor environmental impact (no remediation) • Loss of production less than base level amount • No offsite impact or damage. No public concern or media interest
2	• Serious on-site injuries (temporary disabling worker injuries) • Property damage 1 to 20 times base level • Moderate environmental impact (cleanup or remediation in less than 1 week and no lasting impact on food chain, terrestrial life, or aquatic life) • Loss of production from 1 to 20 times base level • Minor off-site impact (public nuisance, noise, smoke, odor, traffic) • Potential adverse public reaction. Some media awareness
3	• Permanent disabling on-site injuries or possible fatality • Property damage 20 to 50 times base level • Significant environmental impact (cleanup or remediation less than 1 month and minor impact on food chain, terrestrial life or aquatic life) • Loss of production from 20 to 50 times base level • Moderate off-site impact limited to property damage, minor health effects to the public or first aid injuries • Adverse public reaction. Local media concern
4	• Onsite fatality or less than four permanent disabling worker injuries • Property damage 50 to 200 times base level • Serious environmental impact (cleanup or remediation requires 3 to 6 months and moderate impact on food chain, terrestrial life, and/or aquatic life) • Loss of production up from 50 to 200 times base level • Significant off-site impact property damage, short-term health effects to the public or temporary disabling injuries • Significant public concern or reaction. National media concern
5	• Multiple on-site fatalities or four or more permanent disabling on-site injuries • Property damage greater than 200 times base level • Extensive environmental impact (cleanup or remediation exceeding 6 months, significant loss of terrestrial or aquatic life, or damage to food chain uncertain) • Loss of production greater than 200 times base level • Severe off-site impact property damage, off-site fatality, long-term health effect, or disabling injuries • Severe adverse public reaction threatening facility continued operations. International media concern

Consumer Products Safety Act of 1972
  A U.S. law, Title 16 CFR, Parts 1101 to 1406, that established mandatory notification
  by manufacturers of products and the distributors of these products to the Consumer
  Product Safety Commission (CPSC) in the event they become aware of a faulty
  product, or part of that product, that could result in bodily injury or product damage. The act also allows CPSC to ban a product if there is no feasible alternative. See
  also **Consumer Products Safety Improvement Act of 2008**; **Consumer Products
  Safety Commission (CPSC)**.

Consumer Products Safety Commission (CPSC)
  U.S. regulatory agency that protects the public against unreasonable risks of injuries
  or deaths associated with consumer products. The CPSC investigates products to
  determine if they pose a fire, electrical, chemical, or mechanical hazard, or can injure
  children. They have estimated that their evaluations have significantly contributed
  to a 30 percent decline in the rate of death and injury associated with consumer
  products over the last 30 years. See also **Consumer Products Safety Act of 1972**;
  Consumer Products Safety Improvement Act of 2008.

Consumer Products Safety Improvement Act of 2008
  A U.S. law, Title 16 CFR, Part 2051, that established consumer product safety stan-
  dards and other safety requirements for children's products (e.g., lead content, toys
  as choking hazards, etc.) and to reauthorize and modernize the Consumer Product
  Safety Commission. It also provides for "whistleblower" protection, i.e., protection
  of individuals against employers for information on the safety concerns of a product.
  See also **Consumer Products Safety Act of 1972**; **Consumer Products Safety
  Commission (CPSC)**.

Contact Dermatitis
  Dermatitis caused by contact with a substance, i.e., gaseous, liquid, or solid. The
  condition may be due to primary irritation or an allergy. See also **Dermatitis**.

Contact Hazard
  A potential injury source arising from exposure to a machine, equipment, or a process system that contains sharp surfaces, sharp projections, heat, or extreme cold.

Contact Stress
  Exposure of a body part to a hard or sharp surface repetitively or forcefully at a
  workstation or on a tool. Contact stress has been associated with Cumulative Trauma
  Disorders. It may be considered during ergonomic evaluations.

Containment
  Restricting the spreading of fire or toxic or hazardous material.

Contest
  To object to an alleged violation of regulatory standards. An example would be
  disputing a violation alleged by an Occupational Safety and Health Administration
  (OSHA) Area Director, thus placing the dispute before its review commission.

Contingency Plan
  A document setting out an organized, planned, and coordinated course of action to
  be followed in case of a fire, explosion, or other incident that releases toxic chemicals,
  hazardous waste, or radioactive materials that threaten human health or the environment. See also **Emergency Response Plan (ERP)**.

Contingent Business Interruption Insurance
  Contingent business interruption insurance and contingent extra expense insurance
  protect a firm against interruption and extra expense losses resulting from damage caused by an insured peril to property that it does not own, operate, or control.
  There are four situations in which this coverage is used: When the insured depends

FIGURE C.4   Continuous improvement model.

on a single supplier or a few suppliers for materials, the firm on which the insured depends is called a contributing property. When the insured depends on one or a few manufacturers or suppliers for most of its merchandise, the firm upon which the insured depends is called the manufacturing property. When the insured relies on one or a few businesses to purchase the bulk of its (the insured's) products, the firm to which most of the insured's production flows is called the recipient property. When the insured counts on a neighboring business to help attract customers, the neighboring firm is called a leader property. See also **Business Interruption (BI); Business Interruption Insurance**.

Continuous Flow Respirator
An atmosphere-supplying respirator that provides a continuous flow of breathable air to the respirator facepiece.

Continuous Improvement Model, Safety
A concept in safety management that continuous improvement in safety management allows the development of a preventive culture within the company. It is a multi-step process that repeats itself in order to build on previous activities and accomplishments. It is illustrated in Figure C.4.

Continuous Noise
Noise of a constant level as measured over at least one second using the "slow" setting on a sound level meter. Note that a noise that is intermittent, e.g., on for over a second and then off for a period, would be both variable and continuous.

Contractual Liability
Liability as set forth by agreements between people, as distinguished from liability imposed by law (legal liability).

Contributing Cause
Physical conditions, management practices, procedures, or policies that facilitate the occurrence of an incident.

Contributory Negligence
The act or omission amounting to want of ordinary care on the part of a complaining party, which, concurring with the defendant's negligence, is the proximate cause of injury. This is different from assumption of risk, which exists where no fault for injury rests with the plaintiff, but where the plaintiff assumes the consequences of injury occurring through the fault of the defendant, a third person, or through the fault of no one.

Control, Interlocking

A form of electrical interlocking for machinery guards which incorporates an actuating switch operated by the guard and associated electronic devices which control power to the equipment. Failure of any of the elements or their connection wiring can be considered a fail to danger arrangement.

Control Methods or Technology

Engineering measures and techniques designed as a system to eliminate, or reduce to acceptable levels, exposure to harmful agents in the workplace. Includes engineering, administrative controls, work practices, monitoring, and personal protective equipment (PPE). Engineering controls design the hazard out by initial design specifications or by applying substitution methods, isolation, enclosure, or ventilation. In the hierarchy of control methods, engineering control methods should be considered first. Administrative controls reduce employee exposures through methods such as education and training, work reduction, job rotation, maintenance or repairs, housekeeping, personal hygiene, and appropriate work practices. Administrative controls depend on constant employee implementation or intervention. Personal Protective Equipment (PPE) is a device worn by employees to protect them from the environment. PPE includes anything from gloves to full body suits with self-contained breathing apparatus and can be used with engineering and administrative controls. See also **Engineering Controls**.

Controlled Area

A specific area that is restricted due to an identified hazard. See also **ANSI Z535.5, Safety Tags and Barricade Tapes (for Temporary Hazards)**.

Contusion

Injury to the body due to a blunt external force from an object, or a fall or bump. Usually accompanied by swelling and black and blue mark due to rupture of veins. Severity may vary from a small bruise to severe underlying damage to bones, vessels, and nerves. Severe contusions may be accompanied by lacerations as well.

Corrective Action Process

Addressing identified deficiencies, planning, follow-up, and documentation. The corrective action process normally begins with a management review of the audit findings. The purpose of this review is to determine what actions are appropriate, and to establish priorities, timetables, resource allocations and requirements, and responsibilities. In some cases, corrective action may involve a simple change in procedure or minor maintenance effort to remedy the concern. Management of change procedures needs to be used, as appropriate, even for what may seem to be a minor change. Many of the deficiencies can be acted on promptly, while some may require engineering studies or in-depth review of actual procedures and practices. There may be instances where no action is necessary, and this is a valid response to an audit finding. All actions taken, including an explanation where no action is taken on a finding, require documentation as to what was done and why.

Corrective Action Tracking Process

The identification of a deficiency, the status of corrective actions that are to be or were taken, and the audit person or team responsible. Typically it includes periodic update reports provided to management, which highlight specific reports such as completion of an engineering study, hazard analysis, etc., and a final implementation report to provide closure for audit findings that have been through a management of change process safety review, if appropriate, and then shared with affected employees and management. It also provides documentation required to verify that appropriate corrective actions were taken on deficiencies identified in an audit.

Cost-Benefit Analysis

A review that is a determination of the total value of an investment's inputs and out-puts. It is useful to evaluate the justification of safety improvement in the loss preven-tion profession. The technique was first examined by French engineer and economist Jules Dupuit (1804–1866) and later developed by twentieth century economists.

Cost-Effectiveness

A method of economic analysis, where the cost of system changes made to increase safety are compared with either the decreased costs of fewer serious failures, or with the increased effectiveness of the system to perform its task to determine the relative value of the changes.

Coverage, Insurance

An insured risk or liability. That which is insured, as specified in the insurance policy.

CPR (Cardiopulmonary Resuscitation)

An emergency means of providing oxygen and blood circulation through the delivery of rescue breathing and chest compressions to victims of sudden cardiac arrest for adequate life support. This procedure ensures that a critical flow of oxygen-ated blood is maintained to the brain and other vital organs during a resuscitation attempt. It is estimated that more than one million individuals suffer a sudden cardiac arrest every year. Every minute without CPR following sudden cardiac arrest, the probability of survival reduces by 7 to 10 percent. When bystander CPR is delivered, the patient stands a better chance as the probability for survival reduces to 3 to 4 percent per minute.

Cradle-to-Grave or Manifest System

A procedure in which hazardous wastes are identified as they are produced and are followed through further treatment, transportation, and disposal by a series of perma-nent, linkable, descriptive documents.

Crash Safety

A system characteristic that allows the system occupants to survive the impact of a crash and to evacuate the system after potentially survivable incidents.

Crashworthiness

The capacity of a vehicle to act as a protective container and energy absorber during impact conditions.

Credible Scenario

A scenario that has a relatively high probability of occurring for the situation under examination. As part of emergency response planning, credible scenario exercises or pre-planning is undertaken. In order to avoid undue and unrealistic postulation of events that may occur, a collection of credible scenarios is identified and assembled. These can be based on past loss histories or on risk assessments that have been undertaken for the specific aspect being reviewed. See also **Pre-incident Planning**; **Scenario**; **Tabletop Drill**.

Criminal Negligence

Involving or relating to a legal crime due to failure to use a reasonable amount of care when such failure results in injury, illness, or death to another.

Crisis

A situation where something happens that requires major decisions to be under-taken quickly.

Critical Confusion

As defined in American National Standards Institute (ANSI) Z535.3 for the design of safety signs, it is when a safety symbol elicits the opposite interpretation, or prohibited action. As an example, when a safety symbol meaning "No Fires Allowed" is misunderstood to mean "Fires Allowed Here."

Critical Control Point (CCP)

As applied in the food processing industry, it is a point, step, or procedure at which controls can be applied and a food safety hazard can be prevented, eliminated, or reduced to acceptable (critical) levels. The most common CCP is cooking, where food safety managers designate critical limits. See also **Hazard Analysis and Critical Control Point (HACCP).**

Critical Defect

A defect that judgment and experience indicate is likely to result in a hazardous or unsafe condition for individuals using, maintaining, or depending on the product or system or that is likely to prevent adequate performance of the product or system.

Critical Function

An operation or activity that is essential to the continuing survival of a system. Those functions that have a major impact on system performance and safety.

Critical Incident Technique

A set of procedures for collecting direct observations of human behavior in such a way as to facilitate their potential usefulness in solving practical problems and developing broad psychological principles. The critical incident technique outlines procedures for collecting observed incidents having special significance and meeting systematically defined criteria. A randomly selected sample of critical incidents should permit an inference to be made concerning the existence of similar incidents within the population from which the sample was taken.

Crossing Guard Safety

A crossing guard is an individual who guides and controls vehicle and pedestrian traffic at designated locations at streets, schools, railroad crossings, or construction sites. The Bureau of Labor Statistics estimates that there were 67,570 crossing guards employed in the United States in 2007. From 1993 to 2006, 97 fatalities of crossing guards occurred on the job in the United States.

Cryogenic Liquid

A refrigerated liquefied gas having a boiling point less than $-90°C$ ($-130°F$) at 101.3 kPa (14.7 psi). It is a hazard due to its freezing temperature, and the evaporation of the liquid at ambient temperatures may be an asphyxiant, especially in low-lying areas.

Culture

The customs, habits, and traditions that characterize an organization or social group. It includes the attitudes and beliefs that an organization has for profitability, commitment, and health and safety.

Cumulative Injury

An injury caused by repeated events or repeated exposures at work, such as the loss of hearing due to constant loud noise.

Cumulative Risk

The risk of a common toxic effect associated with concurrent exposure by all relevant pathways and routes of exposure to a group of chemicals that share a common mechanism of toxicity.

Cumulative Trauma Disorder (CTD)

A musculoskeletal injury that arises gradually as a result of repeated microtrauma. CTDs are characterized by injuries to the tendons, nerves, or neurovascular system. Muscles and joints are stressed, tendons are inflamed, nerves are pinched, or the flow of blood is restricted. Examples of CTDs include tendinitis, tenosynovitis, carpal tunnel syndrome, thoracic outlet syndrome, and Raynaud's phenomenon (white finger disease).

Cutaneous Hazards
Chemicals that irritate or otherwise damage the skin.

D

Dam Safety

Dam safety is the art and science of ensuring the integrity and viability of dams such that they do not present unacceptable risks to the public, property, and the environment. It requires the collective application of engineering principles and experience, and a philosophy of risk management that recognizes that a dam is a structure whose safe function is not explicitly determined by its original design and construction. It also includes all actions taken to identify or predict deficiencies and consequences related to failure, and to document, publicize, and reduce, eliminate, or remediate to the extent reasonably possible, any unacceptable risks.

Damage

Loss in value, usefulness, etc., to property or things. Harm causing any material loss. Severity of injury or the physical, functional, or monetary loss that could result if hazard is not controlled.

Damage Control

A term used in the maritime industry and navies for the emergency control of situations that may cause the sinking of a vessel or other serious harm to its operation.

Damage Risk Criterion

The suggested base line of noise tolerance, which, if not exceeded, should result in no hearing loss due to noise. A damage risk criterion may include in its statement a specification of such factors as time of exposure, noise intensity, noise frequency, amount of hearing loss that is considered significant, percentage of the population to be protected, and method of measuring the noise level.

Damages

Compensation or indemnity which may be recovered in a judicial or quasi-judicial forum by a party who has suffered loss, damage, or injury, whether to person or property, through the unlawful act, breach of contract, omission, or negligence of another.

Damages, Compensatory

That amount that will compensate the injured party for injury sustained, and nothing more, to make good or replace the loss.

Damages, Punitive

Damages assessed as a punishment to the wrongdoer, or as an example to others, for outrageous conduct or gross, wanton negligence.

Damping Shower

A walk-through decontamination shower specifically designed for the pharmaceutical industry for use in laboratories, production areas, and packaging departments to allow the damping of protective clothing prior to removal in order to avoid the risk of contact with airborne contaminants. It is utilized in environments where there is a potential risk of contamination of protective clothing by airborne particles. It is arranged to start automatically when an individual enters the shower enclosure and breaks the photoelectric beam. It continues to operate until the individual leaves the enclosure. Eliminating manual control prevents the risk of depositing contaminants on operating devices within the cubicle. It delivers a fine water mist to thoroughly

FIGURE D.1   Danger sign.

dampen full-body protective clothing. It also leaves both hands free so that they can be raised for quicker and more efficient damping. The damping of clothing prior to removal prevents the release of airborne contaminants. See also **Safety Shower**.

Danger

A general term denoting liability or potential of injury, illness, damage, loss, or pain.

Danger Analysis

A qualitative safety review methodology, primarily utilized in project management, to identify hazards in activities or systems, their probability of occurrence, and determine if protection measures are adequate. It is similar to a Job Safety Analysis. It is sometimes called a Danger Analysis, Safety Verification, or Preliminary Danger Analysis.

Danger Label or Sign

As defined by American National Standards Institute (ANSI) Z535.4, a sign or label that indicates a hazardous situation that if not avoided, could result in death or serious injury. Danger application is to be limited to the most severe situations. Danger is to be indicated in white letters on red background (Figure D.1). See also **Caution Label or Sign**; **Notice Label or Sign**; **Safety Alert Symbol**; **Safety Sign**; **Warning Label or Sign**.

Danger Tag

As defined by Occupational Safety and Health Administration (OSHA) regulations 29 CFR 1910.145 (f) (5), is a tag used in major hazard situations where an immediate hazard presents a threat of death or serious injury to employees. See also **Caution Tag**; **Warning Tag**.

Danger Tree

In the logging industry it refers to a standing tree that presents a hazard to employees due to certain conditions including, but not limited to, deterioration or physical damage to the root system, trunk, stem or limbs, and the direction and lean of the tree.

Danger Zone

A physical area or location within which a danger exists. It is typically identified with warning signs, painted ground lines, or through barriers either permanent or temporary.

D

Dangerous

Attended with risk; hazardous; unsafe. Something that if in normal use, danger or injury can be anticipated by the user. Something without adequate protection. See also **Dangerous, Imminently; Dangerous, Inherently**.

Dangerous, Imminently

Something, by reason of defective construction, that causes an impending or threatening dangerous situation, which could be expected to cause death or serious injury to persons in the immediate future unless corrective measures are taken.

Dangerous, Inherently

Something that is usually dangerous even in its normal or non-defective state, such as explosives or poisons, and requires special precautions and warnings so as to prevent injury.

Dangerous Failure

A failure of a component in a safety instrumented function that prevents that function from achieving a safe state when it is required to do so. See also **Failure Mode**.

Dangerous Goods

Also referred to as hazardous materials. Any solid, liquid, or gas that can harm people, other living organisms, property, or the environment.

DART Rate (OSHA)

An acronym for Days Away, Restricted work activity, and/or job Transfer (DART) Case Incidence Rate and is defined by the Occupational Safety and Health Administration (OSHA) as the rate of recordable injuries and illness cases per 100 full-time employees resulting in days away from work, restricted work activity, and/or job transfer that a site has experienced in a given time frame. Restricted cases are defined as any occupational injury or illness that results in the limitation of employees' ability to do their job (i.e., no lifting, climbing, etc.) or being transferred to another job (restricted days). The annual DART rate is calculated according to the following formula:

$$\text{DART rate} = \frac{\text{\# of recordable injuries and illnesses} \times 200,000}{\text{\# employee hours worked}}$$

DAW Rate (OSHA)

As defined by the Occupational Safety and Health Administration (OSHA), Days Away from Work (DAW), are the number of days away from work and the number of days of restricted work resulting from the work-related injury or illness. Restricted days are the number of days of restricted work or of being transferred to a different job resulting from the work-related injury or illness.

DAWC Rate (OSHA)

As defined by the Occupational Safety and Health Administration (OSHA), occupational injury or illness cases that result in an employee being unable to work a full assigned work shift. That is, the employee is off from work (lost workday). As defined by OSHA, a fatality is not considered a lost time case.

Days of Disability

Total full calendar days on which an injured person was unable to work as the result of an injury. The total does not include the day of the injury or the day of return to work. See also **Lost Workdays**.

De Minimus Citation

Violation of a regulatory standard that does not involve an immediate or direct relationship to the safety or health of an employee, or when the intent of the regulation is clearly complied with but deviates in a minor, technical, or trivial fashion, or the location is considered "state of the art," which is technically beyond the requirements of the applicable standard and provides equivalent or more effective employee protection of health and safety. It is considered a minor citation. An example of this is a technical violation such as the incorrect height of letters on an exit sign. See also **Citation**.

De Quervain's Disease

Inflammation of the tendon sheath of the thumb attributed to excessive friction between two thumb tendons and their common sheath. Usually caused by twisting and forceful gripping motions with the hands and is considered an ergonomic hazard.

Deadman Switch

A switch that is automatically operated in case the human operator becomes incapacitated. The switch usually stops a machine, and is a form of a fail-safe arrangement that is typically set up so that if the force being applied by an individual to operate the device is removed, the device becomes inactive. It may also be called a Kill Switch.

Deaf

A term used to describe a person who has lost hearing before speech patterns were established.

Deafened

Refers to a person who has lost the ability to hear after normal speech patterns were established.

Death (Accidental)

An injury that terminates fatally and is causally related to an incident. Death resulting from work injuries is assigned a time charge of 6000 days each according to the ANSI Z16.2 Standard. See also **Occupational Injury and Illness Classification System (OIICS)**.

Death Benefits

Benefits paid to dependants when a work injury or illness results in death.

Death Certificate

A vital record, maintained by local governments, signed by a licensed physician that includes cause of death; decedents name, sex, date of birth; date of death; place of residence of death; and usually occupation.

Decibel (dB)

The unit used to express the intensity of sound. The decibel was named after Alexander Graham Bell. The decibel scale is a logarithmic scale in which 0 dB approximates the threshold of hearing in the mid frequencies for young adults and in which the threshold of discomfort is between 85 and 95 dB sound pressure level (SPL) and the threshold for pain is between 120 and 140 dB SPL. The decibel is equal to 10 times the logarithm of the signal power ratio as expressed by the following equation: $n(dB) = 10 \log10 [(P1)/(P2)]$. dBA is the sound level in decibels read on the A scale of the sound level meter. The A scale discriminates against the very low frequencies (as does the human ear) and is therefore superior for measuring general sound levels.

DECIDE Process

A management system used to organize the response to a chemical incident. The factors of DECIDE are Detect, Estimate, Choose, Identify, Do the best, and Evaluate.

Decompression Sickness (Bends, Caisson Disease)

A condition caused by the formation and growth of bubbles in the blood or tissue resulting from a state of supersaturation with gas. This occurs when the sum of partial pressures of gases dissolved in a tissue exceeds the ambient pressure. It occurs in divers and workers in compressed-air environments on return from hyperbaric pressures to surface pressure, or in aviators going from surface pressure to hypobaric pressures at altitude. Several specific clinical syndromes are described: Serious Symptom or Type II includes cerebral, spinal cord, vestibular (the staggers), and pulmonary (the chokes); Simple or Type I includes pain-only bends, and skin bends (the niggles).

Decontamination

Removal of a polluting or harmful substance from air, water, earth surface, etc. For example, the process of removing hazardous chemical contamination from objects or areas.

Decontamination (Radiation)

The removal of radioactive material from a location where it is not desired. In regard to personnel it would include both removal of external contamination by washing and removal of internal contamination by the use of chelating agents or similar methods.

Deductible, Insurance

The portion of an insured loss to be borne by the insured before he or she is entitled to recovery from the insurer. It is normally quoted as a fixed quantity and is a part of most policies covering losses to the policyholder. The deductible must be paid by the insured before the benefits of the policy can apply. Typically, a general rule is the higher the deductible, the lower the premium, and vice versa.

Deductive Approach

A type of reasoning from the general to the specific. Utilized in the loss prevention industry to analyze a system or process that has failed in certain circumstances, an evaluation is made to determine what modes of the system, component, operator, or organization behavior contributed to the failure. See also **Inductive Approach**; **Morphological Approach**.

De-energized

Free from any electrical connection to a source of potential difference and from electrical charge; not having a potential different from that of the earth.

Defect

Nonconformance of a characteristic with specified requirements, or a deficiency in something necessary for an item's intended purpose and proper use that may lead to an incident. Anything that exceeds specifications or standards.

Defective

Lacking in some particular way that is essential to the completeness or security of the object.

Defensive Driving

Advanced techniques and practices (e.g., awareness, observation, spacing, etc.), used by drivers to protect themselves from the errors of other drivers, unsafe road conditions, and adverse weather conditions that increase driving hazards. The National Safety Council (NSC) launched the first Defensive Driving Course, adapted from techniques used by professional drivers in 1964. Since then, the NSC has trained more than 60 million drivers and today offers more than 25 different defensive driving and fleet safety courses. The NSC offers more courses with the highest success rate in reducing the number and severity of collisions and their related costs than any other driver training provider.

Deflagration

An exothermic reaction that propagates from burning gases to unreacted material by conduction, convection, and radiation. The combustion zone progresses through the material at a rate that is less than the velocity of sound in the unreacted material.

Degrees of Negligence

Ordinary negligence is based on the fact that one ought to have known the results of unsafe acts. Gross negligence rests on the assumption that one knew the results of acts but was recklessly or wantonly indifferent to the results. All negligence below that called gross, or ordinary, by the courts is slight negligence.

Delayed Hazard

An adverse effect that has the potential to occur after an extended period of time. Particularly featured in harmful health effects.

Deluge

The immediate release of a commodity; normally refers to the water spray release for fire suppression application.

Demand Respirator

An atmosphere-supplying respirator that admits breathing air to the facepiece only when a negative pressure is created inside the facepiece by inhalation.

Demolition Hazards

Demolition is considered a high-risk dangerous construction operation. Demolition consists of the dismantling or destruction of a building or structure. It may be undertaken by manual techniques or mechanical applications, which may expose individuals to a large number of hazards. Demolition hazards include structural instability, i.e., unexpected collapse or subsidence; falling debris; fire occurrence; exposure to dust, toxic contaminants, or hazardous materials; live electrical circuits; restricted access or egress; or unprotected openings and close proximity to moving cranes and heavy equipment. Occupational Safety and Health Administration (OSHA) requirements for demolition are addressed in 29 CFR 1926, Subpart T.

Denied Claim

An injury case in which the insurance company believes an injury or illness is not covered by workers' compensation.

Department of Labor (DOL)

A U.S. cabinet-level department responsible for agencies that oversee safety and health, including the Occupational Safety and Health Administration (OSHA) and the Mine Safety and Health Administration (MSHA).

Deposition

The testimony of a witness taken upon interrogatories, not in open court, under oath, in writing, and duly authenticated and intended to be used as evidence in court.

Depressurization

The release of unwanted gas pressure from a vessel or piping system to an effective disposal system to prevent the rupture of equipment or for the quick managed disposal of gas to prevent its uncontrolled release.

Dermal Exposure

Skin exposure to chemicals in the workplace. Most chemicals are readily absorbed through the skin and can cause other health effects or contribute to the dose absorbed by inhalation of the chemical from the air. Many studies indicate that absorption of chemicals through the skin can occur without being noticed by the worker. In many cases, skin is a more significant route of exposure than the lung. This is particularly true for non-volatile chemicals that are relatively toxic and that remain on work surfaces for long periods of time. Dermal exposure is considered a significant problem in the United States. Both the number of cases and the rate of skin disease in the United States exceed recordable respiratory illnesses. In 2006, 41,400 recordable skin diseases were reported by the Bureau of Labor Statistics (BLS) at a rate of 4.5

injuries per 10,000 employees, compared to 17,700 respiratory illnesses with a rate of 1.9 illnesses per 10,000 employees.

Dermal Toxicity

Adverse effects resulting from skin exposure to a substance. Dermal toxicity ratings corresponding to the following definitions are derived from data that are obtained from the test methods as described in 16 CFR 1500.40 and categories of toxicity as described in 16 CFR 1500.3.

D

Non-Toxic: The probable lethal dose of undiluted product to 50 percent of the test animals determined from dermal toxicity studies (LD50) is greater than 2 g/kg of body weight.

Toxic: The probable lethal dose of undiluted product to 50 percent of the test animals determined from dermal toxicity studies (LD50) is greater than 200 mg and less than or equal to 2 g/kg of body weight.

Highly Toxic: The probable lethal dose of undiluted product to 50 percent of the test animals determined from dermal toxicity studies (LD50) is less than or equal to 200 mg/kg of body weight.

Dermatitis

Inflammation of the skin. Dermatitis appears as a redness of the skin and progresses to swelling, blistering, cracking, scaling, and crusting. It may be caused by primary irritants (e.g., solvents, strong acids) if permitted to contact skin and act for a sufficient length of time and sufficient concentration. Secondary irritants may be cutaneous sensitizers such as plants, rubber, and many chemical compounds, which do not initially cause a skin change at first contact but effect a specific sensitization, resulting in a dermatitis effect at a future exposure. It is a symptom of an unsatisfactory exposure to a toxic material. It is the most common occupational disease. It can be prevented by meticulous attention to personal hygiene, avoiding contact with all potentially harmful substances (i.e., use of personal protective equipment), and washing thoroughly and immediately after any inadvertent or unavoidable contact. See also **Contact Dermatitis**.

Design (Safety)

The planning of environments, structures, and equipment, and the establishment of procedures for performing tasks, so that human exposure to injury or illness potential will be reduced or eliminated. In product safety, the design of the product for safe use.

Design Speed

The optimal safe speed for travel on a road based on its physical characteristics, such as width, grade, surface, curvature, etc., based on accepted highway design standards.

Detector Tube

A glass tube containing specific chemicals that have been impregnated on inert material granules and that will change color when chemicals in air are drawn through the tube. See also **Draeger Tube®**.

Detonation

An exothermic reaction that is characterized by the presence of a shock wave in the material that establishes and maintains the reaction. A distinctive feature is that the reaction zone propagates at a rate greater than sound velocity in the unreacted material.

Dig Safely

A national campaign to enhance safety, environmental protection, and service reliability by reducing underground facility damage. This damage prevention education

D

FIGURE D.2   Tank diking in a petroleum refinery.

and awareness program is used by pipeline companies, one-call centers, and others throughout the country. Dig Safely was developed through the joint efforts of the Office of Pipeline Safety and various damage prevention stakeholder organizations.

Diking

Temporary or permanent barriers that prevent liquids from flowing away from designated containment areas or direct flow to a disposal area. Figure D.2 provides an example of tank diking for spill containment in a petroleum refinery.

Dilution Ventilation

Airflow designed to dilute contaminants to an acceptable level. It is achieved by introducing large volumes of air to flow through the contaminated region, where the quantity of contaminant is small and of low toxicity. It is typically applied to control vapors from low toxicity solvents. It is also referred to as general ventilation or exhaust.

Direct Cause

Unsafe behaviors or unsafe conditions that contribute sequentially or concurrently in a chain of events leading to an incident.

Direct Damage

Damage caused by the direct action of a peril as distinguished from damage done contingently.

Direct Injury Costs

The sum of compensation payments and medical expenses for an injury.

Disability

Any injury or illness, temporary or permanent, or physiological or psychological condition that prevents an individual from performing his or her usual job functions

or activities. See also **Permanent Disability; Permanent Partial Disability;
Permanent Total Disability**.

Disability Insurance

Insurance that includes paid sick leave, short-term disability benefits, and long-term
disability benefits.

Disabling Injury

Per American National Standards Institute (ANSI) Standard Z16.2, an injury that
prevents a person from performing a regularly established job for one full day (24
hours) beyond the day of the incident. See also **Occupational Injury and Illness
Classification System (OIICS)**.

Disabling Injury Frequency Rate (DIFR)

The number of disabling (lost time) injuries per million employee-hours of exposure:
DIFR = (Disabling Injuries × 1,000,000)/Employee-Hours of Exposure. See also
Incidence Rate.

Disabling Injury Index (DII)

An index computed by multiplying the disabling injury frequency rate by the
disabling injury severity rate and dividing the product by 1000: DII = (DIFR ×
DISR)/1000. This measure reflects both frequency and severity, yielding a combined
index of total disabling injury (ANSI Z16.2). See also **Incidence Rate; Occupational
Injury and Illness Classification System (OIICS)**.

Disabling Injury Severity Rate (DISR)

The total number of days charged per million employee-hours of exposure: DISR =
(Total Days Charged × 1,000,000)/Employee-Hours of Exposure. See also **Incidence
Rate; Occupational Injury and Illness Classification System (OIICS)**.

Disaster

The most serious form of emergency, where the resources, personnel, and/or materials at the site are insufficient to control the situation. Whether an emergency becomes
a disaster can depend on such things as the following:

    Type of site and the hazards in the site operations
    Proximity of neighboring communities or facilities
    Capabilities of emergency response equipment and personnel

Disaster Control

Advanced planning and established procedures for handling emergency situations.

Disaster Recovery Institute International (DRII)

Oversees training and education for professionals in the area of business continuity,
including testing criteria and national tests.

Disaster Recovery Site, Cold

A disaster recovery service that provides space, but the customer provides and
installs all the equipment needed to continue operations. A cold site is less expensive
than a hot or a warm site, but it takes longer to get an enterprise in full operation after
the disaster.

Disaster Recovery Site, Hot

A redundant facility or a commercial disaster recovery service that allows a business or a facility to continue its operations in the event of a disaster. For example,
if an enterprise's data processing center becomes inoperable, that enterprise can
move all data processing operations to a hot site. A hot site has all the equipment
and operations needed for the enterprise to continue its operation, including equip-

ment, machinery, storage space, office space and furniture, telephone jacks, and computer equipment.

Disaster Recovery Site, Warm

A location that can provide partial capabilities with equipment, operation, storage, and computer equipment such as servers, mainframes, and network connectivity. The key concept to consider is the time required to restore a level of service. The closer to "real time" this is, the "hotter" is the recovery site. However, this is rarely the case in manufacturing recovery activity. Warm sites are most typical.

Disclaimer

A legal clause utilized in contracts to specify that an entity does not warrant at all, or that it warrants only against, specified consequences or costs. Disclaimers do not release the manufacturer or defendant from liability for negligence, nor are disclaimers a defense to statutory violations.

Disconnect

A physical separation in a system to introduce an air gap break to positively prevent flow through a system.

Dispersion

The mixing and dilution of contaminate in the ambient environment.

Disposable Respirator

A respirator that is discarded after the end of its recommended period of use, after excessive resistance or physical damage, or when odor breakthrough or other warning indicators render the respirator unsuitable for further use.

Distant Guard

A guard that does not enclose a hazard but is arranged so that it cannot be reached by an individual.

Distraction Theory

A theory of incident causation that states that incidents are caused when workers are distracted when they are performing their work tasks. There are two types of distractions. Job site hazards can be a source of distraction. Workers will try to avoid being injured so they naturally focus on the hazard, but this occurs as they are trying to do work. Trying to get the task done may cause the worker to be distracted and to ignore the hazard, resulting in an injury. Mental worries can be a source of distraction. Workers will try to focus on the work to be done, but may be distracted by worries caused by personal or job-related concerns. Failure to be able to focus on the work increases the likelihood of being injured. See also **Accident Prone Theory**; **Adjustment Stress Theory**; **Goals-Freedom-Alertness Theory**.

Distress

As defined for excavations it is a situation where a cave-in is imminent or likely to occur. Distress is evidenced by such phenomena as the development of fissures in the face of or adjacent to an open excavation; the slumping of material from the face or the bulging or heaving of material from the bottom of an excavation; the spalling of material from the face of an excavation; and raveling, i.e., small amounts of material such as pebbles or little clumps of material suddenly separating from the face and trickling or rolling down into an excavation. See also **Cave-In**.

Do Not Operate Tag

A temporary notice for isolation provided on equipment, processes, or devices, which outlines specific information, reasons, and contacts for the notice in accordance with an organization's work procedures to prevent their unauthorized operation or to warn individual of potential hazards. They are required by Occupational Safety and Health Administration (OSHA) regulation 29 CFR 1910.145 as a means to prevent injury or illness to employees who are exposed to hazardous or potentially hazardous conditions, equipment, or operations that are out of the ordinary, unexpected, or not readily apparent unless other signs, guarding, or other positive means of protection are being used. See also **Accident Prevention Tag**; **Lockout/Tagout (LOTO)**.

Domino Theory

A theory of incident prevention that compares the sequence of events in an incident to five dominos. If the first domino falls the remaining dominos will also fall in a particular sequence. If a domino is removed the sequence of falling is broken, and the result is the last domino cannot fall. It is compared to a preventable incident, which is composed of five factors in a sequence that results in an incident. If a factor is removed the incident is prevented. The five factors or events are:

Ancestry and social environment
Fault of the individual
Unsafe act or mechanical or physical hazard
Incident
Personal injury

The theory was later modified to account for influence of management:

Allowing lack of control by management
Basic cause, i.e., personal and job factors
Immediate cause, e.g., substandard practices, conditions, or errors
Incident
Loss, whether negligible, minor, serious, or catastrophic

The origin of the Domino Theory is credited to Herbert W. Heinrich, circa 1931, who worked for Travelers Insurance. Mr. Heinrich undertook an analysis of 75,000 accident reports by companies insured with Travelers. This resulted in the research report titled "The Origins of Accidents," which concluded that 88 percent of all accidents are caused by the unsafe acts of persons, 10 percent by unsafe physical conditions, and 2 percent are "Acts of God." His analysis of 50,000 accidents showed that, in the average case, an accident resulting in the occurrence of a lost-time work injury was preceded by 329 similar accidents caused by the same unsafe act or mechanical exposure, 300 of which produced no injury and 29 resulted in minor injuries. This is sometimes referred to as "Heinrich's Law." Mr. Heinrich then defined the five factors in the accident sequence, which he identified as the "Domino Theory." Heinrich's work is the basis for the theory of behavior-based safety, which holds that as many as 95 percent of all workplace incidents are caused by unsafe acts. See also **Accident Chain**; **Behavior-Based Safety**.

Dose

The amount of exposure to a bioactive chemical agent, chemical, or ionizing radiation energy that is received by an organism or individual.

Dosimeter

A device (instrument or material) used to measure an individual's exposure to a hazardous environment, particularly when the hazard is cumulative over long intervals of time or a selected period of time. They are typically employed to measure sound, radiation, ultraviolet light, or electromagnetic fields to determine if recommended safe exposure limits are being maintained or exceeded.

D

FIGURE D.3   DOT placard with UN/NA number.

DOT (Department of Transport)

An agency of the federal government that is responsible for the regulation of the transport of hazardous materials.

DOT Hazard Class

The Federal Department of Transportation (DOT) groups chemicals into nine classes, depending on specific hazardous material properties. Some of these classes are also broken up into divisions to further clarify groups within each class. The nine classes include DOT 1, Explosives; DOT 2, Gases; DOT 3, Flammable Liquids; DOT 4, Flammable Solids; DOT 5, Oxidizing Substances; DOT 6, Poisons; DOT 7, Radioactive Materials; DOT 8, Corrosive Materials; and DOT 9, Miscellaneous Hazardous Materials. DOT placards must be displayed on vehicles and freight containers transporting hazardous materials in amounts of 1001 pounds or more, and on vehicles transporting any amount of explosives, poison gases, poisonous liquids that pose an inhalation hazard, water-reactive substances, and certain radioactive materials. Typically, the substance's UN/NA (United Nations/North American) number is shown on the placard along with the hazard class, division symbols, and numbers. The DOT placard in Figure D.3 displays 1203, the UN/NA number for gasoline. See also **NFPA 704, Standard System for the Identification of the Hazards of Materials for Emergency Response**; **UN Number**.

Double Block and Bleed

A three-valve configuration common in shut-off applications to prevent release of a gas or fluid to the area of isolation by providing a positive means of isolation. Two main shut-off valves (block valves) operate on the main process line to stop flow. A third bleed valve in a T-section between the two block valves fitted with a vent or drain can be opened to relieve pressure of the process fluid from the region between the two block valves. In this fashion, passing of the isolation valve on the active side will not allow entry into the isolated equipment. Typically considered as a "one out of two" (1oo2) voting shut-off system provided the bleed valve opening is not critical to achieving the safe state.

Double Contingency Principle

The simultaneous failure of two components. Most risk analyses do not consider the simultaneous failure of two components a high probability, unless the system under review is of very high criticality, and therefore its reliability is of concern.

Double Indemnity

A clause or provision in a life insurance or accident policy whereby the company agrees to pay the stated multiple (i.e., double, even triple indemnity is available) of the face amount in the contract in cases of accidental death. An accidental death is considered a death that is neither intentionally caused by a human being, such as murder or suicide, nor from natural causes, such as cancer or heart disease. In 2004, 4.67 percent of all deaths in the United States were declared accidental by the Centers for Disease Control and Prevention (CDC). Double-indemnity clauses are therefore usually relatively economical and often highly marketed, especially to people over 45 and under 60. Children and people in dangerous jobs, such as heavy construction, are the exceptions.

Double Insulated

Double-insulated or class 2 electrical appliances are devices that have been designed in a way so as not to require a safety connection to electrical ground. They are required to prevent any failure from resulting in dangerous voltage levels becoming exposed, causing a shock, etc. This must be done without the aid of a grounded metal casing. Ways of achieving this include double layers of insulating material or reinforced insulation protecting any live parts of the fitting. There are also strict requirements relating to the maximum insulation resistance and leakage to any functional ground or signal connections of such appliances. Devices of this type are required to be labeled "Class II," "double insulated," or bear the double insulation symbol. See Figure D.4 for Double Insulated symbol.

Draeger Tube®

Commonly supplies trade as a detector device from a specific vendor used in many industries. It consists of a glass vial filled with a chemical reagent that reacts to a specific chemical or family of chemicals. A calibrated 100 mL sample of air is drawn through the tube with the bellows pump. If the targeted chemical is present, the reagent in the tube changes color, and the length of the color change typically indicates the measured concentration. The Draeger-Tube System is the world's most popular form of gas detection. It was first introduced in 1937, and currently there are over 200 different Draeger tubes available that can measure over 500 gases and vapors.

Drowning

Death from acute asphyxia while submerged, whether or not liquid has entered the lungs.

Drug-Free Workforce Program

A comprehensive approach to prevent the use of drugs in the workforce to prevent incidents from occurring and maintain employees' health. It includes a policy, supervisor

FIGURE D.4   Double insulated symbol.

training, employee education, employee assistance, and drug screening/testing. Such programs, especially when drug testing is included, must be reasonable and take into consideration employee rights to privacy. Drug-free workplace programs are not required by the Occupational Safety and Health Administration (OSHA). OSHA states that the industries with the highest rates of drug use are the same as those at a high risk for occupational injuries, such as construction, mining, manufacturing, and wholesale.

D

Drug Safety

Refers to the process by which the Food and Drug Administration (FDA) learns about the safety profile of a drug before approving it to be marketed to the public and monitors its safety once it is in the marketplace. All medicines have risks. According to the FDA, injuries from approved medicines are one of the top 10 causes of death in the United States. FDA tools and techniques to detect rare and unexpected risks include testing and surveillance of drugs, and developing policies, guidance, and standards for drug labeling, current good manufacturing practices, clinical and good laboratory practices, and industry practices that demonstrate that drugs are safe. Drug risks include product quality defects, known side effects, medication errors, and uncertainties. Product quality defects are controlled through good manufacturing practices, monitoring, and surveillance. Most injuries and deaths are due to known side effects that are identified in the drug's labeling; these can be avoidable (by appropriate use of the drug) or unavoidable (e.g., some drugs have side effects even when used appropriately, such as nausea from antibiotics). Medication errors occur when the drug is given incorrectly or the wrong drug or dose is given. Risks from uncertainties include unexpected side effects, long-term effects, and uses of the drugs in groups of people (e.g., children or seniors) who were not studied in the clinical trials (e.g., a rare event which occurs in fewer than 1 in 10,000 people would not be identified by normal pre-market testing).

Drug Screening

A medical examination screening process, required by some employers, to determine if prospective or current employees have or currently utilize drugs that may contribute negatively to a healthy and safe work environment due to their influence on the individual.

Dry Chemical Extinguisher

An extinguisher containing a chemical that extinguishes fire by interrupting the chain reaction, wherein the chemicals used prevent the union of free radical particles in the combustion process so that combustion does not continue when the flame front is completely covered with the agent.

Dry Powder Extinguisher

A fire extinguisher designed for use on combustible metals fires, such as sodium, titanium, uranium, zirconium, lithium, magnesium, and sodium-potassium alloys.

Dual Modular Redundant (1oo2) ESD System

An emergency shutdown (ESD) system that uses two separate processors, each with its own separate input/output modules, bus structure, chassis, software, and power supplies to vote input signals in a 1 out of 2 arrangement. Sensor signals are separated into two isolated paths to two separate input modules where signals are conditioned and communicated by separate busses to separate processors. A valid input signal on either leg of the system will initiate the desired logic response via two separate, fail-safe, output modules. See also **Emergency Shutdown System (ESD): Triple Modular Redundant (TMR) ESD System**.

Ductile Failure

A metal failure as a result of being stretched to the point where the yield stress has been exceeded over a large area. The metal reaches the plastic stage and failure when ductile fracture occurs.

Due Diligence
> A level of judgment, care, prudence, determination, and activity that a person would reasonably be expected to do under particular circumstances.

Dust
> Small solid particles generated by the breaking up of larger particles by processes such as crushing, grinding, drilling, explosions, etc. Dust particles already in existence in a mixture of materials may escape into the air through such operations as shoveling, conveying, screening, sweeping, etc. Dust is a term used in industry to describe airborne sold particles that range in size from 0.1 to 25 microns (1 micron = 1/10,000 cm = 0.001 mm = 1/25,000 in.). See also **Nuisance Dust**.

Dust Collector
> An air-cleaning device to remove particulate loadings from exhaust systems before discharge to the outdoor environment.

Dust Explosion
> Dust explosions occur when finely divided combustible particles are dispersed in air in sufficient concentration in the presence of an ignition source strong enough to cause ignition. Combustion dust explosions have a slower rate of pressure rise and final lower pressure than combustible vapor explosions. Combustible dusts have a flammability range and minimum ignition temperature. Typically for combustible dust explosions it takes a higher ignition energy source than for a combustible vapor explosion. When dust particle size is above 400 micrometers, even a high-energy source cannot ignite the dust cloud. Controlled dust explosions by limited particle size is not an effective explosion control method since even 5 to 10 percent of fine dust within a material above 400 micrometers can develop into an explosive mixture. A dust explosion usually occurs with a small explosion that then causes additional dust to become airborne, resulting in a large dust explosion.

Dye-Penetrant Testing
> A form of nondestructive testing in which the surface of a material is saturated with dye or fluorescent penetrant and a developer is applied. The dyes bleed visibly to the surface indicating defects, while a fluorescent penetrant indicates the defects when an ultraviolet light is shown on them. See also **Nondestructive Testing**.

Dynamic Event Logic Analytical Methodology (DYLAM)
> The dynamic event logic analytical methodology provides an integrated framework to explicitly treat time, process variables, and system behaviors. A DYLAM is usually comprised of the following procedures: (a) component modeling, (b) system equation resolution algorithms, (c) setting of top conditions, and (d) event sequence generation and analysis. DYLAM is useful for the description of dynamic incident scenarios and for reliability assessment of systems whose mission is defined in terms of values of process variables to be kept within certain limits in time. This technique can also be used for identification of system behavior, and thus as a design tool for implementing protections and operator procedures. It is important to note that a system-specific DYLAM simulator must be created to analyze each particular problem. Furthermore, input data such as probabilities of a component being in a certain state at transient initiation, independency of such probabilities, transition rates between different states, and conditional probability matrices for dependencies among states and process variables need to be provided to run the DYLAM package.

Dynamic Event Tree Analysis Method (DETAM)
> Dynamic event tree analysis method is a quantifiable risk analysis approach that treats time-dependent evolution of plant hardware states, process variable values, and

operator states over the course of a scenario. In general, a dynamic event tree is an event tree in which branching is allowed at different points in time. This approach is defined by five characteristic sets: (a) branching set, (b) set of variables defining the system state, (c) branching rules, (d) sequence expansion rule, and (e) quantification tools. The branching set refers to the set of variables that determine the space of possible branches at any node in the tree. Branching rules, on the other hand, refer to rules used to determine when a branching should take place (a constant time step). The sequence expansion rules are used to limit the number of sequences. This approach can be used to represent a wide variety of operator behaviors, model the consequences of operator actions, and also serves as a framework for the analyst to employ a causal model for errors of commission. Thus it allows the testing of emergency procedures and identifies where and how changes can be made to improve their effectiveness. See also **Event Tree Analysis (ETA)**.

E

Es of Safety, the Three (3Es)

The supervisory duties of maintaining and improving safety through Engineering, Education, and Enforcement. Engineer hazards out of the workplace, educate employees in safe work practices and procedures, and enforce all safety rules and policies. It is advocated by the National Safety Council (NSC).

Ear Bands

Hearing protectors that are semi-aural devices fitted with small pods made of a soft material or flexible tips that seal off the ear canal entrances. Ear bands only require minimal insertion to be effective. The band itself, usually made of plastic, is designed to be worn under the chin or behind the neck. Ear bands are easy to keep track of because of higher visibility; compliance is also easier to verify for this same reason. With easy storage around the neck, they are well suited to applications that require repeated on-and-off use.

Ear Muffs

A type of hearing protector that is designed to fit around the head using a band, noise-attenuating cups with soft cushions surround the outer ear to seal out noise. Most workers can use the same size ear muffs, which usually have three different band positions: over the head, behind the neck, and under the chin. Many ear muffs have all three positions available through rotating headbands (the position worn may have an effect on the muffs' noise attenuation, indicated by the noise protection rating [NPR]). Most models are available as hard-hat mountable units, and some may require adapters. Applications requiring frequent removal of hearing protection are well suited to ear muffs. Ear muffs, when well cared for, can last for many years; ear cushions and other wearable parts should be replaced often. See also **Noise Reduction Rating (NRR)**.

Ear Plugs

A type of hearing protector that is designed to fit snugly into the ear canal, ear plugs are made of a soft, pliable material to seal out unwanted noise. They are the least expensive type of hearing protection and can be purchased with or without attached cords. The attached cords will prevent loss of the ear plugs, and the cords also stand out so supervisors can easily verify compliance. Usually the molded foam style plugs are categorized as disposable (good for several usages), while the plastic style are more durable and can be used for much longer periods of time.

Ear Protectors, Hearing

Plugs, muffs, or helmets designed to keep excessive noise from the ear to preserve hearing acuity.

Early Cause

An act on the part of some person or organization or a condition that causes or permits approximate or immediate cause to exist. It is also referred to as Distal Cause.

Early Suppression, Fast Response (ESFR)

A system for extinguishing a hazard upon or shortly after its inception, such as sprinklers and fire protection systems.

Effective Protection Factor (EPF) Study

A study, conducted in the workplace, that measures the protection provided by a properly selected, fit-tested, and functioning respirator when used intermittently for only some fraction of the total workplace exposure time (i.e., sampling is conducted during periods when respirators are worn and not worn). EPFs are not directly comparable to Workplace Protection Factor (WPF) values because the determinations include the time spent in contaminated atmospheres both with and without respiratory protection; therefore, EPFs usually underestimate the protection afforded by a respirator that is used continuously in the workplace. See also **Protection Factor Study**; **Workplace Protection Factor (WPF) Study**.

Efficient Cause

The cause that originates and sets in motion a chain of causation through other causes to the result. The cause of injury is attached to legal liability. The use of this term is not as popular as proximate cause.

Ejection Hazard

A form of machinery exposure, whereby particles or items are emitted or thrown off by a machine, which may lead to an injury. A shield or barrier is a protective device to guard against this immediate hazard.

Electric Shock

The effect produced on the human body by an electric current passing through it as a result of contact with an electrical energy source. Adolescents and adults are prone to high voltage shock caused by mischievous exploration and exposure at work. About 1000 people in the United States die each year as a result of electrocution. Most of these fatalities are related to on-the-job injuries. Many variables determine the type of injuries that may occur, if any. These variables include the type of current (AC or DC), the amount of current (determined by the voltage of the source and the resistance of the tissues involved), and the pathway the electricity takes through the body. Low voltage electricity (less than 500 volts) does not normally cause significant injury to humans. Exposure to high voltage electricity (greater than 500 volts) has the potential to result in serious damage. See also **Shock**.

Electrical Hazard

A condition that exists where contact or equipment failure can result in electric shock, arc flash burn, thermal burn, or blast.

Electrical Lockout

Procedure for the minimum requirements for lockout (tagout) of electrical energy sources. It is used to ensure that conductors and circuit parts are disconnected from sources of electrical energy, locked (tagged), and tested before work begins where employees could be exposed to dangerous conditions. Sources of stored energy, such as capacitors or springs, shall be relieved of their energy, and a mechanism shall be engaged to prevent the re-accumulation of energy. Lockout is the preferred method of controlling personnel exposure to electrical energy hazards. National Fire Protection Association (NFPA) 70E, Standard for Electrical Safety in the Workplace, provides guidance in the application of electrical lockout. See also **Lockout/Tagout (LOTO)**.

Electrical Safety

Measures, procedures, and devices and used to prevent exposure to hazards caused by electricity to prevent injury or death. Electrical current exposes workers to a serious, widespread occupational hazard; practically all members of the workforce are exposed to electrical energy during the performance of their daily duties, and electrocutions occur to workers in various job categories (Figure E.1). Many workers are

FIGURE E.1 Electrical safety.

unaware of the potential electrical hazards present in their work environment, which makes them more vulnerable to the danger of electrocution. Electrical injuries consist of four main types: electrocution, electric shock, burns, and falls caused as a result of contact with electrical energy. Contact with live electricity results, on average, in 4000 non-disabling injuries, 3600 disabling injuries, and 365 fatalities each year, according to a 2008 presentation given at the Institute of Electrical and Electronic Engineers (IEEE), Electrical Safety Workshop. See also **Electric Shock**.

Electrical Safety Foundation International (ESFI)
An organization dedicated exclusively to preventing electrical fires, injuries, and fatalities in the home and the workplace. Its goal is to reduce the number of injuries, fatalities, and fires caused by electrical incidents each year. To accomplish this it promotes electrical safety education and awareness in schools, communities, and workplace environments where hazards are commonly overlooked; encourages and supports educators, employers, retailers, utilities, and other community and professional organizations in their efforts to promote electrical safety; provides active leadership and annual sponsorship of the National Electrical Safety Month campaign each May through proactive media and community outreach; and serves as a national resource and a leading authority for electrical safety information. It was founded in 1994 as a cooperative effort by the National Electrical Manufacturers Association (NEMA), Underwriters Laboratories (UL), and the U.S. Consumer Product Safety Commission (CPSC). ESFI is funded by electrical manufacturers, distributors, independent testing laboratories, retailers, insurers, utilities, safety organizations, and trade and labor associations.

Electrical Tagout
See **Electrical Lockout**; **Lockout/Tagout (LOTO)**.

Electrically Safe Work Condition
A state where an electrical conductor or circuit part has to be disconnected from energized parts, locked and tagged in accordance with established standards, tested to ensure the absence of voltage, and, if necessary, grounded.

Electrocution
See **Electric Shock**.

Electronic Gas Detector
A device for detecting and measuring flammable gases based on their reaction with an electric filament. The resulting combustion and rise in temperature can be detected electronically and displayed as a concentration in air.

Electrostatic Precipitator
An air-cleaning device that involves the following steps: electrical charging of suspended particulate matter; collection of charged particles on a grounded surface; and removal of particulates from the collecting surface by mechanical vibration or flushing with liquid.

Electrostatic Sensitive Device (ESD)
See **Static Sensitive Device**.

Emergency
A non-routine situation that necessitates prompt remedial action, primarily to mitigate a hazard or adverse consequences for human health and safety, quality of life, property, or the environment. This includes conventional emergencies such as fires, release of hazardous chemicals, storms, or earthquakes. It also includes situations for which prompt action is warranted to mitigate the effects of a perceived hazard. See also **Disaster**.

Emergency Action
An action performed to mitigate the impact of an emergency on human health and safety, property, or the environment.

Emergency Alarm
A warning device, usually visual or auditory, that indicates the existence of an emergency situation requiring immediate action.

Emergency Alert System (EAS)
A public alert and warning system in the United States. It provides the President and other authorized federal, state, and local officials the capability to transmit an emergency message to the public during disasters or crises. The national EAS, regulated by the Federal Communications Commission (FCC), is administered by the Department of Homeland Security through the Federal Emergency Management Agency. The President and authorized federal government officials originate national alerts and warnings. During non-federal emergencies, the EAS gives state and local government and emergency management officials the capability to alert and warn their local populations.

Emergency Brake
A separate back-up (secondary) brake system in a transportation vehicle for use in case of failure of the regular (hydraulic or air) brakes. Commonly used as a parking brake in automobiles.

Emergency Depressuring System
A system of valves, piping, actuating devices, and emergency shutdown (ESD) logic used during an emergency to rapidly and safely reduce pressure in process equipment by controlled venting to a disposal system such as a flare, burn pit, or storage. Logic for automated emergency depressuring systems resides within an ESD system. Design guidelines for depressuring systems are provided in American Petroleum Institute (API) Recommended Practice (RP) 521, Guide for Pressure-Relieving and Depressuring Systems.

Emergency Disaster Plan
An organizational plan for swift, efficient, and cost-effective responses to medical, fire, health care, shelter, and communications needs after disasters.

Emergency Drill

Training during which an emergency is simulated and the participants go through the steps of responding as if it were a real emergency for the purposes of familiarization, practice, preparedness, and improvement. See also **Fire Drill; Tabletop Drill.**

Emergency Evacuation

See **Evacuation.**

Emergency Exposure Limit (EEL)

The maximum amount of a toxic agent to which an individual can be exposed for a very brief (emergency) period of time and still maintain physical safety. See also **Acute Exposure Guideline Level (AEGL) Value; Emergency Response Planning Guideline (ERPG) Value; Temporary Emergency Exposure Limit (TEEL) Value.**

Emergency Isolation Valve (EIV)

A valve that, in event of fire, rupture, or loss of containment, is used to stop the release of flammable or combustible liquids, combustible gas, or potentially toxic material. An EIV can be either hand operated or power operated (air, hydraulic, or electrical actuation). EIVs can be actuated either by an emergency shutdown (ESD) system or by a local or remote actuating button, depending on the design of the facility.

Emergency Lighting

Lighting provided to aid in emergency evacuation in case normal lighting provisions fail. Codes for emergency evacuation normally specify the level and duration of emergency exit lighting that is required. National Fire Protection Association (NFPA) 101, Life Safety Code specifies 1 foot-candle (10 lux) at the centerline measured at the floor level of the evacuation route. Power for emergency lighting provision is not to be affected by the incident requiring evacuation; therefore, most emergency lighting equipment is provided with self-contained batteries. The duration specified by NFPA 101, Life Safety Code is 1.5 hours.

Emergency Management Agency

The state and local agencies responsible for emergency operations, planning, mitigation, preparedness, response, and recovery for all hazards. Names of emergency management agencies may vary, such as Division of Emergency Management, Comprehensive Emergency Management, Disaster Emergency Services, Civil Defense Agency, Emergency and Disaster Services.

Emergency Operations Center (EOC)

The location or facility where responsible officials gather during an emergency to direct and coordinate emergency operations, communicate with other jurisdictions and with field emergency forces, and formulate protective action decisions and recommendations during an emergency. An EOC may be a temporary facility or be located in a more central or permanently designated facility. They may be organized by major functional disciplines (e.g., fire, law enforcement, and medical services) or by jurisdiction (e.g., federal, state, regional, county, city, tribal). Large industrial facilities usually have their own designated EOCs, organized by their respective management personnel. They may also be called Emergency Operating Center or Emergency Control Center (ECC).

Emergency Operations Plan

The "steady-state" plan maintained by various jurisdictional levels for responding to a wide variety of potential hazards.

Emergency Planning and Community Right-to-Know Act (EPCRA)

Environmental Protection Agency (EPA) chemical reporting requirements, where facilities must report the storage, use, and release of certain hazardous chemicals. It was created to help communities plan for emergencies involving hazardous substances. The EPCRA has four major provisions: one addresses emergency planning and the three others deal with chemical reporting.

Emergency Preparedness

The capability to take actions that will effectively mitigate the consequences of an emergency for human health and safety, quality of life, property, and the environment.

Emergency Procedure

A plan for action in case of emergency by response personnel.

Emergency Response

The performance of actions to mitigate the consequences of an emergency for human health and safety, quality of life, property, and the environment. It may also provide a basis for the resumption of normal social and economic activity. See also **Response**.

Emergency Response Guidebook (ERG), DOT

A guideline provided by the Department of Transport (DOT) that assists first responders in making decisions at transportation-related chemical incidents. The Guidebook was developed jointly by the U.S. DOT, Transport Canada, and the Secretariat of Communications and Transportation of Mexico for use by firefighters, police, and other emergency services personnel who may be the first to arrive at the scene of a transportation incident involving a hazardous material. It is primarily a guide to aid first responders in quickly identifying the specific or generic classification of the material involved in the incident, and protecting themselves and the general public during the initial response phase of the incident. The ERG is updated every 3 to 4 years to accommodate new products and technology. Copies of the ERG are available free of charge to public emergency responders through State Coordinators.

Emergency Response Personnel

Emergency response personnel are persons engaged in the immediate response to incidents and emergencies. Emergency response personnel can include firefighters, police or sheriffs, medical personnel, civil defense and emergency management personnel, and, sometimes, military, manufacturing, and transportation personnel. Emergency response personnel are sometimes referred to as "emergency responders" and "first responders."

Emergency Response Plan (ERP)

See **Incident Action Plan**.

Emergency Response Planning Guideline (ERPG) Value

Acute exposure limits developed by the American Industrial Hygiene Association to describe the risk to humans resulting from once-in-a-lifetime, or rare, exposure to airborne chemicals. Each material is assigned a value for each chemical. Starting with "0," each successive number is associated with an increasingly severe effect that involves a higher level of exposure. The four values represent threshold levels for:

0—no adverse health effects
1—mild, transient health effects
2—irreversible or other serious health effects that could impair ability to take protective action
3—life-threatening health effects

See also **Acute Exposure Guideline Level (AEGL) Value**; **Temporary Emergency Exposure Limit (TEEL) Value**.

Emergency Response Provider

Includes federal, state, local, and tribal emergency public safety, law enforcement, emergency response, emergency medical (including hospital emergency facilities), and related personnel, agencies, and authorities. Addressed in Section 2 (6), of the Homeland Security Act of 2002, Pub. L. 107-296, 116 Stat. 2135 (2002). May also be called Emergency Responder.

Emergency Services

The local off-site response organizations that are generally available and that perform emergency response functions. These may include police, firefighters and rescue brigades, ambulance services, and control teams for hazardous materials.

Emergency Shutdown (ESD)

A control feature to safely stop a process. An emergency shutdown generally consists of stopping equipment, closing isolation valves on the supply or discharge lines from the process, or causing the system to be depressurized. The emergency shutdown features chosen for a particular process are dependent on the hazards of the process materials, quantities involved, arrangement of equipment, and exposures.

Emergency Shutdown Button

An operator control, usually a push-button with a large, red, protruding mushroom head, that when actuated initiates an emergency stop.

Emergency Shutdown System

A system composed of sensors, logic solvers, and final control elements for the purpose of taking the process, or specific equipment in the process, to a safe state when predetermined conditions are violated. The system is designed to isolate, de-energize, shut down, or depressure equipment in a process unit.

Emergency Shut-Off

A switch placed in a convenient position for cutting off the supply of electricity to a piece of equipment or to a building, in case of emergency.

Emergency Stop

Arrest of dangerous machine motion resulting from actuation of an emergency stop switch. The switch may be in the form of a safety switch, button, trip cable, or foot bar or other mechanical device used in conjunction with an emergency stop safety module.

Emission Control

Engineering measures, including devices, used to prevent worker exposure to contaminants that are released within the workplace. The term also refers to measures used on internal combustion engines, exhaust stacks, and other emission sources that are used to protect the general public.

Employee Assistance Program (EAP)

A program contracted by the employer and staffed by psychologists and other health professionals to which employees under stress or exhibiting behaviors that may create an unsafe work environment can be referred.

Employee Exposure Records

Information, results, or records concerning employee exposures to harmful substances or agents in the workplace, such as inventories of chemicals, material safety data sheets, and work area sampling results.

Employee Medical Records

Documentation of workers' health status by physicians, nurses, or other health professionals; includes exams, first aid records, diagnoses, and treatments.

Employer's Liability

Legal liability imposed on an employer making the employer responsible for paying damages to an employee injured by the employer's negligence. Generally replaced by Workers' Compensation, which pays the employee whether the employer has been negligent or not.

Enclosed Space

Per Occupational Safety and Health Administration (OSHA) regulation 29 CFR 1910.269 (e) and the Maritime Safety Standard, any space, other than a confined space, that is enclosed by bulkheads and overhead that workers may find otherwise ordinary job hazards aggravated or intensified. These may include cargo holds, tanks, quarters, and machinery and boiler spaces. See also **Confined Space**.

E

End of Service Life Indicator (ELSI)

A device that is used on a chemical cartridge for a breathing mask respirator to indicate when the cartridge should be discarded and replaced by showing a colorimetric change on the cartridge. This color change takes place before the contaminant breaks through the cartridge.

Endorsement, Insurance

A written or printed form attached to an insurance policy that alters provisions of the contract.

Energy Isolating Device

A mechanical device that physically prevents the transmission or release of energy. They typically include, but are not limited to, the following: A manually operated electrical circuit breaker; a disconnect switch; a manually operated switch by which the conductors of a circuit can be disconnected from all ungrounded supply conductors, and, in addition, no pole can be operated independently; a line valve; a block; and any similar device used to block or isolate energy. Push buttons, selector switches, and other control circuit type devices are not considered energy isolating devices.

Enforcement

The application by a regulatory body of sanctions against an entity, intended to correct and, as appropriate, penalize non-compliance with conditions of an authorization.

Engineered Failure

In a disaster management system under extreme stress, the identification and selection of priority services and activities that should be preserved, while allowing less critical activities to degrade. The principle is to avoid catastrophic or random failure of the emergency response system, when system capacity or capability is limited. Its intent is to preserve the functions most vital to achieving the organizational goals. May also be referred to as engineered system failure or managed degradation of incident response.

Engineering Controls

Risks are avoided, eliminated, or minimized through good engineering design. Basic methods used to prevent worker exposure to harmful chemical, physical, or biological agents by means of material substitution, equipment isolation, and material removal. Traditionally, a hierarchy of controls has been used as a means of determining how to implement feasible and effective controls. One representation of this hierarchy can be summarized as follows:

Elimination
Substitution
Engineering Controls
Administrative Controls
Personal Protective Equipment (PPE)

The idea behind this hierarchy is that the control methods at the top of the list are potentially more effective and protective than those at the bottom. Following the hierarchy normally leads to the implementation of inherently safer systems, ones where the risk of illness or injury has been substantially reduced. Elimination and substitution, while most effective at reducing hazards, also tend to be the most difficult to implement in an existing process. If the process is still at the design or development stage, elimination and substitution of hazards may be inexpensive and simple to implement. For an existing process, major changes in equipment and procedures may be required to eliminate or substitute for a hazard. Administrative controls and personal protective equipment are frequently used with existing processes where hazards are not particularly well controlled. Administrative controls and personal protective equipment programs may be relatively inexpensive to establish but, over the long term, can be very costly to sustain. These methods for protecting workers have also proven to be less effective than other measures, requiring significant effort by the affected workers. Engineering controls are used to remove a hazard or place a barrier between the worker and the hazard. Well-designed engineering controls can be highly effective in protecting workers and will typically be independent of worker interactions to provide this high level of protection. The initial cost of engineering controls can be higher than the cost of administrative controls or personal protective equipment, but over the longer term, operating costs are frequently lower, and in some instances, can provide a cost savings in other areas of the process. See also **Administrative Controls**.

Engulfment

Engulfment means the surrounding and effective capture of a person by a liquid or finely divided (flowable) solid substance that can be aspirated to cause death by filling or plugging the respiratory system or that can exert enough force on the body to cause death by strangulation, constriction, or crushing.

Episodic Event

An incident of limited duration, e.g., explosion, spill, gas release.

Equipment Breakdown Insurance

Insurance provided against the sudden and accidental breakdown or malfunction of boilers, machinery, electrical equipment, and a vast array of other equipment including air conditioners, heating, electrical, telephone, and computer systems. Most property insurance policies exclude theses losses, which is why separate equipment breakdown insurance is offered. Coverage is usually provided for damage to the equipment, expediting expenses, property damage to the property of others, and supplementary payments and automatic coverage on additional objects. The coverage can normally be extended to cover consequential losses and loss for business interruption. Previously referred to as Boiler and Machinery Insurance.

Ergonomic Chairs

An adjustable office chair that allows variation for seat height, seat depth, seat width, lumbar support, backrest, armrests, and the ability to swivel. These adjustments increase comfort and positioning in supported, neutral positions. To reduce the risk of work-related injuries from long periods of time spent at a desk or on a computer. See also **Ergonomics**.

Ergonomic Hazard Prevention and Control

Elimination or minimizing the ergonomic hazards identified in a worksite analysis through reduction of the frequency, duration, and severity of the exposure to the hazard. It includes work methods training, job rotation, gradual work introduction, and changes in the workstation, tools, or environment to fit the individual. See also **Ergonomics Program**; **Human Factors Engineering (HFE)**.

Ergonomic Worksite Analysis
> A safety and health review that addresses muskoskeletal hazards, the risk factors that pose the hazards, and the cause of the risk factors. See also **Ergonomics Program**.

Ergonomics
> The study of the design requirements of work in relation to the physical and psychological capabilities and limitations of human beings. Ergonomists contribute to the design and evaluation of tasks, jobs, products, environments, and systems in order to make them compatible with the needs, abilities, and limitations of people. Ergonomic disorders are the fastest growing category of work-related illness. According to recent statistics from the U.S. Bureau of Labor Statistics, they account for 56 percent of illnesses reported to the Occupational Safety and Health Administration. See also **Ergonomic Chairs**; **Ergonomics Program**; **Ergonomic Worksite Analysis**; **Human Factors Engineering (HFE)**.

Ergonomics Program
> The application of ergonomics in a system that includes the following components: health and risk factor (job or worksite) surveillance, job analysis and design (hazard prevention and control), medical management, and education and training.

Error
> A mistake or miscalculation of judgment leading to an action resulting in unacceptable or undesirable consequences such as an incident and its subsequent effects. See also **Human Error**.

Error Rate Prediction
> A forecast of the possibility of error based on statistical data.

Escape
> An immediate evacuation. See also **Evacuation**.

Escape Hatch
> An emergency exit door used for evacuation in a confined location such as in an aircraft, boat, or submarine. Usually considered a secondary means of evacuation to be used when the normal egress facility is unavailable.

ESD Safe Bag
> A bag for shipping electronic components, which may be subjected to damage caused by an electrostatic discharge (ESD). See also **Static Sensitive Device**.

Evacuation
> The departure of occupants from a building or area due to an emergency (dangerous or potentially dangerous areas) to an area free of risk from the emergency. Evacuations can be spontaneous (usually unorganized and unsupervised due to readily observed hazards or perceived threats), voluntary and orderly (typically used for potential future threats), or mandatory and directed (individuals must leave as directed by authorities due to imminent dangers). Failure to provide available exit facilities has been a contributing factor in the evacuation of individuals from several major building fire incidents resulting in loss of life. Disorderly or unfamiliar evacuation can lead to evacuee injuries due to congestion and confusion. The term evacuation does not include relocation of personnel to other areas with the same toxic, fire, or explosion risk. See also **Assembly Area**; **Personnel Accountability**; **Safe Refuge**; **Shelter-in-Place**.

Evacuation Capability
> The ability of individuals to evacuate a building or area due to an emergency. Evacuation capability is usually categorized into three levels (especially for healthcare facilities): Prompt, Slow, and Impractical. A prompt evacuation rating indicates the ability of individuals to move reliably in a timely manner. A slow evacuation is

indicative of a rapid move in a timely manner but not as rapidly as the general population. The inability of individuals to move reliably in a timely manner is considered an impractical evacuation.

Evacuation Plan

A prearranged set of instructions for the orderly departure of individuals from a building or area due to an emergency condition, to an area that is free of risk from the emergency. The evacuation plan should include methods of alerting or communication, exit routes, personnel accountability, alternative methods, contingency issues, and areas designated as safe refuge. See also **Assembly Area**; **Evacuation**; **Personnel Accountability**; **Safe Refuge**; **Shelter-in-Place**.

Event Tree Analysis (ETA)

A method of fault propagation modeling. Consists of an analysis of possible causes starting at a system level and working down through the system, sub-system, equipment, and component, identifying all possible causes (e.g., What faults might we expect? How may they be arrived at?). A tree-shaped picture of the chains of events leading from an initiating event to various potential outcomes is drawn to relate the event visually, hence the identification of this method. The tree expands from the initiating event in branches of intermediate propagating events. Each branch represents a situation where a different outcome is possible. After including all of the appropriate branches, the event tree ends with multiple possible outcomes. It begins with a specified event and looks forward in time. It is an inductive approach identifying and evaluating potential outcomes from a designated set of conditions and options. This assessment method allows for quantifying the probability of an incident and the risk associated with plant operation based on the graphic description of incident sequences. Event Tree Analysis is a logical method of analyzing how and why a disaster could occur. It is a useful technique for determining the overall probability of a catastrophic event occurring, and where a substantial economic impact may occur. These methods are used to carry out a mathematical analysis of the incident sequences. It is widely used in the nuclear industry. See also **Bow Tie Analysis**; **Fault Tree Analysis (FTA)**.

Excavation

Any man-made cut, cavity, trench, or depression in an earth surface formed by earth removal. This definition includes environmental characterization (for example, core drilling), jack hammering, and indoor drilling or digging operations that may contact soil. In general, excavations are operations where contact with soil is expected, such as trenching and removing soil to install foundation footings or exposing underground pipes for repair or replacement. Occupational Safety and Health Administration (OSHA) standards on excavations are included in 29 CFR 1910.120 (b) (1) (iii), 29 CFR 1926.650, 29 CFR 1926.651, and 29 CFR 1926.652. The most common hazards associated with excavations, in addition to those posed by working with heavy and mechanical equipment, are the following:

Unidentified or misidentified utilities: Workers may be exposed to hazards such as electric shock, suffocation, or explosions if they unexpectedly come in contact with utility lines.

Hazardous atmospheres: Workers may be exposed to hazards such as suffocation, chemical exposure, or explosions if they enter excavations with hazardous atmospheres.

Cave-ins: Worker injury or structural damage may result from sidewalls of excavations caving in.

Structural instability: Structures may become unstable if excavation occurs below the base of building or equipment pad foundations, or below retaining wall footings.

Water accumulation: Water accumulation in excavations can cause sloughing of excavation sidewalls, resulting in unsafe conditions for those entering the excavation, particularly if the use of electrical equipment is required.

Lack of egress: Workers may become injured while exiting an excavation if egress is not adequate, particularly if an emergency evacuation is required.

Falls: Workers or passersby may fall into an open, unprotected excavation, or vehicles may accidentally be driven into an uncovered or inadequately barricaded pit.

See also **Benching; Shoring; Trenching Hazards**.

Excavation Protective System

A method of protecting employees from cave-ins, from material that could fall or roll from an excavation face or into an excavation, or from the collapse of adjacent structures. Protective systems include support systems, sloping and benching systems, shield systems, and other systems that provide the necessary protection. See also **Benching; Shoring**.

Excavation Support System

Structures such as underpinning, bracing, and shoring that provide support to an adjacent structure or underground installation during an excavation.

Exclusions

Provisions in an insurance policy that indicate denied coverage. Common exclusions are hazards that are deemed so catastrophic in nature that they are uninsurable.

Executive Action

The control process performed to initiate critical instructions or signals to safety devices.

Exhaust (General)

Diluting the general room atmosphere with outdoor air fast enough to keep the concentration of toxic vapor in the room air within safe limits. It may also be known as General Ventilation or Dilution Ventilation.

Exhaust (Local)

A local exhaust system is used to collect air contaminants at the source, as contrasted with general ventilation, which allows the contaminant to spread throughout the workroom, later to be diluted by exhausting quantities of air from the room. Local exhaust may be achieved using an enclosure, a receiving hood, or an exterior hood.

Exhaust Ventilation

The removal of air or other gas from any work space, usually by mechanical means.

Exit

The portion of an exit route that is generally separated from other areas to provide a protected way of travel to the exit discharge from the effects of an incident. An example of an exit is a 2-hour fire-resistance-rated enclosed stairway that leads from the fifth floor of an office building to the outside of the building. Exits include exterior exit doors, exit passageways, horizontal exits, and separated exit stairs or ramps. See also **Fire Escape**.

Exit Access

The portion of an exit route that leads to an exit. An example of an exit access is a corridor on the fifth floor of an office building that leads to a two-hour fire-resistance-rated enclosed stairway (the Exit).

Exit Discharge
  The part of the exit route that leads directly outside or to a street, walkway, refuge area, public way, or open space with access to the outside. An example of an exit discharge is a door at the bottom of a two-hour fire-resistance-rated enclosed stairway that discharges to a place of safety outside the building.

Exit Marking
  The identification provided for an exit from a building, structure, or area to an area free from the risk exposure. See also **Exit Sign**.

Exit Route
  A continuous and unobstructed path of exit travel from any point within a workplace to a place of safety (including refuge areas). An exit route consists of three parts: the exit access; the exit; and, the exit discharge (an exit route includes all vertical and horizontal areas along the route) (see Figure E.2).

E

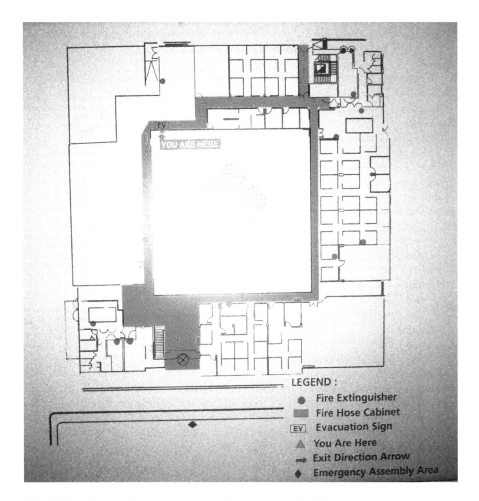

LEGEND :
- ● Fire Extinguisher
- ■ Fire Hose Cabinet
- EV Evacuation Sign
- ▲ You Are Here
- → Exit Direction Arrow
- ◆ Emergency Assembly Area

FIGURE E.2   Placard of building exit routes. (Photo courtesy of D. P. Nolan.)

Exit Sign

A designated identification label provided at or near an exit that is clearly recognizable and visible and identifies the exit or the path to an exit. Some codes require red exit signs and others green. Red is usually associated with fire and exit lights and green for safety. The Life Safety Code (NFPA 101) of the National Fire Protection Association (NFPA) recognizes either color as acceptable.

Experience Period

The time period to which a company refers when evaluating an insurance policy.

Expert Testimony

The opinion of a witness skilled in a particular art, trade, or profession or possessed of special knowledge derived from education or experience not within the range of common experience, education, or knowledge.

Expert Witness

A person possessing particular knowledge, wisdom, skill, or information acquired by study, investigation, observation, practice, or experience regarding the subject matter under consideration and not likely to be possessed by ordinary or inexperienced persons. It is up to the judge to determine if a person is qualified as an expert and if the testimony is pertinent to, or would shed light on, the case.

Explode

To undergo a rapid chemical or nuclear reaction with the production of noise, heat, and violent expansion of gases, causing destructive forces to occur based on the magnitude of the reaction.

Explosion

The sudden conversion of potential energy (chemical or mechanical) into kinetic energy with the production and release of gases under pressure, or the release of gas under pressure. These high-pressure gases then do mechanical work such as moving, changing, or damaging nearby materials. A study by the Industrial Risk Insurers has estimated the average cost of an industrial explosion at $1,000,000. See also **Blast**.

Explosion Detection

Typically applied where dust explosion potentials exist in enclosures from potential overpressures generated from dust or vapor. Spark detectors (i.e., infrared detectors) or pressure detectors are typically used that instantaneously initiate a suppression system to inhibit or extinguish the spark and prevent an explosion. See also **Explosion Isolation**.

Explosion Door

A door in a furnace or boiler setting designed to be opened by a predetermined gas pressure.

Explosion Isolation

A system, e.g., detection and isolation valves, to provide a positive mechanical barrier and block potential flame paths in process piping or ductwork that lead to other process equipment or operator occupied areas. Used where there is the potential for flame and pressure to travel to other areas resulting in additional fire or explosion hazards. An isolation valve system utilizes detection and control functions that initiate valve closure within milliseconds and operates at the earliest possible moment of a deflagration.

Explosion Venting

A means provided for the release of high pressures caused by explosion. Dust and vapor explosions create a shock wave. Explosion venting systems are commonly installed where there may a risk of a dust explosion. They typically incorporate a predetermined "blow out" panel, which, if an explosion occurs, allows the shock wave to vent in a safe location, usually to the external area of the facility. National Fire Protection Association (NFPA) 68, Explosion Protection by Deflagration Venting, provides guidelines for the design, sizing, and application of explosion vents.

Explosionproof

A common industry term characterizing an electrical apparatus that is designed so that an explosion of flammable gas inside the enclosure will not ignite flammable gas outside the enclosure. It is also used to mean a device that has the capability of preventing the ignition of a specified vapor or gas by such an explosion and operating at an external temperature too low to ignite a surrounding atmosphere. Nothing is technically explosionproof, as all equipment has a relative resistance or vulnerability to the effects of explosions depending on the force of the explosion exposure and strength of construction.

Explosive Atmosphere

Atmosphere containing a mixture of vapor gas or particulate matter that is within the explosive or flammable range.

Explosive Decompression

A sudden rapid decrease in barometric pressure. This may occur from loss of integrity of a pressurized aircraft cabin or recompression chamber, or in a diver who surfaces rapidly from depth due to a loss of buoyancy control. It may result in decompression sickness or pulmonary hyperinflation and air embolism.

Explosive Limits

The minimum (lower) and maximum (upper) concentration of vapor or gas in air or oxygen below or above which explosion or propagation of flame does not occur in the presence of a source of ignition. The explosive or flammable limits are usually expressed in terms of percentage by volume of vapor or gas in air. In reality, explosive limits for a material vary since they depend on many factors such as air temperature. Therefore, the values given on a Material Safety Data Sheet (MSDS) are approximate. The difference between the lower and upper flammable (explosive) limits is the "range," expressed in terms of percentage by volume of vapor or gas in air. See also **Lower Explosive Limit (LEL)**; **Upper Explosive Limit (UEL)**.

Explosive Materials

Typically are intended to include explosives, blasting agents, and detonators, including dynamites, slurries, emulsions and water gels, black powder, smokeless powder, detonators and safety fuses, squibs, detonating cord, and other materials whose primary function is to function by explosion.

Explosive Mixture

A mixture of flammable vapor or gas and air within the lower and upper limits of the explosive range.

Explosivemeter

An instrument used to determine whether an atmosphere has sufficient gas and oxygen in a mixture to be explosive.

Explosives

Any chemical compound or mechanical mixture that is used or intended for the purpose of producing an explosion and that causes a sudden release of pressure, gas, and heat. Contains any oxidizing and combustive units or other ingredients in such proportions, quantities, or packing that an ignition by fire, by friction, by concussion, by percussion, or by detonation of any part of the compound or mixture may cause such a sudden generation of highly heated gases that the resultant gaseous pressures are capable of producing destructive effects on contiguous objects or of destroying life or limb.

Exposure

Contact between a chemical, physical, or biological agent and the outer surfaces of an organism. Exposure to an agent does not imply that it will be absorbed or that it will produce an effect.

Exposure, Casualty

Proximity to a condition that may produce injury, death, or damage from dusts, chemicals, high pressure, explosives, etc.

Exposure, Occupational

The quantity of time involved, and the level (quantity) and nature (quality) of employee involvement with certain types of environments possessing various degrees or types of hazards in the course of their work.

Exposure Assessment

The qualitative or quantitative determination or estimation of the magnitude, frequency, duration, and route of exposure.

Exposure Hours

Total number of employee hours worked by all employees including those in operating, production, maintenance, transportation, clerical, administrative, sales, and other activities. See also **Occupational Injury and Illness Classification System (OIICS)**.

Exposure Level

The level or concentration of a physical or chemical hazard to which an employee is exposed.

Exposure Limits

Concentration of substances (and conditions) under which it is believed that nearly all workers may be repeatedly exposed day after day without adverse effects. American Conference of Governmental Industrial Hygienists (ACGIH) limits are called threshold limit value and Occupational Safety and Health Administration (OSHA) exposure limits are referred to as permissible exposure limits (PEL). See also **Permissible Exposure Limit (PEL)**.

Exposure Records

The records kept by an employer, or company doctor or nurse of an employee's exposure to a hazardous material or physical agent in the workplace. These records show the time, level, and length of exposure for each substance or agent involved.

Extinguisher

See **Fire Extinguishers**.

Extinguishing Agent

Material or substance that performs a fire extinguishing function. Common extinguishing agents are water, carbon dioxide, dry chemical, "alcohol" foam, and halogenated gases (Halons). It is important to know which extinguishers can be used so they can be made available at the worksite. It is also important to know which agents cannot be used since an incorrect extinguisher may not work or may create a more hazardous situation. If several materials are involved in a fire, an extinguisher effective for all of the materials should be used. Sometimes referred to as Extinguishing Media.

Eye Protection

A device that safeguards the eye in an eye-hazard environment. The devices include safety glasses, chemical splash goggles, face shields, etc. Occupational Safety and Health Administration (OSHA) standard 29 CR 1910.133 require eye protection for exposure to eye hazards. Employers must assess the workplace and determine if

TABLE E.1
Eye and Face Hazard Assessment

Hazard	Examples of Hazard	Typical Tasks
Impact	Flying objects such as large chips, fragments, particles, sand, and dirt	Chipping, grinding, machining, masonry work, wood working, sawing, drilling, chiseling, powered fastening, riveting, and sanding
Heat	Anything emitting extreme heat	Furnace operations, pouring, casting, hot dipping, and welding
Chemical	Splash, fumes, vapors, and irritating mists	Acid and chemical handling, degreasing, plating, and working with blood
Dust	Dust and small particulates suspended in the air	Woodworking, buffing, and general dusty conditions
Radiation	Radiant energy, glare, and intense light	Welding, torch-cutting, brazing, soldering, and laser activities

hazards that necessitate the use of eye and face protection are present or are likely to be present for assigning eye protection to workers. A hazard assessment should determine the risk of exposure to eye and face hazards, including those that may be encountered in an emergency (see Table E.1).

Eye Safety

National Institute for Occupational Safety and Health (NIOSH) records indicate that each day approximately 2000 U.S. workers have a job-related eye injury that requires medical treatment. About one-third of the injuries are treated in hospital emergency departments, and more than 100 of these injuries result in one or more days of lost work. The Occupational Safety and Health Administration (OSHA) estimates that eye injuries alone cost more than $300 million per year in lost production time, medical expenses, and worker compensation. The majority of these injuries result from small particles or objects striking or abrading the eye. Examples include metal slivers, wood chips, dust, and cement chips that are ejected by tools, wind blown, or fall from above a worker. Some of these objects, such as nails, staples, or slivers of wood or metal penetrate the eyeball and result in a permanent loss of vision. Large objects may also strike the eye or face, or a worker may run into an object causing blunt force trauma to the eyeball or eye socket. Chemical burns to one or both eyes from splashes of industrial chemicals or cleaning products are common. Thermal burns to the eye occur as well. Among welders, their assistants, and nearby workers, ultraviolet (UV) radiation burns (welder's flash) routinely damage workers' eyes and surrounding tissue. In addition to common eye injuries, health care workers, laboratory staff, janitorial workers, animal handlers, and other workers may be at risk of acquiring infectious diseases via ocular exposure. Infectious diseases can be transmitted through the mucous membranes of the eye as a result of direct exposure (e.g., blood splashes, respiratory droplets generated during coughing or suctioning) or from touching the eyes with contaminated fingers or other objects. The infections may result in relatively minor conjunctivitis or reddening and soreness of the eye or in a life-threatening disease.

Engineering controls should be used to reduce eye injuries and to protect against ocular infection exposures. Personal protective eyewear, such as goggles, face

FIGURE E.3 Eye protection.

shields, safety glasses, or full face respirators must also be used when an eye hazard exists (Figure E.3). The eye protection chosen for specific work situations depends upon the nature and extent of the hazard, the circumstances of exposure, other protective equipment used, and personal vision needs. Eye protection should be fit to an individual or adjustable to provide appropriate coverage. It should be comfortable and allow for sufficient peripheral vision. Selection of protective eyewear appropriate for a given task should be made based on a hazard assessment of each activity, including regulatory requirements when applicable.

Eye Wash, Emergency

The flushing and irrigation of eyes with water from an exposure to a hazardous material. It can be used for dilution, cooling or warming, and irrigation. Dilution: the water reduces the concentration of chemical in the eyes to an acceptable level. Cooling or warming: water warms or cools the eyes due to chemical reaction exposure which has caused a temperature hazard, Irrigation: the water flushes the chemical away from the eyes. It is a requirement of Occupational Safety and Health Administration (OSHA) regulation 29 CFR 1910.151 (c), which states: "Where the eyes or body of any person may be exposed to injurious corrosive materials, suitable facilities for quick drenching or flushing of the eyes and body shall be provided within the work area for immediate emergency use." See also **ANSI Z358.1, Standard for Emergency Eye Wash and Shower Equipment; Safety Shower**.

Eyeglass Side Shields
Eyeglass side shields are incorporated into the frames of safety spectacles when workplace operations expose workers to angular impact hazards and heat hazards. See also **ANSI Z87.1-2003, Standard for Occupational and Educational Eye and Face Protection Devices: Safety Glasses**.

E

F

Face Shield
A protective device designed to prevent hazardous substances, dust particles, sharp objects, and other materials from contacting the face or eyes. It may be worn over safety glasses or goggles. Face shields are required to meet American National Standards Institute (ANSI) Z87 and be clearly marked with the manufacturer's name. See also **ANSI Z87.1-2003, Standard for Occupational and Educational Eye and Face Protection Devices; Eye Protection.**

Facepiece
The portion of a respirator that covers the user's nose and mouth in a half-mask facepiece, or the nose, mouth, and eyes in a full facepiece. It is designed to make a gas-tight or dust-tight fit with the face and includes the headband, exhalation valves, and connections for an air purifying device or respirable gas source or both. See also **Eye Protection.**

Facility Emergency Coordinator
A representative of a facility covered by environmental law (e.g., a chemical plant) who participates in the emergency reporting process with the Local Emergency Planning Committee.

Facility Layout
The act or process of laying out, or planning in detail, to show the arrangement of equipment and other elements of a facility to establish a safe working environment.

Factor of Safety (FS)
The ratio of ultimate strength of a material or structure to the allowable stress. The factor of safety is used for the safe design of structures and the materials used in such structures. It varies immensely for different materials, e.g., steel, concrete.

Factory Mutual (FM)
A loss prevention and control service organization maintained for the policyholders of three major industrial and commercial property insurance companies: Allendale Insurance, Arkwright, and Protection Mutual Insurance. Factory Mutual provides loss control engineering, loss adjustment, insurance appraisals, building plan review, research, and education services. Fire and loss prevention equipment that has passed specific FM testing standards are considered acceptable for fire protection service and are provided with an FM listing or label.

Fail Close
See **Failure Mode.**

Fail Open
See **Failure Mode.**

Fail-Safe
A system design or condition such that the failure of a component, subsystem, or system, or input to it, will automatically revert to a predetermined safe static condition or state of least critical consequence. The opposite of fail-safe is fail to danger. See also **Failure Mode; Failure Mode and Effects Analysis (FMEA/FMECA).**

Fail Steady (FS)

A condition wherein the component stays in its last position when the actuating energy source fails. It may also be called Fail-in-Place.

Fail to Danger

A system design or condition such that the failure of a component, subsystem, or system or input to it, will automatically revert to an unsafe condition or state of highest critical consequence for the component, subsystem, or system.

Failure

An inability to perform an intended function. It is a condition in which a human, structure, component, device, or system fails to adequately perform its intended purpose.

Failure, Critical

A failure that could result in major injury or fatality to people or major damage to any system or loss of a critical function.

Failure, Dependent

See **Failure, Secondary**.

Failure, Independent

See **Failure, Primary**.

Failure, Primary

The failure that is responsible for a system malfunction.

Failure, Secondary

A failure that occurs as the consequence of another failure.

Failure Analysis

The logical systematic examination of an item to identify and analyze the cause, mode, and consequence of a real failure.

Failure Assessment

The process by which the cause, effect, responsibility, and cost of any reported problem in the system is determined and reported.

Failure Management

Decisions, policies, and planning that identify and eliminate or control potential failures and implement corrective or control procedures following real failures.

Failure Mechanism

The physics or chemistry of the failure event, i.e., the cause of the failure.

Failure Mode

The action of a device or system to revert to a specified state or operating condition upon failure. These ways are generally grouped into one of four failure modes: Safe Detected (SD), Dangerous Detected (DD), Safe Undetected (SU), and Dangerous Undetected (DU) per Instrument Society of America (ISA), TR84.0.02. Failure modes for valves are normally specified as fail open (FO), fail close (FC), or fail steady (FS), i.e., in the last operating position, which will result in a fail-safe or fail to danger arrangement. Valves that are required to shut off fuel supplies are normally specified as fail close, and those that are used to release gases to a vent or flare are specified as fail open. Electrical circuit devices either fail open or fail close. The failure mode of all the devices in a system must be known before a Failure Mode and Effects Analysis (FMEA) is undertaken. See also **Fail-Safe**.

Failure Mode and Effects Analysis (FMEA/FMECA)

A qualitative investigative safety review technique. It was developed in the 1950s by reliability engineers to determine problems that could arise from malfunctions of military systems. Failure mode and effects analysis is a procedure by which each potential failure mode in a system is analyzed to determine its effect on the system and to classify it according to its severity (Figure F.1). When the FMEA is extended

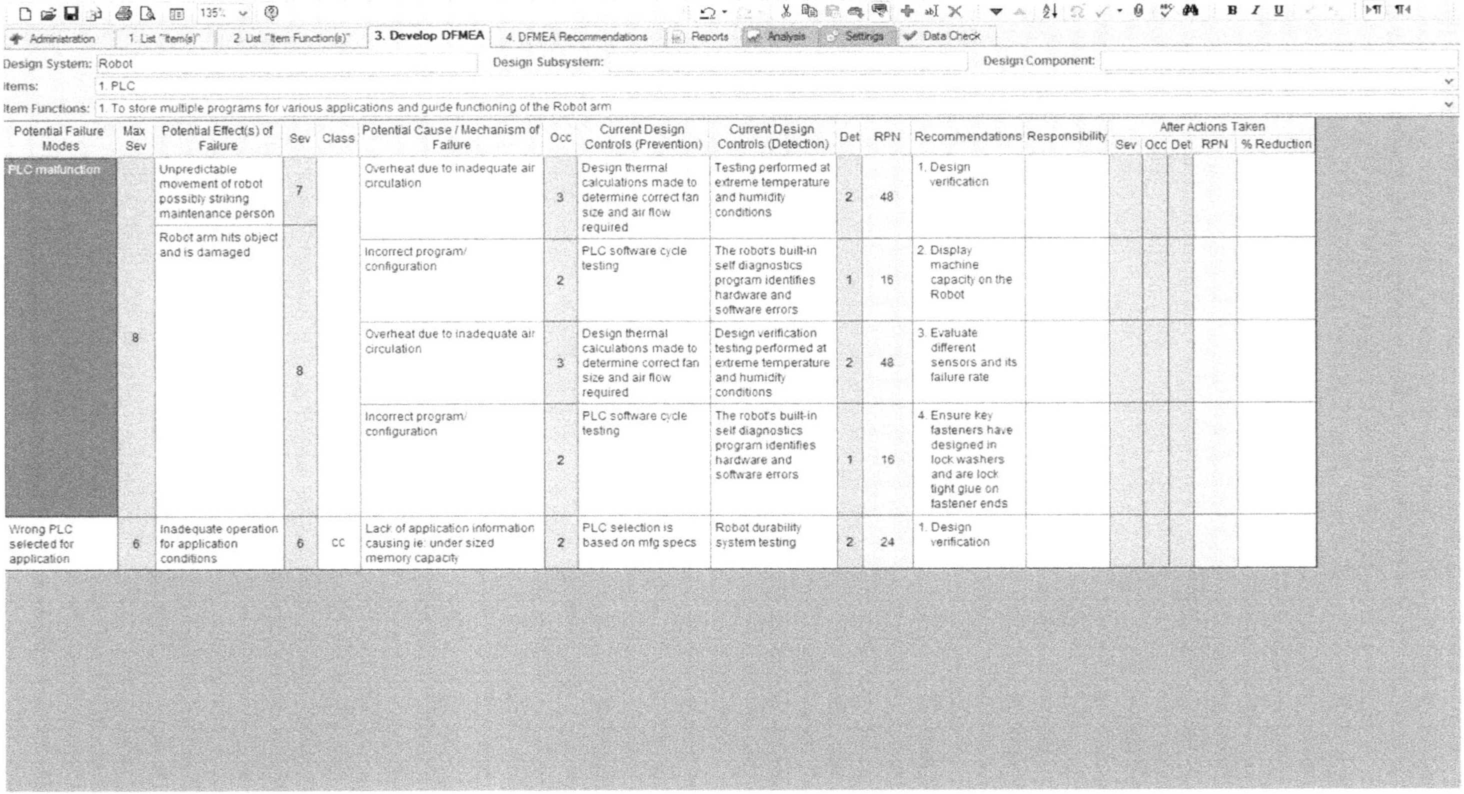

Potential Failure Modes	Potential Effect(s) of Failure	Max Sev	Sev	Class	Potential Cause / Mechanism of Failure	Occ	Current Design Controls (Prevention)	Current Design Controls (Detection)	Det	RPN	Recommendations	Responsibility	After Actions Taken				
													Sev	Occ	Det	RPN	% Reduction
PLC malfunction	Unpredictable movement of robot possibly striking maintenance person	8	7		Overheat due to inadequate air circulation	3	Design thermal calculations made to determine correct fan size and air flow required	Testing performed at extreme temperature and humidity conditions	2	48	1 Design verification						
	Robot arm hits object and is damaged		8		Incorrect program/ configuration	2	PLC software cycle testing	The robots built-in self diagnostics program identifies hardware and software errors	1	16	2 Display machine capacity on the Robot						
					Overheat due to inadequate air circulation	3	Design thermal calculations made to determine correct fan size and air flow required	Design verification testing performed at extreme temperature and humidity conditions	2	48	3 Evaluate different sensors and its failure rate						
					Incorrect program/ configuration	2	PLC software cycle testing	The robots built-in self diagnostics program identifies hardware and software errors	1	16	4 Ensure key fasteners have designed in lock washers and are lock tight glue on fastener ends						
Wrong PLC selected for application	Inadequate operation for application conditions	6	6	CC	Lack of application information causing ie under sized memory capacity	2	PLC selection is based on mfg specs	Robot durability system testing	2	24	1 Design verification						

FIGURE F.1   Failure mode and effects analysis example. (Software example courtesy of Dyadem.)

by a criticality analysis, the technique is then called failure mode and effects critical-ity analysis (FMECA). Failure mode and effects analysis has gained wide acceptance by the aerospace and military industries. In fact, the technique has adapted itself in other forms such as misuse mode and effects analysis. An FMEA review tends to be more labor intensive, as the failure of each individual component in the system has to be considered. See also **Action Error Analysis (AEA)**.

Failure Modes Effects and Diagnostics Analysis (FMEDA)

This is a detailed analysis of the different failure modes and diagnostic capabilities for a piece of equipment. It is an effective method for determining failure modes and failure rates, a requirement for certification against IEC 61508 in most certification agencies.

Failure Pattern

The relationship between the conditional probability of failure of an item and its age. Failure patterns are generally applied to failure modes. Research in the airline indus-try established that there are six distinct failure patterns. The type of failure pattern that applies to any given failure mode is of vital importance in determining the most appropriate equipment maintenance strategy. This fact is one of the key principles underlying Reliability-Centered Maintenance.

Failure Rate

The number of failures of an item per unit time (cycles, hours, miles, events, etc., as applicable for the item).

Fall Arrest System

A fall arrest system is an assembly of components and subsystems, including the necessary connectors, used to arrest the user in a fall from a working height and sus-pend the user until rescue can be affected. A fall arrest system must always include a full body harness and connecting means between the harness and an anchorage or anchorage connector. Such connecting means may consist of a lanyard, energy (shock) absorber, fall arrester (rope grab), lifeline, self-retracting lanyard, or qualified combinations of these. See also **Body Harness, Full**; **Lifeline**.

Fall Protection

Devices (e.g., fall arrest system) or barriers (e.g., handrails) provided to prevent an individual from falling from a height and sustaining injuries. Any time a worker is at a height of four feet or more, the worker is at risk and needs to be protected. Under Occupational Safety and Health Administration (OSHA) regulations, fall protection must be provided at four feet in general industry, five feet in maritime, and six feet in construction. Additionally, regardless of the fall distance, fall protection must be pro-vided when working over dangerous equipment and machinery. The U.S. Department of Labor (DOL) lists falls as one of the leading causes of traumatic occupational death, accounting for 8 percent of all occupational fatalities from trauma. A total of 809 fatal work-related falls occurred in 2006, according to the Bureau of Labor Statistics. Fall protection is considered either passive or active. Passive includes guardrails, railings, parapet walls, safety nets, and covers for walkway openings. Active includes full body harness, connected to a lanyard or self-retracting lifeline to an anchorage point. OSHA regulations that may apply for general industry are 29 CFR 1910 Subparts D, F, I, and R. The following American National Standards Institute (ANSI) standards may be used as guidelines: A1264.1, Safety Requirements for Workplace Walking/Working Surfaces and Their Access; A10.32, Fall Protection Systems for Construction and Demolitions Operations; ANSI/IWCA 1-14.1, Window Cleaning Safety; and Z359.1, Safety Requirements for Personal Fall Arrest Systems, Subsystems, and Components.

Fall Protection Program

Occupational Safety and Health Administration (OSHA) regulations for construction, 29 CFR 1926.502, require a fall protection program where fall hazard exposures exist for employees. The program should include addressing safety and hazards in project planning; identification of fall hazards at worksites; employee training; job hazard analysis for tasks to be performed; provision, training, inspection, and use of fall protection systems; worksite safety inspections; inclusions of environmental considerations; multi-language differences; alternative methods for tasks; medical and rescue programs; and employee participation in safety.

Fall Restraint System

A fall protection system that prevents the user from falling any distance. The system is comprised of either a body belt or body harness, along with an anchorage, connectors, and other necessary equipment. The other components typically include a lanyard, and may also include a lifeline and other devices. See also **Anchor Point**; **Body Harness, Full**; **Lanyard**; **Personal Fall Arrest System**.

Fatal Accident

An incident resulting in the death of one or more persons.

Fatality

A death resulting from an incident.

Fatality Rate

Represents the number of fatal injuries per 100,000 workers as used by the Bureau of Labor Statistics. It is calculated by the following equation $(N/W) \times 100,000$, where N = number of fatal injuries, W = number of workers employed, and 100,000 = base to express the fatality rate per 100,000 workers.

Fatigue

The physical or mental responses to a detrimental activity (e.g., prolonged work hours, lack of sleep), which show themselves in individuals as a diminished capacity for work.

Fault Hazard Analysis

The analysis of hazards or hazard potential situations using fault tree methodology. See also **Fault Tree Analysis (FTA)**.

Fault Propagation Modeling

The analysis of the chain of events that leads to an incident. By analyzing what events initiate that chain, which events contribute to it or allow the incident to propagate, and establishing how they are logically related, the event frequency can be determined. Fault propagation modeling techniques use the failure rates of individual components to determine the failure rate of the overall system.

Fault Tolerance

Ability of a functional unit to continue to perform a required function in the presence of random faults or errors. For example, a one out of two (1oo2) voting system can tolerate one random component failure and still perform its function. Fault tolerance is one of the specific requirements for safety integrity level (SIL) and is described in more detail in International Electrotechnical Commission (IEC) 61508 Part 2, Tables 2 and 3, and in IEC 61511 (ISA 84.01 2004) in Clause 11.4.

Fault-Tolerant System

A system incorporating design features that enable the system to detect and log transient or steady-state fault conditions and take appropriate corrective action while remaining on-line and performing its specified function.

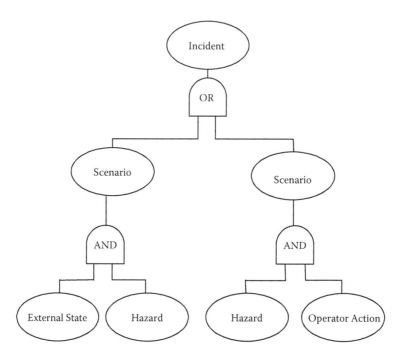

FIGURE F.2 Fault tree analysis.

Fault Tree Analysis (FTA)

A qualitative or quantitative risk analysis technique. The concept of fault tree analysis was originated by Bell Telephone Laboratories in 1962 as a technique with which to perform a safety evaluation of the Minutemen Intercontinental Ballistic Missile Launch Control System. A fault tree is a logical diagram that shows the relation between system failure, i.e., a specific undesirable event in the system, and failures of the components of the system. It is a technique based on deductive logic. An undesirable event is first defined and causal relationships of the failures leading to that event are then identified. It focuses on a designated "top event" and looks backwards in time to identify those specific factors, conditions, and events that could, in combination, result in that specified top event. From a time progression perspective, the fault tree ends with the top event. Fault tree can be used in qualitative or quantitative risk analysis. The difference in them is that the qualitative fault tree is looser in structure and does not require use of the same rigorous logic as the formal fault tree. Fault Tree Analysis is used in a wide range of industries, and there is extensive support in the form of published literature and software packages (Figure F.2). See also **Action Error Analysis (AEA); Bow Tie Analysis; Failure Mode and Effects Analysis (FMEA/FMECA); Hazard and Operability Study (HAZOP); Management Oversight and Risk Tree (MORT).**

Federal Emergency Management Association (FEMA)

A U.S. government agency that maintains the United States Fire Administration. Its main goal is to reduce the loss of life and property and protect the United States from all hazards, including natural disasters, acts of terrorism, and other man-made disasters, by leading and supporting the nation in risk-based, comprehensive emergency management system of preparedness, protection, response, recovery, and mitigation.

If requested they respond to emergencies outside the United States to coordinate and assist in rescue and aid efforts. In 2003 FEMA became part of the Department of Homeland Security (DHS).

Federal Insecticide, Fungicide, and Rodenticide Act (FIFRA)
A U.S. law that requires that certain poisons, such as chemical pesticides, sold to the public contain labels that carry health hazard warnings to protect users. It is administered by the Environmental Protection Agency (EPA). See also **Toxic Substance Control Act (TSCA) of 1976**.

Federal Railway Safety Act (FRSA)
Protects railroad employees from hazardous and unsafe working conditions as well as discrimination and retaliation. The FRSA also ensures that injured railroad workers will receive prompt medical attention. See also **Railroad Safety**.

F

Fibrosis
A medical condition that occurs to individuals which is marked by an increase of interstitial fibrous tissue. Exposures to contaminants through inhalation may lead to fibrosis or scarring of the lung, a particular concern in industrial hygiene.

Field Calibration
Field calibration means verifying whether an instrument is functioning properly and giving the correct readout within the limits specified by the manufacturer of the span (calibration or cal) gas.

Film Badge
A piece of masked photographic film worn by employees in the nuclear industry. It is darkened by nuclear radiation. Radiation exposure is checked by examining the film.

Filter
A straining device used to remove solid or liquid aerosols from the air.

Filter, HEPA
A High Efficiency Particulate Air dry type filter, with at least 99.97 percent efficiency in removing thermally generated monodisperse dioctyl phthalate smoke particles with a diameter of 0.0003 mm. The equivalent National Institute for Occupational Safety and Health (NIOSH) 42 CFR Part 84 particulate filters are the N100, R100, and P100 filters.

Filter, Respirator
A fibrous medium used in respirators to remove solid or liquid particles from the airstream entering the respirator.

Filtering Facepiece (Dust Mask)
A negative pressure particulate respirator with a filter as an integral part of the facepiece or with the entire facepiece composed of the filtering medium.

Final Position (After an Incident)
The place where objects or persons involved in the incident finally come to rest without application of power. This is the position before anything is moved to help the injured or remove vehicles or equipment.

Fire
Rapid oxidation with the evolution of heat and light.

Fire Alarm
A device or system (visual, auditory, local, or transmitted to other locations, etc.) that signals the presence of a fire to occupants and those who will provide assistance.

Fire Alarm Control Panel (FACP)
A control system for receiving fire alarm signals and initiating actions to highlight conditions (alarms and beacons) or institute actions to automatically activate fire protective systems (i.e., fire pump startup, HVAC shutdown, etc.). The fire alarm

control panel also provides an indication of the fire detection activation point through an annunciator panel or area and zone indicator lights, which highlight the specific location in a facility from where an alarm has been initiated. The FACP is required to meet specific performance requirements for reliability. National Fire Protection Association (NFPA) 72, National Fire Alarm Code (NFAC), provides guidance in the provision and features of a fire alarm control panel.

Fire Class

A simple letter designation given to a particular fire category for the purposes of generally classifying (severity and hazard) the fire according to the type of fuel and possible spread of the fire and type of extinguisher agent to use for it. A is used for ordinary combustibles (wood, cloth, paper, rubber, and many plastics), B for flammable liquids and gases (oils, liquid fuels, lubrication oils, hydraulic fluids, greases, tars, oil-based paints, lacquers, aerosols, cleaning compounds, and cutting or fuel gases), C for energized electrical fires, and D for combustible metal fires (magnesium, titanium, zirconium, sodium, or potassium).

Fire Code

A regulatory document for the implementation of measures to prevent the occurrence, spread, and institute suppression capability for unwanted fires. The National Fire Protection Association (NFPA) is a major source of fire codes that are used for adoption by legal agencies (e.g., city or state ordinances or laws). Enforcement is maintained by state Fire Inspectors and Fire Marshals. Most codes are determined by consensus agreement and therefore may not represent the highest or lowest level of protection that may be available, but what has been determined as appropriate by the experienced members determining the fire code specifications. Most fire codes have been prescriptive in nature, i.e., specifying exact requirements; however, the trend has been to provide performance-based codes that require a specific outcome and allow the detailed requirements to be determined by the premises owner as long as they meet the performance requirements.

Fire Detection

A device for the detection and notification of a fire event. Fire alarms can be activated by people or automatic devices that can detect the presence of fire. These include heat-sensitive devices, which are activated if a specific temperature is reached; a rate-of-rise heat detector, which is triggered either by a quick or a gradual escalation of temperature; and smoke detectors, which sense changes caused by the presence of smoke, in the intensity of light, in the refraction of light, or in the ionization of air. The arrangement and type of fire detectors for optimum performance is usually specified by fire codes or industry guidelines.

Fire Door

Doors that are rated and tested for resistance to various degrees of fire exposure and utilized to prevent the spread of fire through horizontal and vertical openings. The doors must remain closed normally or be closed automatically in the presence of fire. The degree of resistance required is determined by the type of occupancy, the anticipated fire exposure, and the resistance of the structure in which it is installed.

Fire Drill

A planned or unplanned evacuation of a building, facility, or location in order to familiarize and train the occupants in the means of escape to a safe location.

Fire Escape

A means of rapid egress from a building, primarily intended for use in case of fire. Several types have been used: a knotted rope or rope ladder secured to an inside wall; an open iron stairway on the building's exterior; an iron balcony; a chute; and an enclosed fire- and smokeproof stairway. The iron stairway is the most common because it can be added to the outside of nearly any building of modest height, although it has certain drawbacks; unless built against a blank wall it may be rendered useless by smoke from windows, and a means must be provided for keeping it in a state of readiness while denying its use to thieves and prowlers. The iron balcony extends around the exterior of a building to provide a corridor along which persons can flee from fire-imperiled rooms to safety behind a firewall or in an adjacent building. The chute, or slide escape, is either a curved or a straight incline and may be open or enclosed; it is well suited to such buildings as hospitals, from which patients can be evacuated on their mattresses. The best fire escape, however, is a fully enclosed fireproof stairway in the building or in an adjoining tower. Elevators are not considered safe because fire damage may cause them to fail and heat-sensitive call buttons may stop the car where the fire is hottest. See also **Exit.**

F

Fire Extinguisher, Portable

Portable or movable apparatus used to put out a small fire by directing onto it a substance that cools the burning material, deprives the flame of oxygen, or interferes with the chemical reactions occurring in the flame. They are intended as the first line of defense against fires of limited size. Portable extinguishers may be of water-based, gaseous, or dry chemical type. Most portable fire extinguishers are small tanks provided with an expelling gas that has been compressed (e.g., compressed air or carbon dioxide) to propel the extinguishing agent through a nozzle and onto the fire. This method supersedes the previous method used in the soda-acid fire extinguisher whereby carbon dioxide (CO_2) is generated by mixing sulfuric acid with a solution of sodium bicarbonate. The type of portable fire extinguisher selected depends primarily on the nature of the materials that are burning. Secondary considerations include cost, stability, toxicity, ease of cleanup, and the presence of an electrical hazard. Small fires are classified according to the nature of the burning material.

Class A fires involve wood, paper, and similar cellulosic materials. Class B fires involve flammable liquids, such as cooking fats and paint thinners. Class C fires are those in electrical equipment, and Class D fires involve highly reactive metals, such as sodium and magnesium. Water is suitable for putting out fires of only one of these classes (A), though this is the most common because it can cool and protect exposures as well. Water converts to steam when it absorbs heat, and the steam displaces the air from the vicinity of the flame. The water may contain a wetting agent to make it more effective against fires in upholstery, an additive to produce stable foam that acts as a barrier against oxygen or antifreeze to prevent freezing in cold ambient temperatures.

Fires of classes A, B, and C can be controlled by carbon dioxide (CO_2), halogenated hydrocarbons such as environmental friendly Halon substitutes, or dry chemicals such as sodium bicarbonate or ammonium dihydrogen phosphate. Class D fires are ordinarily combated with dry chemicals.

The CO_2 extinguisher is a steel cylinder filled with liquid carbon dioxide, which, when released, expands suddenly and causes so great a lowering of temperature that it solidifies into powdery "snow." This snow volatilizes (vaporizes) on contact with the burning substance, producing a blanket of gas that cools and smothers the flame (Figure F.3).

Early fire extinguishers in the 1730s were just glass balls of water or saline solution that were thrown on fires. They were invented by a German physician, M. Fuches, in 1734. Although they were widely advertised and sold, in general they were not really used (primarily because they were too small to be effective). In 1816, George Manby, an English Army Captain, invented the first practical extinguisher

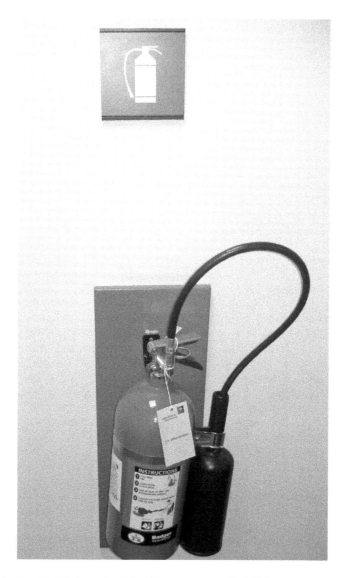

FIGURE F.3   Portable CO_2 fire extinguisher. (Photo courtesy of D. P. Nolan.)

similar to modern models. It used compressed air to force water (i.e., pressurized water) out of a cylinder through a control valve. It delivered 11.4 liters (3 gallons) of water from a cylinder that was pressurized with compressed air and was three-quarters full of water. A more efficient portable extinguisher was invented by Fraccoi Carier, a French doctor, in about 1866. He mixed sodium bicarbonate with water and fixed a glass bottle of sulfuric acid inside the extinguisher near the neck. The bottle was broken by striking a pin and the chemical mixed. This produced carbon dioxide (CO_2) gas that forced out the water. In 1909, Mr. Edward Davidson of New York patented the use of carbon tetrachloride (CCl_4). It was ejected out of the extinguisher by pressurized carbon dioxide. It vaporized immediately to form a heavy noncombustible gas that smothered the fire. Four years before this, foam extinguishers had been invented in St. Petersburg, Russia, by Professor Alexander Laurent. He mixed a solution of aluminum sulfate and sodium bicarbonate with a stabilizing agent. The foam bubbles that were formed contained carbon dioxide gas. They were able to float on burning oil, paint, or petrol and smother a fire. The National Fire Protection Association (NFPA) standard on portable fire extinguishers was developed in 1921.

Fire extinguishers are also grouped by the means of expelling the agent. Five methods are commonly employed. These include self-expelling, gas cartridge or cylinder, stored pressure, mechanically pumped, and hand propelled or applied (Table F.1). See also **Fire Extinguisher Rating**.

Fire Extinguisher Rating

A rating of relative extinguishing effectiveness of a portable fire extinguisher along with the class of fire it is rated for, as specified by National Fire Protection Association (NFPA) 10, Standard for Portable Fire Extinguishers. Fire extinguisher ratings consist of a relative number followed by a fire class letter, i.e., A, B, C, or D. Color coding is also a part of the identification scheme. A green triangle is used for Class A, a red square for Class B, a blue circle for Class C, and a yellow five-pointed star for Class D. Fire extinguishers classified for use on Class C or D hazards are not required to have a number preceding those specific classification letters. A Class C rating is not provided unless a Class A or B rating has been established. Fire extinguisher ratings began around 1948. They were developed in order to classify the appropriate extinguishing agent to the fire hazard and quantify the relative effectiveness of portable fire extinguishers. See also **Fire Extinguisher, Portable**.

Fire Extinguishers

Devices having characteristics essential to extinguish flame. Fire extinguishers may contain either liquid or dry chemicals, or gases (water, dry chemicals, carbon dioxide, etc.). They are tested and rated to indicate their ability to handle specific classes and sizes of fires. Class A extinguishers are for ordinary combustibles, such as wood, paper, and textiles, where a quenching-cooling effect is required. Class B extinguishers are for flammable liquid and gas fires, such as oil, gasoline, paint, and grease, where oxygen exclusion or a flame interruption effect is essential. Class C extinguishers are for fires involving energized electrical wiring and equipment where the nonconductive property of the extinguishing agent is of prime importance. Class D extinguishers are for fires in combustible metals such as magnesium, potassium, powdered aluminum, zinc, sodium, titanium, zirconium, and lithium.

Fire Hazard

Any situation, process, material, or condition that, on the basis of applicable data, can cause a fire or explosion or provide a ready fuel supply to augment the spread or intensity of a fire or explosion and that poses a threat to life, property, continued business operation, or the environment. The relative degree of hazard can be evaluated and appropriate safeguards provided. According to the Bureau of Labor Statistics' Census of Fatal Occupational Injuries Charts, 1992–2007, fires and explosions accounted for 3 percent of workplace fatalities in 2007. See also **Fire Safety, Workplace**.

Fire Hazard Identification

A system of labeling established by the National Fire Protection Association (NFPA) to provided a readily identifiable means to ascertain material hazards. The system identifies fire hazards in three main areas: health, flammability, and reaction or instability. The relative ranking in severity of each hazard category is indicated

F

TABLE F.1
Fire Extinguisher Comparisons

Extinguishing Material	Water and Antifreeze	Wetting Agent	AFFF and FFFP	Load-d Stream	Multi Dry Chemical	CO$_2$	Dry Chemical	Inert Gas Agents	Dry Powder
Self-expelling						X		X	
Cartridge or GN$_2$ cylinder			X	X	X		X		X
Stored pressure	X	X	X	X	X		X	X	X
Pump	X								
Hand applied	X								X

with a numerical value from 0 to 4 (no risk to severe risk). The system is primarily provided for emergency situations and is not intended to apply to normal hazard evaluations. The specific features of the system are outlined in NFPA 704, Standard System for the Identification of the Fire Hazards of Materials. Generally it consists of a diamond-shaped placard, divided into four smaller diamonds or quadrants. Each quadrant is color coded and specifically arranged for the three main hazard areas. The health rating is provided on the left at the 9 o'clock position and is colored blue. The flammability rating is provided at the top or 12 o'clock position and is colored red. The reactivity hazard is provided on the right at the 3 o'clock position and is colored yellow. The relative rankings for each hazard are indicated in each quadrant. Special hazard identifiers are provided in the bottom quadrant at the 6 o'clock position, which is usually white. Special hazard qualifiers generally include radioactivity, explosives, corrosive, water reactive, oxidizer, etc. The NFPA fire hazard identification scheme is somewhat limited as it only identifies relative potential hazards with the individual material. It does not identify the material itself or all of its potential reactions with other materials. See Figure C.3 for a depiction of this placard. See also **Chemical Hazard Label; Hazardous Materials Identification System (HMIS®); NFPA 704, Standard System for the Identification of the Hazards of Materials for Emergency Response**.

Fire Precautions
See **Fire Protection**.

Fire Prevention
Measures directed toward avoiding the inception of fire. Fire prevention and control is the prevention, detection, and extinguishment of fires, including such secondary activities as research into the causes of fire, education of the public about fire hazards, and the maintenance and improvement of firefighting equipment. Up until the 1920s, little official attention was given to fire prevention, because most fire departments were concerned only with extinguishing fires. Since then most urban areas have established some form of a fire-prevention unit. The staff of this unit concentrates on such measures as heightening public awareness; incorporating fire-prevention measures in building design and in the design of machinery and the execution of industrial activity; reducing the potential sources of fire; and outfitting structures with such equipment as extinguishers and sprinkler systems to minimize the effects of fire. The importance of increasing public understanding of the causes of fire and of learning effective reactions in case of fire is essential to a successful fire-prevention program. To reduce the impact and possibility of fire, the building codes of most cities include fire safety regulations. Buildings are designed to separate and enclose areas, so that a fire will not spread. Building fire codes incorporate fire-prevention devices, alarms, and exit signs; to isolate equipment and materials that could cause a fire or explode if exposed to fire; and to install fire-extinguishing equipment at regular intervals throughout a structure. Fire-retardant building materials have also been developed, such as paints and chemicals. They are used to coat and impregnate combustible materials, such as wood and fabric. Perhaps more important than firefighting itself in many modern industrial countries is fire prevention. In Russia and Japan, for example, fire prevention is treated as a responsibility of citizenship. Fire departments are charged with enforcement of the local fire-prevention code and of state fire laws and regulations.

Fire (Flame) Proof
Material incapable of burning. The term *fire proof* is considered false. No material is immune to the effects of a fire possessing sufficient intensity and duration. It is com-

monly, although erroneously, used synonymously with the term "fire resistive." Use of the term is discouraged since it is misleading. See also **Fire Resistive**.

Fire Protection

In general terminology this refers to the prevention, detection, and extinguishment of fire and reduction or avoidance of losses in human terms, assets, business activities, environmental impact, and prestige. In a specific application, it is the providing of fire control or extinguishment. It may also be used to signify the degree to which protection from fire is applied. See also **Fire Prevention**.

Fire Protection Engineering

The discipline of engineering that applies scientific and technical principles to safeguard life, property, loss of income, and threat to the environment from the effects of fires, explosions, and related hazards. It is associated with the design and layout of buildings, industrial properties, structures, equipment, processes, and supporting systems. It is concerned with fire prevention, control, suppression, and extinguishment and provides for consideration of functional, operational, economic, aesthetic, and regulatory requirements.

Fire Protection System

An integrated system that affords protection against fire and its effects. It may be composed of either active or passive fire protection measures. The fire protection system should be commensurate with the level of hazard it is protecting.

Fire Pump

A pump specifically designed, designated, and installed to provide adequate and sufficient water supplies for controlling and suppressing unwanted fires. Firewater pumps are required to be constructed and installed to recognized standards, such as National Fire Protection Association (NFPA) 20, Installation of Centrifugal Firewater Pumps, for reliability in emergency situations and levels of performance. A firewater pump may be driven by an electric motor, diesel engine, or steam turbine. Gasoline engines are not recommended due to their reduced reliability and inherent fire hazard. Fire pumps may be mobile or stationary. Mobile pumps used by fire departments and mounted on trucks are referred to as fire pumpers.

Fire Resistive

Refers to properties of materials or designs to resist the effects of any fire to which the material or structure may be expected to be subject. A building constructed of fire-resistive materials can withstand a burnout of its contents without subsequent structural collapse. Fire resistive implies a higher degree of fire resistance than noncombustible.

Fire Retardant

In general this denotes a substantially lower degree of fire resistance than fire resistive. The term is frequently used to refer to materials or structures that are combustible but have been subjected to treatments or surface coverages to prevent or retard ignition or the spread of fire.

Fire Risk Indexing

A simplified fire safety evaluation of a building (see Table F.2). It consists of analyzing and scoring hazard and other related risk parameters to produce a rapid and simple estimate of relative fire risk. A detailed fire risk evaluation may not include attributes such as human behavior and attitudes. The structure of a risk index system facilitates quantification and inclusion of such factors. Where a quantitative fire safety evaluation is desirable, detailed fire risk assessment may not be cost-effective or appropriate. Fire risk indexing may provide a cost-effective means of fire safety

TABLE F.2
Fire Risk Indexing Table

HISTORIC FIRE RISK INDEX
Fire Safety for Historic House Museums
National Center for Preservation Technology and Training Fire Safety Institute

Location:
Date:

Attribute	Grade	Weight	Score
Fire prevention		15%	
Exits/evacuation		13%	
Significance		13%	
Vertical openings		12%	
Automatic suppression		8%	
Building height and construction		8%	
Compartmentation		8%	
Fuel (contents and finishes)		8%	
Detection and alarm		5%	
Emergency response		5%	
Smoke control		5%	
	Total	100%	

Source: Fire Safety Institute, Reproduced with permission, Ref. *SFPE Handbook of Fire Protection Engineering.* See http://firesafetyinstitute.org/fireriskindex.html

F

evaluation that is sufficient for a large proportion buildings and processes. The Fire Safety Institute was a primary developer of the technique.

Fire-Safe Cigarette
A cigarette that has been manufactured in accordance with the American Society for Testing Materials (ASTM) Standard E 2187, Standard Test Method for Measuring the Ignition Strength of Cigarettes, whereby no more than 25 percent of 40 cigarettes placed on 10 layers of filter paper are to indicate full-length burns. Essentially they have a reduced propensity to burn when left unattended than normal cigarettes. This is commonly achieved by the manufacturer placing two or three thin bands of less porous paper wrapped around the cigarette at points along the cigarette, which slows the burning action of the cigarette. See also **Smoking Safety**.

Fire Safety
The National Fire Protection Association (NFPA) reports that there were 1,451,000 fire incidents in the United States in 2008, which resulted in 3320 fatalities, 16,705 injuries, and $15.5 billion in property losses. Home structure fires caused 83 percent of the fire deaths and 79 percent of the fire injuries. Cooking is the leading cause of fires and injuries for incidents in the home, while smoking is the leading cause of fatalities in home fires. See Figure F.4 for recent causes of home structural fires. More fires are reported on Independence Day (July 4th) than on any other day in the

F

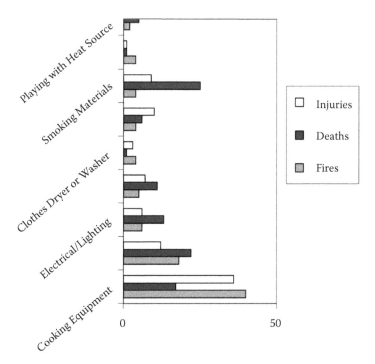

FIGURE F.4   Leading causes of home structural fires 2003–2006 (percentages). (Reproduced from Ahrens, Marty. *Leading Causes of Home Structural Fires 2003–2006.* Copyright © 2009, National Fire Protection Association. With permission.)

year. Overall, statistics from the NFPA for the last 30 years have shown a consistent downward trend in the number of fire incidents, fatalities, and injuries. Occupational Safety and Health Administration (OSHA) regulations cited for fire safety in general industry are 29 CFR 1910 Subparts E, G, H, L, N, O, R, and Z. For the construction industry, the regulations are 29 CFR 1926 Subparts C, D, F, H, J, K, R, S, T, and U. See also **Fire Safety, Home**; **Fire Safety, Vehicles**; **Fire Safety, Workplace**; **Fireworks Safety**.

Fire Safety, Home

The National Fire Protection Association reports that cooking is the leading cause of fires and injuries for incidents in the home, while smoking is the leading cause of fatalities in home fires. Heating equipment contributed to 22 percent of home fire deaths, while electrical equipment was the third leading cause of home fires. Homes that have older wiring are considered at a higher electrical fire risk according to a study by the Consumer Product Safety Commission. Intentionally set fires (arson) are reported as the fourth leading cause of home deaths. See also **Smoking Safety**.

Fire Safety, Vehicles

The National Fire Protection Association (NFPA) reports that Fire Departments in the United States responded to an estimated 278,000 vehicle incident fires in 2006. These incidents resulted in 490 fatalities, 1200 injuries to the occupants, and $1.3 billion in property losses. Performing vehicle maintenance is considered a major factor in preventing vehicle fire incidents.

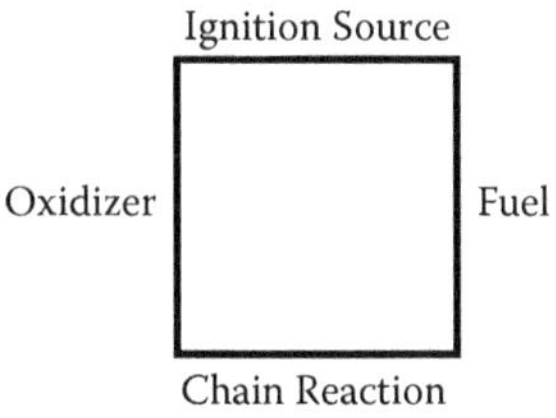

FIGURE F.5   Fire square.

Fire Safety, Workplace

Each year in the United States it is estimated that 70,000–80,000 workplaces experience a serious fire. About 200 employee fatalities per year occur in these fires and another 5000 employees are injured. Property losses from workplace fires exceed $2 billion annually. About 15 percent of workplace fires result from a catastrophic failure of equipment, and approximately 85 percent are caused by factors related to human behavior. In Occupational Safety and Health Administration (OSHA) regulation 29 CFR 1910, Subpart L states regulations for fire safety. This regulation covers the requirements of employer fire brigades, portable fire extinguishers, various fixed fire suppression equipment, fire detection systems, and employee alarm systems. Occupational Safety and Health Administration (OSHA) regulation 29 CFR 1910.38-39 contains regulations for Emergency Action Plans and Fire Prevention Plans. See also **Smoking Safety**.

Fire Safety Concepts Tree

A logic diagram that spans the entire realm of fire safety measures. It is defined in National Fire Protection Association (NFPA) 550, Guide to the Fire Safety Concepts Tree, which describes its structure, application, and limitations.

Fire Safety Institute (FSI)

A nonprofit organization founded in 1981 to encourage an integrated approach to the reduction of loss from fire through informed fire safety decision making. The institute pursues this goal by application of information science to collect and organize current and developing fire safety concepts; research methods of decision analysis to develop improved techniques to utilize fire safety technology; and education of engineers, architects, and other professionals on fire safety principles and practices.

Fire Square

A graphical symbolic representation of the four factors needed for the propagation of combustion or fire. Each side of the square is representative of a factor. The four factors include fuel, oxidizer, ignition source, and chain reaction. Removal or blockage of one of the elements prevents the combustion process from occurring or continuing (Figure F.5). See also **Combustion**; **Fire Tetrahedron**; **Fire Triangle**.

Fire Tetrahedron

A graphical symbolic representation of the four factors needed for the propagation of combustion or fire. Each side of the tetrahedron is representative of a factor. The four factors include fuel, oxidizer, ignition source, and chain reaction. Removal or blockage of one of the elements prevents the combustion process from occurring or continuing. See also **Combustion**.

Fire Training

Where an individual is expected to undertake fire suppression activities, this involves basic education on fire types and classes, the general principles of fire

FIGURE F.6   Fire training. (Photo courtesy of NASA.)

extinguishment, and the hazards involved with firefighting (Figure F.6). Training should include reporting fires, evaluating fire size, practice using the provided portable fire extinguishers or facility fire hoses, and maintaining an exit route. Where additional hazards and special extinguishing methods are required for an assignment, the appropriate higher level of training should be provided. Occupational Safety and Health Administration (OSHA) regulations for firefighting in workplaces is provided in 29 CFR 1910, Subpart L, App A.

Fire Triangle

Geometric symbolic representation of the combustion process whereby each side of the triangle is one independent element of the process, namely fuel (usually in vapor form), oxidizer, and ignition source (of sufficient energy and high temperature to initiate a combustion process). Removal of one element of the triangle stops the combustion process. The chain reaction of fire is sometimes indicated in the middle of the triangle (Figure F.7). Fuel must generally be in a vapor form for combustion. Liquid mist and finely divided particles that are readily converted to vapor have combustion characteristics much the same as vapors. Carbon and some metals and dusts are an exception in that fuel must be in a vapor form for a combustion process. See also **Combustion**; **Fire Square**; **Fire Tetrahedron**.

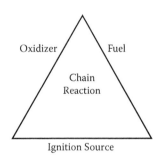

FIGURE F.7   Fire triangle.

Fire Wall

A fire-resistant wall designed to prevent the horizontal spread of fire into adjacent areas that is generally self-supporting and designed to maintain its integrity if the structure on either side completely collapses. If a wood roof is involved, the wall must extend through and above the roof.

Fire Watch

An individual assigned for temporary standby operations where there is a possibility for a fire incident to occur and no other protective means is readily available. A qualified individual is provided to ensure continuous and systematic surveillance of ongoing operations to prevent the evolution of a fire incident. Fire watches are commonly provided for industrial "hot work" activities, e.g., welding, torch cutting, etc., or where a fire protection system has been impaired, e.g., building fire detection and alarm system. The prime objective is to identify and control fire hazards, detect undesirable fires, highlight the occurrence of a fire by use of local means, and notifying the fire department. Fire watches are required as part of the International Fire Code (IFC) 901.7, Occupational Safety and Health Administration (OSHA) regulations 29 CFR 1910.252 (a) (2) (iii) and 29 CFR 1915.504, National Fire Protection Association (NFPA) 51B, and American National Standards Institute (ANSI) Z49.1.

Fireball

An expanding cloud of flaming vapor, usually caused by a sudden release of combustible vapor under high pressure from a vessel or tank. The inner core of the cloud consists mostly of fuel that has been released, while the outer layers consist of a flammable fuel-air mixture. As the buoyancy force of the hot gases increases, the cloud will rise, expand, and assume a spherical shape, resulting in the term fireball.

Fireworks Safety

The National Fire Protection Association (NFPA) reports that in 2007, approximately 9800 individuals were treated for fireworks-related injuries. Forty-nine percent of these injuries were the result of burns. It was reported that injuries were two and half times more likely for children aged 5 to 9 or 10 to 14 than for the general population. The NFPA states that the risk of fire death relative to exposure indicates that fireworks are the riskiest consumer product. Additionally, for 2006, fireworks were estimated to have caused 32,600 reported fires. In a typical year, more fires are reported on Independence Day (July 4th) than on any other day of the year, and fireworks account for half of the reported fire incidents. See also **Pyrotechnics**.

First Aid

The skilled emergency care or treatment of a person who is injured or ill to prevent death or further injury, to relieve pain, and to counteract shock with available materials until medical aid can be obtained. The purpose of first aid is to minimize injury and future disability. In serious cases, first aid may be necessary to keep the victim alive.

First Aid Injury

An injury requiring first aid treatment only.

First Aid Kit

A collection of basic medical supplies or equipment for use in giving first aid. First aid kits may be made up of different contents depending on who has assembled the kit and for what purpose. They may also vary by region due to varying advice or legislation between particular governmental authorities or organizations. The International Organization for Standardization (ISO) identifies first aid kits as having a green color with a white cross, in order to provide easy recognition for individuals requiring first aid.

First Aid Treatment

The following is defined by the Occupational Safety and Health Administration (OSHA) to be first aid treatment: using nonprescription medication at nonprescription strength; administering tetanus immunizations; cleaning, flushing, or soaking surface wounds; applying wound coverings, butterfly bandages, steri-strips; applying hot or cold therapy; providing non-rigid means of support; applying a temporary immobilization device used to transport accident victims; drilling of fingernail or toenail, draining fluid from blister; applying eye patches; removing foreign bodies from eye using irrigation or cotton swab; removing splinters or foreign material from areas other than the eye by irrigation, tweezers, cotton swabs, or other simple means; applying finger guards; providing massages; providing drinking fluids for relief of heat stress.

First Out Alarm

Computer emergency shut down logic that discriminates from a group of inputs the input that tripped first to cause a shutdown. It may also be called First-Up Alarm.

First Responder

The first trained personnel to arrive on the scene of a motor vehicle accident, hazardous material incident, or drowning. Usually refers to officials from local emergency services, firefighters, and police.

Fit Factor

A quantitative estimate of the fit of a particular respirator to a specific individual, which typically estimates the ratio of the concentration of a substance in ambient air to its concentration inside the respirator when worn.

Fit Test

The qualitative or quantitative evaluation of the fit of a respirator on an individual.

Five Whys Technique

An investigative procedure for identifying root causes by determining why each causal factor or item of note occurred or existed. Available evidence is used and Who, What, When, Where, and How are asked until the root cause is determined, i.e., management system failure(s). See also **Root Cause**.

Fixed Guard

A shield or barrier that has no moving parts associated with it or is not dependent upon the mechanism of any machinery, which when in position, prevents access to a hazard or area. See also **Automatic Guard**; **Distant Guard**; **Machine Guarding**.

Flagman

An individual who directs traffic through a construction site or other temporary traffic control zone passing area using signs or flags (Figure F.8). They are responsible for maintaining the safety and efficiency of traffic, as well as the safety of road workers, while allowing construction, incident recovery, or other tasks to proceed. They are commonly used to control traffic when two-way roads are reduced to one lane, and traffic must alternate in direction. Their duties are to direct traffic to safer areas where construction, incidents, and severe traffic are taking place. They guide motorists to follow the traffic laws, but may not be able to enforce the law.

Flame

The visible heat rays that appear when the ignition of a material is reached. Hydrogen is one of the exceptions since the heat rays are not visible.

FIGURE F.8   Flagman.

Flame (Flash) Arrestor
Devices utilized on vents for flammable liquid or gas tanks, storage containers, cans, gas lines, or flammable liquid pipelines to prevent flash-back (movement of flame) through the line or into the container when a flammable or explosive mixture is ignited.

Flame Propagation
The spread of flame throughout a combustible vapor area, which may be in a container or across a surface, independent of the ignition source. Generally used in connection with the capability and rate of such movement.

Flameout
The unexpected loss of a burner flame during furnace operations, which can create potentially explosive conditions due to buildup of combustible vapors.

Flammable
In a general sense, this refers to any material that is easily ignited and burns rapidly. Numerous testing method are available for testing the flammability of materials; therefore, one has to know or specify the test standard that should be required or applied to determine the specific flammability of a material. A universal test standard has not been adopted to define the flammability of materials. It is synonymous with the term inflammable, which is generally considered obsolete due to its prefix "in" which may be incorrectly misunderstood as not flammable (e.g., incomplete is not complete).

Flammable Limits
See **Explosive Limits**.

Flammable Liquid
As defined by the National Fire Protection Association (NFPA), 30, Flammable and Combustible Liquids Code, any liquid having a flash point below 37.8°C (100°F), except any mixture having components with flash points of 37.8°C (100°F) or higher, the total of which make up 99 percent or more of the total volume of the mixture. Flammable liquids are known as Class I liquids. Class I liquids are divided into three classes: IA, IB, and IC. Class IA includes liquids having flash points below 22.8°C (73°F) and a boiling point below 37.8°C (100°F). Class IB includes liquids having flash points below 22.8°C (73°F) and a boiling point at or above 37.8°C (100°F).

Class IC includes liquids having flash points at or above 22.8°C (73°F) and a boiling point below 37.8°C (100°F). See also **Combustible Liquid.**

Flammable Vapor

A concentration, by volume, of vapors in air from a flammable liquid within the lower and upper flammable limits.

Flare

The flame condition of a fire in which burning occurs with an unsteady flame. In the process industries (i.e., chemical and petroleum) a flare refers to a primary fire safety system used for the safe remote disposal of gases by burning from normal processes or emergency conditions. Process gases that cannot be safely disposed of may contribute to fire destructiveness or cause a vessel rupture or BLEVE (Boiling Liquid Expanding Vapor Explosion). American Petroleum Institute (API) Recommended Practice 521, Guide for Pressure Relieving and Depressurizing Systems, provides information on the design and permissible heat radiation levels for flares.

Flash Burn

Injury or destruction of body tissue caused by exposure to a flash or sudden release of intense radiant heat.

Flash Point

The lowest temperature of a liquid at which it gives off sufficient vapors to form an ignitable mixture with the air near the surface of the liquid or within the vessel used. The flash point can be determined by the open cup or the closed cup method. The latter is commonly used to determine the classification of liquids that flash in the ordinary temperature range.

Flotation Device

See **Personal Flotation Device (PFD).**

Flux

Usually refers to a substance used to clean surfaces and promote fusion in soldering. However, fluxes of various chemical natures are used in the smelting of ores in the ceramic industry, in assaying silver and gold ores, and in other endeavors. The most common fluxes are silica, various silicates, lime, sodium and potassium carbonate, and litharge and red lead in the ceramic industry.

FM Approved

The phrase ''FM Approved'' is used to describe a product or service that has satisfied the criteria for approval by Factory Mutual (FM) Approvals. The FM Approval Guide has a complete list of products and services that are FM Approved. See also **Factory Mutual (FM).**

FM Data Sheets

Engineering guidelines produced by Factory Mutual (FM) to help reduce the risk of property loss due to fire, weather, and/or electrical or mechanical equipment failure. They are based on input from loss experience, research results, consensus standards committees, equipment manufacturers, and other interested participants. The subjects covered include construction, sprinklers, water supply, extinguishing equipment, electrical equipment, boilers and industrial heating equipment, hazards, storage, miscellaneous, human factors, systems instrumentation and control, pressure vessels, mechanical, welding, and boiler and machinery. They may also be referred to as FM Global Property Loss Prevention Data Sheets. See also **Factory Mutual (FM).**

Foam, Firefighting

A fluid aggregate of gas- or air-filled bubbles formed by chemical or mechanical means that will float on the surface of flammable liquids or flow over solid surfaces. The foam functions to blanket and extinguish fires and/or to prevent ignition of material.

F

Fog Index

A measure of readability of documents in the English language developed by Robert Gunning in 1952. It is based on a combination of two criteria: (1) the average number of words per sentence, and (2) the percentage of words containing three or more syllables. A great number of studies have concluded that most health information is too difficult for the average adult reader. Much is written at the post-graduate level, while the average adult in the United States reads at the 7th-grade level. This means that less than 15 percent of the adult population will read it. The fog index is generally used by people who want their writing to be read easily by a large segment of the population. The resulting number is an indication of the number of years of formal education that a person requires in order to easily understand the text on the first reading. That is, if a passage has a fog index of 12, it is at the reading level of a U.S. high school senior. Texts that are designed for a wide audience generally require a fog index of less than 12. Texts that require a close-to-universal understanding generally require an index of less than 8. It has been estimated that the fog index has 80 percent accuracy in predicting the difficulty of a written passage. It may also be called Gunning Fog Index.

The Gunning Fog Index can be calculated as follows:

Take a full passage that is around 100 words (do not omit any sentences).

Find the average sentence length (divide the number of words by the number of sentences).

Count words with three or more syllables (complex words), not including proper nouns (for example, Djibouti), familiar jargon or compound words, or common suffixes such as -es, -ed, or -ing as a syllable.

Add the average sentence length and the percentage of complex words (e.g., +13.37%, not simply +0.1337).

Multiply the result by 0.4.

The steps may be written by the formula:

$$0.4 \times [(\text{Words/Sentence}) + 100 \, (\text{Complex Words/Words})]$$

Foolproof

So plain, simple, obvious, and reliable as to leave no opportunity for error, misuse, or failure to implement the correct action. Safety systems or devices sometimes are designed to be foolproof to reduce the possibility of failure or error.

Foot Protection

Per Occupational Safety and Health Administration (OSHA) regulations, employers shall ensure that each affected employee wears protective footwear when working in areas where there is a danger of foot injuries due to falling or rolling objects or objects piercing the sole. Such equipment shall comply with American National Standards Institute (ANSI) Z-41-1991, "American National Standard for Personal Protection-Protective Footwear." Protective footwear is a part of personal protective equipment (PPE) that is typically worn in industry. See also **ASTM F 2412, Test Methods for Foot Protection, and ASTM F 2413, Standard Specification for Performance Requirements for Foot Protection; Safety Shoe.**

Footcandle (fc)

A measure of illuminance produced on a surface all points of which are 1 foot from a directionally uniform point of 1 candela. That illuminance is 1 lumen/ft^2 or 1 footcandle. Under SI, illuminance is measured in terms of lux units, where 1 lux = 1 $lm/m^2 \times cd \times sr$, where m = meter, cd = candela, and sr = steradian for solid angle. To convert: 1 footcandle = 10.76391 lux. The footcandle (or lumen per square foot) is a non-SI unit of illuminance. Its use is common mainly in the United States, particularly in construction-related engineering and in building codes. Because lux and footcandles are different units of the same quantity, it is perfectly valid to convert footcandles to lux and vice versa. See also **LUX**.

Foreseeability

The legal theory that a person may be held liable for actions that result in injury or damage only where the person was able to foresee dangers and risks that could reasonably be anticipated.

Foreseeable Emergency

Potential occurrence such as equipment failure, rupture of containers, or failure of control equipment, which could result in an uncontrolled release of a hazardous chemical.

Forklift Safety

Each year, the Occupational Safety and Health Administration (OSHA) records that about 100 warehouse employees die and 95,000 are injured every year in forklift incidents. The majority of fatalities are caused by forklift turnovers. Being crushed between a forklift and another surface makes up the second highest percentage, followed by getting struck by a forklift and then getting hit by falling material from a dropped load. Some employees are injured when forklifts are inadvertently driven off loading docks, forklifts fall between docks and unsecured trailers, employees are struck by a forklift, or when employees fall while on elevated pallets. Most incidents also involve property damage, including damage to overhead sprinklers, racking, pipes, walls, and machinery. Most employee injuries and property damage can be attributed to lack of safe operating procedures, lack of safety-rule enforcement, and insufficient or inadequate training. OSHA standard 29 CFR 1910.178 covers the construction and operations of powered industrial trucks (i.e., forklifts) in general industrial locations.

Four-to-One Ratio

An arbitrary ratio frequently used in the comparison of the indirect costs of an incident to the direct costs. Generally considered obsolete since no fixed ratio exists among various types of exposures.

Frequency (in Cycles per Second [CPS], Hertz [Hz])

The time rate of repetition of a periodic phenomenon. The frequency is the reciprocal of the period. It defines pitch or the highness or lowness of sound.

Frostbite

A freezing injury of the skin due to exposure to extreme cold. It may be recognized by whitening of the skin and loss of sensation. It is treated by rapid thawing. Deep frostbite with freezing of tissues deep to the skin generally results in dry gangrene.

Full Body Harness

See **Body Harness, Full**.

Fume

Solid particles generated by condensation from the gaseous state, generally after volatilization from a solid, usually from molten metals. A fume is formed when a volatilized solid, such as metal, condenses in cool air. For example, a hot zinc vapor may form when zinc-coated steel is welded. The vapor then condenses to form fine zinc fume as soon as it contacts the cool surrounding air. The solid particles that make up a fume are extremely fine, and therefore are more easily breathed into the lungs. In most cases, the hot material reacts with the air to form an oxide.

Fume Fever

An acute condition caused by a brief high exposure to the freshly generated fumes of metals.

Function Check

A function check, or bump test, means using simple tests (such as exposing sensors to calibration gas or exhaling into the oxygen sensor) to show that the instrument will respond to the chemical(s) of concern and that all alarms operate as they were designed.

Functional Safety Assessment

Activity performed by a competent senior engineer to determine if the safety system meets the specification and actually achieves functional safety (freedom from unacceptable risk). This assessment is an important part of reducing systematic failures. It must be performed at least after commissioning and validation but before the hazard is present.

Fuse, Electrical

A form of electrical circuit protection that comprises a strip of metal of such size to melt at a predetermined value of current flow. It is placed in the electrical circuit, and upon melting, due to excessive current flow, prevents the flow of electricity supply to the circuit.

Fusible Link

A release device that is activated by the heat effects of a fire. It typically consists of two pieces of metal joined by a low-melting-point solder. They are manufactured in various temperatures and with varying amounts of applied tension. When installed and the rated temperature is reached, the solder melts and the two metal parts separate, initiating the desired fire safety function.

Fusible Plug

A hollowed threaded plug having the hollowed portion filled with a low-melting-point material. This element is often used to provide a mechanical relief device triggered by temperature causing the process fluid to vent when the plug material melts. It is typically used in a pneumatic fire detection system or sprinkler system to indicate the presence of a fire condition and disperse the gas or fluid in the contained system.

F

G

Gas Detector

A device that detects the presence of various gases within an area, usually as part of a system to warn about gases that might be harmful to humans or animals. Various types of gas detectors are used to detect primarily combustible, toxic, and oxygen levels (i.e., deficiency), and carbon monoxide and carbon dioxide gases. They are typically used in confined spaces, utilities and chemical plants, industrial hygiene applications, and air quality testing. They can be fixed devices or portable devices worn by an individual. See also **Combustible Gas Detector**.

Gas Mask

A face covering connected to its own purifying device, which filters harmful gases or pollutants from the air so uncontaminated air may be inhaled. Gas masks do not add oxygen to air and cannot be used where there is oxygen deficiency. See also **Respirator**.

Gases

Normally formless fluids that occupy a space or enclosure and can be changed to the liquid or solid state only by the combined effect of increased pressure and decreased temperature. Depending on their individual characteristics, gases may be harmful to humans, e.g., carbon monoxide is a toxic gas, but, being colorless, odorless, tasteless, and non-irritating, it is very difficult for individuals to detect.

Gauntlets

Protective shirt arm material, similar to safety cuffs, but extending farther down the arm to provide protection to the lower forearm. See also **Safety Cuffs**.

Geiger Counter

A common trade name for a device for the detection and measurement of radioactivity. It can be used for the monitoring of background radiation at particularly low levels.

General (Dilution) Ventilation

A ventilation system consisting of mechanical air movement to mix with and dilute the contaminants in the space and exhaust them to the outside. This is not the type of ventilation that is designed to maintain normal ventilation and thermal comfort in a space. General ventilation is not designed to control highly toxic materials or large quantities of air contaminants.

General Industry Safety Orders (GISO)

Occupational Safety and Health Administration (OSHA) orders that apply to all businesses and industries.

General Industry Safety Standards

Generally refers to the Occupational Safety and Health Administration (OSHA) standards and requirements defined in 29 CFR 1910. OSHA uses the term "general industry" to refer to all industries not included in agriculture, construction, or maritime. General industries are regulated by OSHA's general industry standards, directives, and standard interpretations.

General Liability Insurance

A form of insurance designed to protect owners and operators of businesses from a wide variety of liability exposures. The exposures may include liability arising out of incidents resulting from the premises or the operations of the insured, products sold by the insured, operations by the insured, and contractual liability. See also **Liability Insurance**.

Glare

The sensation produced by luminance within the visual field that is sufficiently greater than the luminance to which eyes are adapted to cause annoyance, discomfort, or loss in visual performance or ability (i.e., eye strain).

Goals-Freedom-Alertness Theory

A theory of accident proneness that proposes that individuals have incidents due to a lack of alertness brought about by having no choice or freedom in setting or choosing the goals established at work. Where workers have the ability to set and freely pursue attainable goals should result in a "rich" work climate with alertness and fewer incidents, i.e., the workers stay focused and quality improves.

Goggles, Safety

They seal the entire eye area to protect the eyes against dust, impacting objects, chemical splashes, strong light, sparks, or other harmful environmental influences. They are contoured for full facial contact and held in place by a headband or other suitable means, and are required to meet American National Standards Institute (ANSI) Z87 and be clearly marked with the manufacturer's name. See also **ANSI Z87.1-2003, Standard for Occupational and Educational Eye and Face Protection Devices**.

Going Postal

A slang expression that is increasingly being applied to highlight workplace violence. Its origin occurred after several U.S. postal worker incidents that occurred in 1980s, in which managers, coworkers, police, and the public were fatally injured due to incidents of workplace rage. The phrase has now been applied to murders committed by employees in acts of workplace rage, irrespective of the employer. It is generally used to describe any fits of rage, though not necessarily at the level of murder, in or outside the workplace. Researchers have found that the homicide rates for workers at postal facilities were actually lower than at other workplaces. In major industries, the highest rate of homicides for workers was in retail. The next highest rate was in public administration, which includes police officers. Taxi driving has the highest homicide rate. Not all murders on the job are directly comparable to "going postal." Taxi drivers, are much more likely to be murdered by passengers than by their peers. Working in retail exposes individuals to store robberies. See also **Workplace Stress**.

Graduate Safety Practitioner (GSP)

An interim designation awarded to graduates of qualified academic programs to show progress toward the Certified Safety Professional® (CSP®) recognition title awarded by the Board of Certified Safety Professionals (BCSP). See also **Certified Safety Professional® (CSP®)**.

Green Cross

Commonly a symbolic reference to safety or commercially a symbol for pharmacies. It is also the logo of the National Safety Council (NSC). The Japanese Industrial Standard Z9103 designates the symbol as the safety indication sign. The color green normally symbolizes healing and nurturing, while the cross is considered a sign of faith. See also **Blue Cross; National Green Cross for Safety Medal; Red Cross**.

Ground Fault Circuit Interrupter (GFCI)

A fast-acting circuit breaker that is sensitive to very low levels of current leakage to ground. They are designed to provide protection against electrical shock from ground faults, or leakage currents, which occur when the electrical current flows outside of the circuit conductors. The GFCI is designed to limit the electric shock to a current and time duration value below that which can produce serious injury. See also **Arc Fault Circuit Interrupter (AFCI)**.

Grounding

The procedure or method used to carry an electrical charge to ground through a conductive path to avoid an electrical buildup that may eventually harm individuals from a shock hazard or initiate a fire or explosion as an ignition source. A typical ground may be connected directly to a conductive pipe or to a grounding bus and ground rod. See also **Ground Fault Circuit Interrupter (GFCI)**.

Guard

An enclosure that prevents entry into the point of operation of a machine or renders contact with any substance or object harmless. See also **Guarded**; **Machine Guarding**.

Guard, Fixed

See **Fixed Guard**.

Guard, Interlocking Barrier

An enclosure attached to the frame of a machine and interlocked so that the machine cycle cannot be started normally unless the guard, including its hinged or movable sections, are in position. In some situations, movement of the guards will interrupt the machine cycle. See also **Machine Guarding**.

Guarded

Protection afforded by being covered, shielded, fenced, enclosed, or otherwise protected by means of suitable enclosures, covers, casings, shields, troughs, railings, screens, mats, or platforms, or by location, to prevent injury.

Guarding, Proximity

See **Machine Guarding**.

Guarding System

A combination of physical safeguards and safety devices applied to the work environment.

Guardrail System, Scaffold

A protective barrier consisting of top rails, mid-rails, toeboards, and supporting uprights, erected to protect individuals from falling off an elevated work area and to prevent objects from falling onto individuals below.

Guardrails

Top rails and mid-rails secured to uprights/posts, erected at the exposed sides and ends of platforms to prevent workers from falling off the elevated work area or to prevent them from reaching or contacting a potential hazard. See also **Railings**.

H

Habitual Violator

An individual whose record, during a given time period, shows reports of repeated violations of laws or regulations. In traffic safety, a habitual violator is usually any driver whose record during a consecutive 12-month period shows reports of more than three convictions for traffic violations or more than five times the average number of convictions for all drivers in the state, whichever is greater. See also **Accident Prone Theory**.

Haddon Matrix

An analytical tool for evaluating the cause of injuries based on the host (i.e., the person injured), the agent (i.e., what caused the injury, e.g., electrical energy), and the environment (i.e., the physical and social context in which the injury occurred). These aspects are looked at over the time periods leading up to the injury event, the injury event itself, and directly after the event. These are arranged and identified in a matrix named for Dr. William Haddon, a physician and engineer who worked on the design of safer roads in the United States in the late 1950s. It concludes with a three-tiered approach to injury prevention, which includes behavioral, environmental, and policy changes.

Hand Protection

The protection of hands from injuries that may result from cuts, abrasions, burns, and from manual handling operations and extremes of temperature. The protection commonly used includes gloves and gauntlets.

Handrail

A building component located at or near the open sides of elevated walking surfaces, which minimizes the possibility of a fall from the walking surface to a lower level. It typically consists of a single bar or pipe supported on brackets from a wall or partition, as on a stairway or ramp, to furnish persons with a handhold in case of tripping or falling. Handrails are required to meet certain building code strength requirements (e.g., minimum code-prescribed load requirements specified in International Building Code [IBC] Section 1607.7).

Hard Guard

Screens, bars, or other mechanical barriers affixed to the frame of a machine, intended to prevent entry by personnel into the hazardous area(s) of a machine, while allowing the point of operation to be viewed. The maximum size of openings is determined by Table O-10 of Occupational Safety and Health Administration (OSHA) standard 1910.217. It may also be called a fixed barrier guard.

Hard Hat

A protective hat used in industry to safeguard the head of an individual from a certain level of hazards anticipated in industry and in construction activities (Figure H.1). It must meet certain impact tests and may have some electrical insulation as defined in the American National Standards Institute (ANSI) standard for their manufacture. It may also be called a Safety Cap, Safety Hat, or Safety Helmet. See also **ANSI Standard Z89.1-2003, Protective Headware for Industrial Workers**, **Bump Cap**.

FIGURE H.1   Hard hat.

Harm

Injury (physical or mental) or physical damage.

Harmful

Causing or capable of causing harm.

Hazard

A hazard is the potential for adverse or harmful consequences. In practical terms, a hazard is often associated with a condition or activity that, if left uncontrolled, can result in an injury, illness, property damage, business interruption, harm to the environment, or an impact on the reputation of an entity.

Hazard Abatement

See **Abatement**.

Hazard Analysis

An analysis performed to identify hazardous conditions for the purpose of their elimination or control.

Hazard Analysis and Critical Control Point (HACCP)

A systematic, preventive approach to food and pharmaceutical safety that includes physical, chemical, and biological hazards as a means of prevention rather than finished product inspection. It has seven key principles, which are conduct a hazard analysis, identify critical control points, establish critical limits for each critical control point, establish critical control point monitoring requirements, establish corrective actions, establish record-keeping procedures, and establish procedures for ensuring the HACCP system is working as intended. HACCP is used in the food industry to identify potential food safety hazards, so that key actions, known as Critical Control Points (CCPs), can be taken to reduce or eliminate the risk of the hazards being realized. The system is used at all stages of food production and preparation processes including packaging, distribution, etc. The Food and Drug Administration (FDA) and the United States Department of Agriculture (USDA) indicate that HACCP programs for juice and meat are an effective approach to food safety and protecting public health. Meat HACCP systems are regulated by the USDA, while seafood and juice are regulated by the FDA. The use of HACCP is currently voluntary in other food industries. See also **Critical Control Point (CCP)**.

Hazard and Operability Study (HAZOP)

A qualitative investigative safety review technique. The HAZOP technique was developed in the early 1970s by Imperial Chemical Industries Ltd. HAZOP can be defined as the application of a formal systematic critical examination of the process and engineering intentions of new or existing facilities. Its function is to assess the hazard potential that arises from deviation in design specifications and the consequential effects on the facilities as a whole. This technique is usually performed by a qualified team that uses a set of prompting guidewords, i.e., More/No/Reverse Flow, High/Low Pressure, High/Low Temperature, etc., to identify concerns from the intended design. From these guidewords, the team can identify scenarios that may result in a hazard or an operational problem. The consequences of the hazard and measures to reduce the frequency with which the hazard will occur are then discussed. This technique had gained wide acceptance in process industries (i.e., oil, gas, chemical processing) as an effective tool for plant safety and operability improvements. HAZOPs are also used in a wide range of other industries, and there is extensive support in the form of published literature and software packages. See also **What-If Analysis (WIA)**.

Hazard Classification

A designation of relative incident potential based on probability of incident occurrence.

Hazard Communication (HAZCOM)

Primarily refers to evaluating the hazard of chemicals used in the workplace and communicating the information concerning the hazard of chemicals and appropriate protective measures to employees as required by Occupational Safety and Health Administration (OSHA) requirements under 29 CFR 1910.1200.

Hazard Communication Standard (HCS)

Occupational Safety and Health Administration (OSHA) regulation 29 CFR 1910.1200 et seq., Hazard Communication Standard (HCS). It requires the development and dissemination of such information. Chemical manufacturers and importers are required to evaluate the hazards of the chemicals they produce or import, and prepare labels and Material Safety Data Sheets (MSDSs) to convey the hazard information to their downstream customers. All employers with hazardous chemicals in their workplaces must have labels and MSDSs for their exposed workers, and train them to handle the chemicals appropriately. See also **Hazardous Communication (HAZCOM)**.

Hazard Control

That function in an organization directed toward the recognition, evaluation, and reduction or elimination of the destructive effects of hazards emanating from human acts of commission and omission, and from the physical and environmental aspects of the workplace to prevent their re-occurrence.

Hazard Elimination

To remove the probability of an incident occurring, and the chance of individuals being injured or harmed.

Hazard Evaluation
A component of risk evaluation that involves gathering and evaluating data on the types of health injuries or diseases that may be produced by a chemical and on the conditions of exposure under which such health effects are produced.

Hazard Identification (HAZID)
A tool for systematically reviewing and identifying risks for an operation or facility. The results are usually listed in a Hazard Identification List or Hazard Register.

Hazard Identification Plan (HIP)
A thorough, qualitative review of potential hazards for construction work activities that should include a list of all hazards, their assessment (considering severity and probability), and corrective actions to mitigate or prevent worker injuries or illness, and is prepared prior to work activities.

Hazard Information Bulletin (HIB)
See **Safety and Health Information Bulletins (SHIBs)**.

Hazard Label
See **Hazardous Materials Identification System® (HMIS®)**.

Hazard Level
A qualitative measure of hazards used in government or industry stated in relative terms. Category I: Negligible—will not result in personal injury or system damage. Category II: Marginal—can be counteracted or controlled so that no injury to personnel or major system damage will be sustained. Category III: Critical—will cause personal injury or major system damage or both. Category IV: Catastrophic—will cause death to personnel.

Hazard Matrix
A category-based method for assigning a Safety Integrity Level (SIL). The user must create a matrix that assigns defined categories to the consequence (one axis dimension) and likelihood (other axis dimension) components of the risk with a SIL assignment associated for each entry in the matrix. In some cases, quantitative tools, such as Layers of Protection Analysis (LOPA), are used to assist the analyst in determining which category to use, but often the assignment is done qualitatively, using engineering judgment.

Hazard Pay
Extra compensation payments to workers in dangerous occupations or while engaged in work where the probability of an injury is greater than normal.

Hazard Point
The closest reachable point of the hazardous area.

Hazard Recognition
The act or process of identifying, recognizing or condition of being recognized, the probability of being injured or harmed and its acceptance or acknowledgment.

Hazard Sign
A sign with symbols or wording to indicate a risk to individuals. The arrangement and colors of signs that concern hazards are defined by American National Standards Institute (ANSI) Z535.2 and Z535.3 to ensure consistency and understanding by society. See also **Hazard Symbols**.

Hazard Symbols
Easily recognizable icons designed to warn about hazardous materials or locations. The use of hazard symbols is usually regulated by law and directed by standards organizations. Hazard symbols may appear with different colors, backgrounds, borders, and supplemental information in order to signify the type of hazard. Table H.1 provides examples of some common hazard symbols. See also **ANSI Z535.3, Criteria for Safety Symbols; Chemical Hazard Label**.

H

TABLE H.1
Examples of Some Common
Hazard Symbols

Type	Symbol
Danger Warning Caution (for Injuries)	
Toxic	
Radiation	
Biohazard	
High Voltage	
Fire	
Personal Hygiene	
Static Sensitive Device	
Traffic (Road Condition)	

Hazardous Area

An area that poses an immediate or impending physical hazard.

Hazardous Area, Electrical

A U.S. classification for an area in which explosive gas/air mixtures are, or may be expected to be, present in quantities such as to require special precautions for the construction and use of electrical apparatus.

> Division 1 (hazardous). Where concentrations of flammable gases or vapors exist (a) continuously or periodically during normal operations; (b) frequently during repair or maintenance or because of leakage; or (c) due to equipment breakdown or faulty operation, which could cause simultaneous failure of electrical equipment.
>
> Division 2 (normally nonhazardous). Locations in which the atmosphere is normally nonhazardous and may become hazardous only through the failure of the ventilating system, opening of pipe lines, or other unusual situations.

> Areas not classified as Division 1 or Division 2 are considered nonhazardous.

Hazardous Atmosphere

An atmosphere that, by reason of being explosive, flammable, poisonous, corrosive, oxidizing, irritating, oxygen deficient, toxic, or otherwise harmful, may cause death, illness, or injury.

Hazardous Condition

The physical condition or circumstance that is causally related to incident occurrence. The hazardous condition is related directly to both the incident type and the agency of the incident.

Hazardous Liquid

A hazardous liquid is a liquid that is dangerous to human health or safety or the environment if used incorrectly or if not properly stored or contained.

Hazardous Material

A hazardous material is any substance or material that is dangerous to human health or safety or the environment if used incorrectly or if not properly stored or contained. See also **Hazardous Substance**.

Hazardous Material (HAZMAT) Response Team

An organized group of employees, designated by the employer, who are expected to perform work to handle and control actual or potential leaks or spills of hazardous substances requiring possible close approach to the substance. The team members perform responses to releases or potential releases of hazardous substances for the purpose of control or stabilization of the incident. Occupational Safety and Health Administration (OSHA) Regulation 29 CFR 1910.120, Hazardous Waste Operations and Emergency Response, contains information on requirements for HAZMAT responders.

Hazardous Materials Identification System (HMIS®)

A simple identification system used to communicate to workers the hazards of materials. HMIS® is somewhat similar to the National Fire Protection Association (NFPA) 704 Hazardous Material placard, which is used for emergency response incidents. Instead of the NFPA diamond shape, which many feared would be confused with the placarding system, the HMIS® uses a four-color bar system. The top bar indicates the level of health hazard. The second bar from the top is red for Flammability; the third bar from the top is yellow for Reactivity, which is used in the second edition, while Physical Hazard is used in the third edition of the guidance; and white at the bottom is used for Personal Protection. A white square is provided at the right of each of the colored bars to indicate its hazard level from 0 to 4, similar to the NFPA 704 placard. For the personal protection white bar, instead of a hazard ranking a level of protection is indicated by a letter, with each letter specifying a different level of protection (see Table H.2). It was developed by the National Paint & Coatings Association, Inc. (NPCA), to aid employers in the implementation of an effective Hazard Communication Program, as required by the Occupational Safety and Health Administration (OSHA), Hazard Communication Standard (HCS), 29 CFR 1910.1200, and is described in their guideline HMIS® III, Implementation Manual, Third Edition. The system is designed to be flexible and for the user to exercise "professional judgment" in deriving the specific ratings. It is to be used according to a specific set of guidelines to develop ratings for substances and mixtures. The guide also states that the ratings are accurate according to a group of professionals at a point in time. New toxicological data or research information may have been published since the guide's publication date that renders one or more of the ratings outdated. HMIS® is registered trademark of the National Paint & Coatings Association, Inc. See also **Hazard Communication Standard (HCS)**.

H

TABLE H.2
HMIS® Personal Protective Equipment (PPE) Indicators

[Chemical Name]

Health	0–4
Flammability	0–4
Reactivity (Version 2)	0–4
Physical Hazard (Version 3)	0–4
Personal Protection	[Ltr]

Note: 0, minimal hazard; 1, slight hazard; 2, moderate hazard; 3, serious hazard; 4, severe hazard.

In some cases you may see an asterisk (*) by the number in the Health box. This means there is a long-term (chronic) risk.

Personal Protection Indicators

Letter	Personal Protective Equipment (PPE) Required
A	Safety glasses
B	Safety glasses and gloves
C	Safety glasses, gloves, and an apron
D	Face shield, gloves, and an apron
E	Safety glasses, gloves, and a dust respirator
F	Safety glasses, gloves, apron, and a dust respirator
G	Safety glasses and a vapor respirator
H	Splash goggles, gloves, apron, and a vapor respirator
I	Safety glasses, gloves, and a dust/vapor respirator
J	Splash goggles, gloves, apron, and a dust/vapor respirator
K	Air line hood or mask, gloves, full suit, and boots
L–Z	Custom PPE specified by employer

Hazardous Ranking System (HRS)
The principal screening tool used by the Environmental Protection Agency (EPA) to evaluate risks to public health and the environment associated with abandoned or uncontrolled hazardous waste sites. The HRS calculates a score based on the potential of hazardous substances spreading from the site through the air, surface water, or ground water, and on other factors such as density and proximity of human population. This score is the primary factor in deciding if the site should be on the National Priorities List and, if so, what ranking it should have compared to other sites on the list.

Hazardous Substance
Any material that poses a threat to human health and/or the environment. Typical hazardous substances are toxic, corrosive, ignitable, explosive, or chemically reactive. It may also refer to any substance designated by the Environmental Protection Agency (EPA) to be reported if a designated quantity of the substance is spilled in the waters of the United States or is otherwise released into the environment. See also **Hazardous Material**.

Hazardous Waste
As defined under the Resource Conservation and Recovery Act (RCRA), any solid or combination of solid wastes that, because of its physical, chemical, or infectious characteristics, may pose a hazard when improperly disposed of. Possesses at least one of four characteristics (ignitability, corrosivity, reactivity, or toxicity), or appears on special Environmental Protection Agency (EPA) lists. See also **Resource Conservation and Recovery Act (RCRA)**.

Hazardous Waste Manifest
A specific shipping document required by the Department of Transportation (DOT) and the Environmental Protection Agency (EPA) for hazardous waste shipments. Also referred to as the Uniform Hazardous Waste Manifest (UHWM). If all DOT requirements, i.e., the basic description (proper shipping name, hazard class/division, ID No., and packing group) are entered on the UHWM, the manifest may be used as a shipping paper per 49 CFR 172.205.

Hazardous Waste Operations and Emergency Response (HAZWOPER)
Occupational Safety and Health Administration (OSHA) regulations 29 CFR 1910.120, for response to hazardous materials incidents.

HAZCOM
Occupational Safety and Health Administration (OSHA) Hazard Communication Standard 29 CFR 1910.1200. It details requirements on labeling and Material Safety Data Sheets (MSDSs) for materials considered hazardous. Information on material hazards is to be communicated by employers to employees. See also **Hazard Communication (HAZCOM); Hazard Communication Standard (HCS)**.

HAZCOM Label
Identification labels applied to containers of hazardous materials to readily inform employees of their workplace hazards as required under Occupational Safety and Health Administration (OSHA) regulation 29 CFR 1910.1200. They use an easy to understand identification hazard rating. See also **Hazardous Materials Identification System (HMIS®)**.

HAZMAT
See **Hazardous Material**.

HAZOP
See **Hazard and Operability Study (HAZOP)**.

HAZWOPER

Acronym for Hazardous Waste Operations and Emergency Response. It is in reference to the Occupational Safety and Health Administration (OSHA) Standard 29 CFR 1910.120, to protect employees from hazardous waste operations.

Head Injury

Damage to any of the structures of the human head as a result of trauma. While the term "head injury" is most often used to refer to an injury to the brain, head injuries may also involve the bones, muscles, blood vessels, skin, and other organs of the face or head. A head injury does not always mean that there is an associated brain injury. Most head injuries are caused by blows to the head from numerous causes including motor vehicle accidents and falls. Head injuries are one of the most common causes of death and disability in the United States. See also **Trauma**.

Head Protection

Devices used to protect the head from injury. These include safety helmets or hard hats that are primarily used to protect the head from falling objects or overhead hazards; industrial scalp protectors such as bump caps that protect against striking fixed objects, scalping, or entanglement; and caps and hair nets, which are used to prevent the hair from coming in contact with moving machinery or the parts of machinery.

Health and Insurance Plan

A program of providing financial protection to workers and their families against death, illness, accidents, and other risks, in which the costs may be borne in whole or in part by the employer. One or more of the following major benefits may be provided for workers and, frequently, their dependents: life insurance, accidental death and dismemberment benefits, accident, and sickness.

Health and Safety Program

A systematic combination of activities, procedures, and facilities designed to ensure and maintain a safe and healthy workplace.

Health Hazard

A chemical, mixture of chemicals, or pathogen for which there is statistically significant evidence based on at least one study conducted in accordance with established scientific principles that acute or chronic health effects may occur in exposed employees. The term "health hazard" includes chemicals that are carcinogens, toxic or highly toxic agents, reproductive toxins, irritants, corrosives, sensitizers, hepatotoxins, nephrotoxins, neurotoxins, agents which act on the hematopoietic system, and agents which damage the lungs, skin, eyes, or mucous membranes. It also includes stress due to temperature extremes.

Health Risk Appraisal

A generic term applied to methods for describing an individual's chances of becoming ill or dying from selected causes. It is used to indicate risks to health and safety that are influenced by an individual's lifestyle behaviors.

Healthy Worker Effect

A phenomenon observed in studies of occupational diseases. Workers usually exhibit lower overall mortality rates than the general population, due to the fact that the severely ill and disabled are not employed.

Hearing Conservation

Preventing or minimizing noise-induced hearing loss through the use of hearing protection devices, the control of noise through engineering methods, periodic audiometric tests, and employee training.

Hearing Conservation Program (HCP)

An HCP includes (1) noise exposure monitoring; (2) audiometric testing and analysis; (3) provision and effective use of hearing protection; (4) training on the effects of noise, the care and use of hearing protection, etc.; and (5) record keeping. Per Occupational Safety and Health Administration (OSHA) regulation 29 CFR 1910.95, hearing protection must be made available to employees when they are exposed to 8-hour time-weighted average of 85 dB(A) for constant noise.

Hearing Level (Hearing Loss)

The deviation in decibels of an individual's threshold from the zero reference of the audiometer. Formerly called hearing loss.

Hearing Loss, Noise Induced (NIHL)

Slowly progressive inner-ear hearing loss resulting from exposure to continuous noise over a long period of time, as contrasted with acoustic trauma or physical injury to the ear. NIHL is permanent, painless, and progressive, but is preventable if hearing protection is worn 100 percent of the time during noise exposure. See also **Acoustic Trauma**.

Hearing Loss Prevention Program Audit

An assessment performed prior to putting a hearing loss prevention program into place or before changing an existing program. The audit should be a top-down analysis of the strengths and weaknesses of each aspect of the program.

Hearing Protection

Devices or programs used to decrease the intensity of sound that reaches the eardrum. See also Hearing Conservation Program (HCP).

Hearing Protection Device (HPD)

Any device designed to reduce the level of sound reaching the eardrum. Earmuffs, earplugs, and ear canal caps (also called semi-inserts) are the main types of hearing protectors. A wide range of hearing protectors exists within each of these categories. For example, earplugs may be subcategorized into foam, user-formable (such as silicone or spun mineral fiber), pre-molded, and custom-molded earplugs. In addition, some types of helmets (in particular, flight helmets worn in the military) also function as hearing protectors. Hearing protectors are rated by their noise reduction rating (NRR). See also **Ear Bands**; **Ear Muffs**; **Ear Plugs**; **Noise Reduction Rating (NRR)**.

Hearing Protector

A device that is worn to reduce the effect of noise on the auditory system.

Heat Cramps (Mines' or Stokers' Cramps)

Painful spasms of the voluntary muscles due to salt depletion. These occur in healthy, heat acclimatized individuals and are due to excessive sweating without salt replacement rather than directly to high temperature exposure. They are relieved with fluid and salt replacement.

Heat Exhaustion (Heat Collapse; Heat Prostration; Heat Stress)

A state of peripheral vascular collapse in an unacclimatized individual attributable to exposure to a high temperature environment, radiant heat sources, high humidity, direct physical contact with hot objects, or strenuous physical activities. Treatment consists of removal to a cool environment, rest, and salt and water replacement. It is prevented by controlling heat exposures, adequate acclimatization, and maintaining adequate salt and water intake. Exposures include iron and steel foundries, nonferrous foundries, brick-firing and ceramics plants, glass products facilities, rubber products factories, electrical utilities (particularly boiler rooms), bakeries, confectioneries, commercial kitchens, laundries, food canneries, chemical plants, mining sites, smelters, and steam tunnels.

Heat Pyrexia (Heat Stroke)
A very serious and often fatal condition resulting from breakdown of thermoregulatory mechanisms during exposure to high temperature environments. It is characterized by extremely high deep-body temperature and an absence of sweating. Treatment consists of rapid cooling in an ice bath.

Heat Stress
See **Heat Exhaustion (Heat Collapse; Heat Prostration; Heat Stress)**.

Heat Stroke
A form of hyperthermia that occurs due to an abnormally elevated body temperature with accompanying physical and neurological symptoms. Unlike heat cramps and heat exhaustion, two forms of hyperthermia that are less severe, heat stroke is a medical emergency that can be fatal if not properly and promptly treated. The body normally generates heat as a result of metabolism, and is usually able to dissipate the heat by either radiation of heat through the skin or by evaporation of sweat. However, in extreme heat, high humidity, or vigorous exertion under the sun, the body may not be able to dissipate the heat and the body temperature rises, sometimes up to 41.1°C (106°F) or higher. Another cause of heat stroke is dehydration. A dehydrated person may not be able to sweat fast enough to dissipate heat, which causes the body temperature to rise. See also **Heat Exhaustion (Heat Collapse; Heat Prostration; Heat Stress); Hyperthermia**.

Heinrich's Law
See **Domino Theory**.

HEPA
See **Filter, HEPA**.

Hertz (Hz)
The unit measurement for audio frequencies. The frequency range for human hearing lies between 20 Hz and approximately 20,000 Hz. The sensitivity of the human ear drops off sharply below about 500 Hz and above 4000 Hz. The frequency is measured in cycles per second (cps). 1 Hz = 1 cps.

Hidden Failure
A failure that, on its own, does not become evident to the operating organization under normal circumstances.

Hidden Hazard
Typically refers to unseen electrical lines, gas lines, waste lines, water lines, or other lines that, if disturbed during an excavation, may injure personnel or damage equipment.

High Consequence Area (HCA)
A location that is specially defined in pipeline safety regulations as an area where pipeline releases could have greater consequences to health and safety or the environment. For oil pipelines, HCAs include high population areas, other population areas, commercially navigable waterways, and areas unusually sensitive to environmental damage. Regulations require a pipeline operator to take specific steps to ensure the integrity of a pipeline for which a release could affect an HCA, and thereby the protection of the HCA.

High Frequency Hearing Loss
A hearing deficit starting at 2000 Hz and higher.

High-Impact Lenses
As part of American National Standards Institute (ANSI) Z87 two-level classification for impact protection for eye and face protectors, "high impact" lenses and all frames

have to pass more stringent high mass and high velocity impact tests than "basic impact" lenses.

High Integrity Protective Systems (HIPS)

High availability, fail-safe, safety integrity level (SIL)-3 emergency shutdown (ESD) systems, designed to augment safety relief devices or mitigate worst-case relieving loads, or that function in lieu of over-pressure protective devices in well-head, flare, or off-site pipelines. See also **Emergency Shutdown System (ESD)**; **Safety Integrity Level (SIL)**.

Highly Protected Risk (HPR)

Term used within the insurance industry to describe a property risk that has a high degree of care taken for safety and protection (e.g., provision of fire sprinklers in a building) and is considered a superior facility from a loss viewpoint (i.e., low probability of loss); therefore, it has a very low insurance rate compared to other industrial risks.

Historical Incident Data

Information that has been collected and collated from past incidents. Used for regulatory reporting purposes, trend analysis, lagging safety key performance indicators (KPIs), and determining emphasis for safety improvements.

Hoistway-Door Interlock

A hoistway-door interlock is a device on an elevator shaftway, the purpose of which is, first, to prevent the operation of the elevator machine in a direction to move the car away from a landing unless the hoistway door at that landing at which the car is stopping or is at rest is locked in the closed position, and second, to prevent the opening of the hoistway door from the landing side except by special key, unless the car is at rest within the landing zone, or is coasting through the landing zone with its operating device in the stop position.

Hold Harmless Agreement

A contract under which the legal liability of one party for damages is assumed by another party to the contract. The principal in a large construction project will frequently demand hold harmless agreements from all subcontractors in respect to claims made against the principal arising out of the subcontractors' negligence. The principal often stipulates the purchase of a liability policy by the subcontractor to support the hold harmless agreement.

Home Safety

Home-related injuries result in approximately 20,000 deaths and 21 million medical visits on average each year. An average of more than 91,000 individuals dies each year from an unintentional injury, and approximately 20 percent of these deaths occur in the home. Of all nonfatal, unintentional injury events, 42 percent occur in the home, translating to nearly 12 million nonfatal home injuries each year. In addition, emergency departments treat more than 10 million home injuries annually, and an average of 11 million home injuries are seen by private physicians. Since the majority of incidents in the home are relatively very minor, many may go unrecorded. Falls are the leading cause of unintentional home injury death, accounting for about 33 percent. There is an average of 5.1 million injuries and nearly 6000 deaths each year from falls according to a Home Safety Council report of 2004. Poisoning is the second leading cause at 27 percent, third is fires or burns at 19 percent, followed by choking or suffocation at 6 percent and drowning or submersion at 5 percent.

Home Safety Council (HSC)

A national nonprofit charitable organization dedicated to preventing home-related injuries through national programs, partnerships, and the support of volunteers. HSC educates people of all ages to be safer in and around their homes. HSC programs are designed to educate people of all ages to take the steps needed to protect against home injuries. The programs target different age-specific audiences, ranging from children to older adults, and focus on the five leading causes of home injury.

Horseplay

Rough or boisterous play, gay or light-hearted recreational activity, for diversion or amusement. Horseplay among workers can distract them; result in anger, hurt feelings, distrust among workers, and possibly a desire for revenge; and cause an individual to be seriously injured or a fatality to occur. Horseplay activities are prohibited under Occupational Safety and Health Administration (OSHA) regulations 29 CFR 1910.1450 App A, (E) (1) (g) and 29 CFR 1910.178 (n) (9). Horseplay injuries are not covered by workers' compensation since they are not work-related injuries. A horseplay injury could result in civil action or criminal prosecution. The courts have held that these injuries are not the result of an incident but a deliberate act.

Hose Reel

A device for compactly storing a hard rubber fire hose. It also allows the firewater nozzle to be readily used as the hose is being unwound from the reel. The hose is permanently connected to a water supply source at one end, and a nozzle is provided at the free end. A shutoff valve is provided on the pipe supplying water to the hose reel. Primarily installed on mobile fire apparatus vehicles and at industrial facilities. They generally are provided with 1.9-, 2.54-, or 3.8-centimeter (0.75-, 1.0-, or 1.5-inch) hoses that can be up to 60 meters (200 feet) in length and can be adapted for the application of foam water.

Hospitalization Benefits

A plan that provides workers, and in many cases their dependents, with hospital room and board or cash allowances toward the cost of such care for a specified number of days, plus the full cost of specified services. Usually a part of a more inclusive health and insurance program.

Hostage Control Device

A term used in American National Standards Institute (ANSI) standards to describe any actuating control device or mechanism that prevents the operator from reaching the hazard point during normal cycling of the machine. A two-hand-control device is an example of a hostage control device. See also **Two-Hand Controls**.

Hot Tap

A procedure used in the repair, maintenance, and service activities that involves welding on a piece of equipment (pipelines, vessels, or tanks) under pressure in order to install connections or appurtenances. It is commonly used to replace or add sections of pipeline without the interruption of service for air, gas, water, steam, and petrochemical distribution systems, but requires extensive safety precautions.

Hot Work

Any activity that uses a heat producing process, e.g., cutting, welding, grinding, brazing, soldering, torch-applied roofing, and similar work involving an open flame, heat application, or the production of sparks, that constitutes a fire risk because it may act as an ignition source. Suitable fire safety precautions must be instituted where hot work occurs. Numerous major fires have occurred as a result of inadequate precautions undertaken during hot work activities. National Fire Protection Association (NFPA) 51B, Standard for Fire Prevention in Use of Cutting and Welding Processes, provides guidance for safety precautions while undertaking hot work. See also **Cold Work**; **Work Permit**.

Hot Work Permit

A safety process control for hot work activity to ensure proper initiation, review, and execution. See also **Hot Work**; **Work Permit**.

Housekeeping

Housekeeping is the routine care and cleaning that needs to be acted on daily in order for a facility to function safety and properly. In Occupational Safety and Health Administration (OSHA) regulation 29 CFR 1910.38 (b) (3), it states that the employer shall control accumulations of flammable and combustible waste materials and residues so that they do not contribute to a fire emergency. Additionally, housekeeping procedures shall be included in the written fire prevention plan. Additionally, 29 CFR 1926.25 similarly states the need for housekeeping activities at construction locations. Common injuries due to poor housekeeping include slips, trips, and falls.

Human Error

Limitations in human capacity to perceive, attend to, remember, process, and act upon information. It is associated with lapses of attention, mistaken actions, misperceptions, mistaken priorities, and in some cases willfulness.

Human Error Analysis

The systematic identification and evaluation of the possible errors that may be made by individuals. Various analysis methods are used such as Task Analysis, Job Safety Analysis, and Root Cause Map. See also **ATHEANA (A Technique for Human Error Analysis)**; **Job Safety Analysis (JSA)**; **Root Cause Map™ (RCM)**.

Human Factors

Human Factors is that field which is involved in conducting research regarding human capabilities and limitations for psychological, social, physical, and biological characteristics; maintaining the information obtained from that research; and working to apply that information with respect to the design, operation, or use of products, systems, and work environments for optimizing human performance, health, safety, and/or habitability. Areas of interest for human factors practitioners may include workload, fatigue, situational awareness, usability, user interface, learnability, attention, vigilance, human performance, control and display design, stress, visualization of data, individual differences, aging, accessibility, shift work, work in extreme environments, and human error.

Human Factors and Ergonomics Society (HFES)

A nonprofit organization founded in 1957 to promote the discovery and exchange of knowledge concerning the characteristics of human beings that are applicable to the design of systems and devices of all kinds. It is to operate exclusively for charitable, educational, scientific, and literary purposes. The society furthers serious consideration of knowledge about the assignment of appropriate functions for humans and machines, whether people serve as operators, maintainers, or users in the system. It advocates systematic use of such knowledge to achieve compatibility in the design of interactive systems of people, machines, and environments to ensure their effectiveness, safety, and ease of performance.

Human Factors Engineering (HFE)

The application of human factors principles to the design of devices and systems. It is often interchanged with the terms human engineering, usability engineering, or ergonomics. HFE typically focuses on the device-user interface (also called the UI or the man-machine interface). The user interface includes all components and accessories necessary to operate and properly maintain the device, including the controls, displays, software, logic of operation, labels, and instructions.

Human Health Risk

The likelihood that a given exposure or series of exposures may have damaged or will damage the health of individuals.

Human Reliability Analysis

An evaluation method to determine the probability that a system-required human action, task, or job will be successfully completed within the required time period and that no extraneous human actions detrimental to system performance will be performed. It provides quantitative estimates of human error potential due to work environment, human-machine interfaces, and required operational tasks. Such an evaluation can identify weaknesses in operator interfaces with a system, quantitatively demonstrate improvements in human interfaces, improve system evaluations by including human elements, and demonstrate quantitative prediction of human behavior. See also **ATHEANA (A Technique for Human Error Analysis)**; **Human Error Analysis**.

Hydraulic Shoring

A pre-engineered shoring system of aluminum or steel hydraulic cylinders (crossbraces) used with vertical rails (uprights) or horizontal rails (walers) and designed specifically to support side walls of an excavation to prevent cave-in.

Hygiene Practices

A broad term for personal health habits that may reduce or prevent the exposure of a worker to chemical or biological substances.

Hypersensitive

The condition of being reactive to substances that normally would not affect most people.

Hyperthermia

An elevated body temperature due to failed thermoregulation. It is usually defined as a temperature greater than 37.5°C to 38.3°C (100°F to 101°F). Hyperthermia occurs when the body produces or absorbs more heat than it can dissipate. When the elevated body temperatures are sufficiently high, hyperthermia is a medical emergency and requires immediate treatment to prevent disability and death. See also **Heat Exhaustion (Heat Collapse; Heat Prostration; Heat Stress)**.

Hypothermia

Lowered core body temperature, usually considered a drop in core temperature to 35°C (95°F) or lower. It produces shivering and discomfort sufficient to adversely affect performance; at about 25°C (77°F) hypothermia is ordinarily fatal. Cold water immersion produces hypothermia very rapidly, whereas exposure in cold air environments is tolerable for much longer periods. Treatment for hypothermia is rapid re-warming in a warm bath. See also **Frostbite**.

H

I

Iceberg Principle
   A correlation of the costs of an incident, illustrating that most of the impact or cost is initially hidden and that only the direct costs are known at the time of occurrence, similar to an iceberg where only the tip appears out of the water and the bulk of the ice is hidden below the water (see Figure I.1). See also **Tip of the Iceberg**.

Ignition
   The process of initiating a combustion process through the input of energy. Ignition occurs with a material when the temperature is raised to the point at which its molecules react spontaneously with an oxidizer and combustion occurs.

Illumination
   The amount of light flux a surface receives per unit area. It may be expressed in lumens per square foot or in footcandles. The rate at which a source emits light energy, evaluated in terms of its visual effect, is spoken of as light flux, and is expressed in lumens. See also **Brightness**.

Immediate Cause
   The initial factor for why something occurred. Usually there are additional underlying reasons or root causes beyond an immediate cause that have prompted the

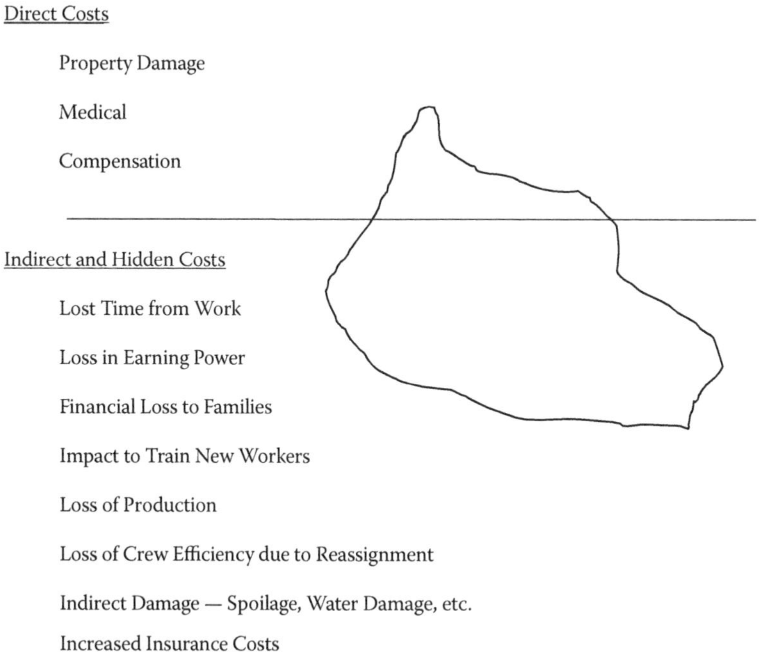

Direct Costs

   Property Damage

   Medical

   Compensation

Indirect and Hidden Costs

   Lost Time from Work

   Loss in Earning Power

   Financial Loss to Families

   Impact to Train New Workers

   Loss of Production

   Loss of Crew Efficiency due to Reassignment

   Indirect Damage — Spoilage, Water Damage, etc.

   Increased Insurance Costs

FIGURE I.1   Iceberg principle.

occurrence. It is not deep enough to be a root cause. There may be several layers of intermediate causes between the causal factor level and the root cause level. See also **Causal Factor (CF)**.

Immediate Hazard

The potential to cause an adverse effect within a short period of time.

Immediate Severe Health Effect

Acute clinical sign of a serious, exposure-related reaction that occurs within 72 hours.

Immediately Dangerous to Life or Health (IDLH)

An atmospheric concentration of any toxic, corrosive, or asphyxiant substance that poses an immediate threat to life or would cause irreversible or delayed adverse health effects or would interfere with an individual's ability to escape from a dangerous atmosphere.

Imminent Danger

An impending or threatening dangerous situation that could be expected to cause death or serious injury to persons in the immediate future unless corrective measures are taken.

Imminent Hazard

A hazard that would likely result in unreasonable adverse effects on humans or the environment or risk unreasonable hazard to an endangered species during the time required for a pesticide registration cancellation proceeding.

Impact Resistance

As applied to eye protection, the ability of a protector to resist the force of an object that comes into contact with the lens or eye protector at the velocity specified in this standard.

Impact (Shock) Loading

The highest load produced during an impact test. Often this point may also correspond to the onset of material damage or complete failure. Also known as Peak Load or Maximum Load.

Impactor (Sampling)

A type of personal air sampling instrument that utilizes bioaerosol inertia to collect the sample onto a solid or semi-solid collection medium. The impactor device forces the air stream to turn a tight corner. If the inertia of the bioaerosol is too great, the bioaerosol will not be able to follow the air flow lines and will instead impact onto the collection medium as the sample. See also **Impinger (Sampling)**.

Impervious

A characteristic of a material that does not allow another substance to pass through it or penetrate it. On a Material Safety Data Sheet (MSDS), impervious is a term used to describe protective gloves and other protective clothing. If a material is impervious to a chemical, then that chemical cannot readily penetrate through the material or damage the material. Different materials are impervious (resistant) to different chemicals. No single material is impervious to all chemicals. If an MSDS recommends wearing impervious gloves, you need to know the type of material from which the gloves should be made. For example, neoprene gloves are impervious to butyl alcohol but not to ethyl alcohol.

Impinger (Sampling)

A type of personal air sampling instrument that incorporates a glass tube containing a liquid medium. This reacts either chemically or physically to dissolve the contaminant. A known volume of air is passed through the impinger and the liquid is analyzed by gas chromatography or photospectrometry. See also **Impactor (Sampling)**.

Implosion

A rapid expenditure of energy producing an inward burst, as opposed to an explosion.

Impulse Noise

See **Noise, Impulse**.

Impurity

The presence of one substance in another, often in such low concentration that it cannot be measured quantitatively by ordinary analytical methods. In the air, trace amounts of sulfur dioxide and carbon monoxide are potentially dangerous impurities in concentrations of 5 ppm of sulfur dioxide and 50 ppm of carbon monoxide.

Incendiary

A substance causing or designed to cause fires. Also, a person who willfully destroys property by fire.

Incidence Rate

The number of injuries, illnesses, or lost workdays related to a common exposure base of 100 full-time workers as used by the U.S. Bureau of Labor Statistics. The common exposure base enables one to make accurate inter-industry comparisons, trend analysis over time, or comparisons among firms regardless of size. This rate is calculated as $IR = (N/EH) \times 200,000$, where N is number of injuries and/or illnesses or lost work days, EH is total hours worked by all employees during the calendar year, and 200,000 is the base for 100 full-time equivalent workers (working 40 hours per week, 50 weeks per year).

Incident

An event or sequence of events or occurrences, natural or man-made, that results in undesirable consequences and requires an emergency response to protect life and property. Incidents can include major disasters, emergencies, terrorist attacks, terrorist threats, fires, explosions, toxic vapor releases, hazardous materials spills, nuclear incidents, transportation mishaps (aircraft, motor vehicles, rail, shipping), structural failures, soil subsidence, earthquakes, hurricanes, tornadoes, floods, tropical or winter storms, war-related disasters, public health and medical emergencies, and other occurrences requiring an emergency response. An incident may lead to ill health, injury (fatal, major, moderate, or minor), property damage, environmental impact, loss of income, increased liabilities, or loss of prestige or self-esteem. All incidents are considered as preventable if suitable precautions are invoked. See also **Emergency**; **Major Disaster**.

Incident Action Plan

A plan containing general objectives reflecting the overall strategy for managing an incident. It includes the identification of operational resources and assignments. It also may include information for management of the incident during one or more operational periods.

Incident Command Post (ICP)

A facility located at a safe distance from an emergency site, where the incident commander, key staff, and technical representatives can make decisions and deploy emergency manpower and equipment, i.e., tactical level on-scene incident command functions. It is typically identified by a green rotating or flashing light.

Incident Command System (ICS)

An organized system of roles, responsibilities, and standard operating procedures used to manage and direct emergency operations. This is also sometimes referred to as an Incident Management System (IMS). ICS is the combination of facilities, equipment, personnel, procedures, and communications operating within a common organizational structure, designed to aid in the management of resources during incidents. It is applicable for all types of emergencies: small, large, simple, and complex incidents. ICS is used by various jurisdictions and functional agencies, both public and private, to organize field-level incident management operations.

Incident Commander (IC)

Individual who is responsible and the authority for the overall management of an incident. This individual's major duties include the following:

Assess the situation (obtain a briefing from the prior Incident Commander).
Determine Incident Objectives and strategy.
Establish the immediate priorities.
Establish an Incident Command Post.
Establish an appropriate organization.
Ensure planning meetings are scheduled as required.
Approve and authorize the implementation of an Incident Action Plan.
Ensure that adequate safety measures are in place.
Coordinate activity for all Command and General Staff.
Coordinate with activated Disaster Operations Centers (DOCs) and Emergency Operations Centers (EOCs) as required.
Coordinate with key people and officials.
Approve requests for additional resources or for the release of resources.
Keep agency administrator informed of incident status.
Approve the use of trainees, volunteers, and auxiliary personnel.
Authorize release of information to the news media.
Order the demobilization of the incident when appropriate.

Incident Investigation

See **Accident Investigation**.

Incident Investigation Team

A group of qualified people (i.e., knowledgeable and experienced in the subject matter) that examine an incident in a manner that is timely (i.e., adequate to gather evidence without disturbances and time-affecting factors), objective, systematic, and technically accurate to determine that factual information pertaining to the event is documented, probable causes are ascertained, and a complete technical understanding of such an event is achieved.

Incident Management System (IMS)

Management actions to direct, control, and coordinate response and recovery operations. Similar to the Incident Command System (ICS).

Incident Management Team (IMT)

The Incident Commander and appropriate command and general staff personnel assigned to an incident.

Incident Objectives

Statements of guidance and direction necessary for selecting appropriate strategies and the tactical direction of resources. Incident objectives are based on realistic expectations of what can be accomplished when all allocated resources have been effectively deployed. Incident objectives must be achievable and measurable, yet flexible enough to allow strategic and tactical alternatives. See also **Credible Scenario**.

Incident Report, Injury and Illness, OSHA

Details concerning harm that has occurred to an employee that is recorded on Occupational Safety and Health Administration (OSHA) Form 301. See Figure I.2A and B. This form lists information about the employee, the physician or health care individual who attended the case, and the incident details. It must be completed

OSHA's Form 300A

Summary of Work-Related Injuries and Illnesses

Year 20____

U.S. Department of Labor
Occupational Safety and Health Administration

Form approved OMB no. 1218-0176

All establishments covered by Part 1904 must complete this Summary page, even if no work-related injuries or illnesses occurred during the year. Remember to review the Log to verify that the entries are complete and accurate before completing this summary.

Using the Log, count the individual entries you made for each category. Then write the totals below, making sure you've added the entries from every page of the Log. If you had no cases, write "0."

Employees, former employees, and their representatives have the right to review the OSHA Form 300 in its entirety. They also have limited access to the OSHA Form 301 or its equivalent. See 29 CFR Part 1904.35, in OSHA's recordkeeping rule, for further details on the access provisions for these forms.

Number of Cases

Total number of deaths

Total number of cases with days away from work

Total number of cases with job transfer or restriction

Total number of other recordable cases

(G)

(H)

(I)

(J)

Number of Days

Total number of days of job transfer or restriction

Total number of days away from work

(K)

(L)

Injury and Illness Types

Total number of . . .

(M)

(1) Injuries

(2) Skin disorders

(3) Respiratory conditions

(4) Poisonings

(5) All other illnesses

Establishment information

Your establishment name _____

Street _____

City _____ State _____ ZIP _____

Industry description (e.g., Manufacture of motor truck trailers) _____

Standard Industrial Classification (SIC), if known (e.g., SIC 3715) _____

Employment information (If you don't have these figures, see the Worksheet on the back of this page to estimate.)

Annual average number of employees _____

Total hours worked by all employees last year _____

Sign here

Knowingly falsifying this document may result in a fine.

I certify that I have examined this document and that to the best of my knowledge the entries are true, accurate, and complete.

Company executive _____ Title _____

Phone _____ / _____ Date _____

Post this Summary page from February 1 to April 30 of the year following the year covered by the form.

Public reporting burden for this collection of information is estimated to average 50 minutes per response, including time to review the instructions, search and gather the data needed, and complete and review the collection of information. Persons are not required to respond to the collection of information unless it displays a currently valid OMB control number. If you have any comments about these estimates or any other aspects of this data collection, contact: US Department of Labor, OSHA Office of Statistics, Room N-3644, 200 Constitution Avenue, NW, Washington, DC 20210. Do not send the completed forms to this office.

Continued

FIGURE I.2 OSHA Form 301. (Source: U.S. Department of Labor.)

OSHA's Form 301

Injury and Illness Incident Report

U.S. Department of Labor
Occupational Safety and Health Administration

Form approved OMB no. 1218-0176

Attention: This form contains information relating to employee health and must be used in a manner that protects the confidentiality of employees to the extent possible while the information is being used for occupational safety and health purposes.

This *Injury and Illness Incident Report* is one of the first forms you must fill out when a recordable work-related injury or illness has occurred. Together with the *Log of Work-Related Injuries and Illnesses* and the accompanying *Summary*, these forms help the employer and OSHA develop a picture of the extent and severity of work-related incidents.

Within 7 calendar days after you receive information that a recordable work-related injury or illness has occurred, you must fill out this form or an equivalent. Some state workers' compensation, insurance, or other reports may be acceptable substitutes. To be considered an equivalent form, any substitute must contain all the information asked for on this form.

According to Public Law 91-596 and 29 CFR 1904, OSHA's recordkeeping rule, you must keep this form on file for 5 years following the year to which it pertains.

If you need additional copies of this form, you may photocopy and use as many as you need.

Information about the employee

1) Full name

2) Street

City ____ State ____ ZIP

3) Date of birth ___ / ___ / ___

4) Date hired ___ / ___ / ___

5) ☐ Male　☐ Female

Information about the physician or other health care professional

6) Name of physician or other health care professional

7) If treatment was given away from the worksite, where was it given?

Facility

Street

City ____ State ____ ZIP

8) Was employee treated in an emergency room?
☐ Yes ☐ No

9) Was employee hospitalized overnight as an in-patient?
☐ Yes ☐ No

Completed by

Title

Phone ( ) ___ - ___　　Date ___ / ___ / ___

Information about the case

10) Case number from the *Log* _____ *(Transfer the case number from the Log after you record the case.)*

11) Date of injury or illness ___ / ___ / ___

12) Time employee began work _____ AM / PM

13) Time of event _____ AM / PM　☐ Check if time cannot be determined

14) **What was the employee doing just before the incident occurred?** Describe the activity, as well as the tools, equipment, or material the employee was using. Be specific. *Examples:* "climbing a ladder while carrying roofing materials"; "spraying chlorine from hand sprayer"; "daily computer key-entry."

15) **What happened?** Tell us how the injury occurred. *Examples:* "When ladder slipped on wet floor, worker fell 20 feet"; "Worker was sprayed with chlorine when gasket broke during replacement"; "Worker developed soreness in wrist over time."

16) **What was the injury or illness?** Tell us the part of the body that was affected and how it was affected; be more specific than "hurt," "pain," or sore." *Examples:* "strained back"; "chemical burn, hand"; "carpal tunnel syndrome."

17) **What object or substance directly harmed the employee?** *Examples:* "concrete floor"; "chlorine"; "radial arm saw." *If this question does not apply to the incident, leave it blank.*

18) **If the employee died, when did death occur?** Date of death ___ / ___ / ___

Public reporting burden for this collection of information is estimated to average 22 minutes per response, including time for reviewing instructions, searching existing data sources, gathering and maintaining the data needed, and completing and reviewing the collection of information. Persons are not required to respond to the collection of information unless it displays a current valid OMB control number. If you have any comments about this estimate or any other aspects of this data collection, including suggestions for reducing this burden, contact: US Department of Labor, OSHA Office of Statistics, Room N-3644, 200 Constitution Avenue, NW, Washington, DC 20210. Do not send the completed forms to this office.

FIGURE I.2 (Continued) OSHA Form 301. (Source: U.S. Department of Labor.)

within 7 days for an injury or illness occurring in the workplace. Required by federal regulation 29 CFR 1904.29.

Incident Response

Activities that address the short-term, direct effects of an emergency. A response includes immediate actions to save lives, protect property, and meet basic human needs. Response also includes the execution of emergency operations plans and of mitigation activities designed to limit the loss of life, personal injury, property damage, and other unfavorable outcomes.

Incompatible Materials

Materials that can react with a product or with components of the product and may destroy the structure or function of a product; cause a fire, explosion, or violent reaction; or cause the release of hazardous chemicals.

Independent Adjuster

An individual who charges a fee to the insurance company to adjust claims.

Independent Protection Layer (IPL)

Protection measures that reduce the level of risk or a serious event by 100 times, have a high degree of availability (greater than 0.99), or have specificity, independence, dependability, and auditability.

Indirect Costs

Monetary losses resulting from an incident other than medical costs and workers' compensation payments. See also **Accident Costs**.

Indirect Damage

Loss resulting from a hazardous condition or incident but not caused directly thereby.

Inductive Approach

A type of reasoning from individual cases to a general conclusion. In the loss prevention industry, it postulates that a system element has failed in specific circumstances. An investigation is then undertaken to determine what occurs to the whole system or process. See also **Deductive Approach**; **Morphological Approach**.

Industrial Accident

See **Occupational Injury**.

Industrial Fire Brigade

A fully trained and fully equipped firefighting team that is developed by a facility, utilizing employees whose full-time job may or may not be as firefighters. The brigade typically does not respond to fire beyond the limits of its employer's property.

Industrial Hygiene (IH)

Industrial hygiene is that science and art devoted to the anticipation, recognition, evaluation, and control of those environmental factors or stresses arising in or from the workplace that may cause sickness, impaired health and well-being, or significant discomfort and inefficiency among workers or among the citizens of the community.

Industrial Hygiene Program

Environmental factors or stresses (i.e., chemical, physical, ergonomic, or biological hazards) in the workplace that can cause illness or discomfort among employees and the adjacent public near a facility. An industrial hygiene program should address these issues. It includes the identification of health hazards, evaluation of their magnitude, and development of corrective measures. The goals of an industrial hygiene program are to:

Protect employees and the public from health hazards occurring in the workplace.
Ensure that individuals are physically, mentally, and emotionally capable of performing their jobs efficiently and safely.

Ensure proper medical attention and rehabilitation of individuals who have suffered an injury or illness on the job.

Educate and encourage employees on health maintenance and its benefits.

Industrial Hygienist

A person with the training and ability to recognize the environmental factors and stresses associated with work and work operations to understand their effect on workers and their well-being. Evaluate, on the basis of experience and with the aid of quantitative measurement techniques, the magnitude of these stresses in terms of the ability to impair worker health and well-being. Prescribe methods to eliminate, control, or reduce such stresses when necessary to alleviate their effects.

Industrial Safety

See **Occupational Safety**.

Industrial Ventilation

An integral part of a system to condition air, which may be used in combination with heating, cooling, and humidifying. When used alone, it may be used to remove contaminated air from a workspace and for heat control, and includes a supply system and an exhaust system. A well-designed supply system consists of an air inlet section, filters, heating and/or cooling equipment, and registers/grilles for air distribution within the workspace. The exhaust system may include a general exhaust system and a local exhaust system. See also **Exhaust (General)**; **Exhaust (Local)**.

Industrial Waste

Unwanted materials produced in or eliminated from an industrial operation and categorized under a variety of headings, such as liquid wastes, sludge, solid wastes, and hazardous wastes.

Inerting

Rendering the atmosphere in a confined or enclosed space or volume non-flammable, non-explosive, or otherwise chemically non-reactive by displacing or diluting the original atmosphere with an inert gas. It is normally accomplished by purging with a gas that will not support combustion such as nitrogen or carbon dioxide (CO_2). The normal oxygen level in the atmosphere is 21 percent (approximately 20.9 percent oxygen, 78.1 percent nitrogen, 1 percent argon, carbon dioxide, and other gases). Combustion of stable hydrocarbon gases and vapors will usually not continue when the ambient oxygen level is below 15 percent. Acetylene is an unstable gas and requires an oxygen level below 4 percent before extinguishment will occur. For ordinary combustibles (e.g., wood, paper, cotton, etc.), the oxygen concentration level must be lowered to 4 or 5 percent for total fire extinguishment. Inerting for fire extinguishment results in an asphyxicant safety hazard for personnel. Inerting, in effect, reduces the oxygen content of the air in the enclosed space below the lowest point at which combustion can occur by replacing the oxygen in the enclosure with an inert gas. The inert gas chosen should not react with the medium contained within the enclosure. It produces an immediately dangerous to life and health (IDLH) oxygen-deficient atmosphere. See also **Purging**.

Inflammable

A general term once used to describe combustible gases, liquids, or solids. Now obsolete. See also **Flammable**.

Inflammation

A tissue reaction to infection or trauma produced by an injury, irritant, corrosive, or other chemical or physical agent. An inflammation is characterized by redness, heat, and swelling.

Infrared (IR)

A form of non-ionizing radiation emitted by all hot bodies. Long-term exposure may cause damage to the eyes. It is descriptive of invisible heat rays located beyond the long wavelength end of the visible spectrum (beyond 7600 angstroms) and possessing high penetrating power. Wavelengths of the electromagnetic spectrum are no longer than those of visible light and are shorter than radio waves (104 to 101 centimeters). IR radiation is emitted by the combustion process, and fire detection devices are available to detect these emissions.

Inhalation

The act of breathing in, or taking into the lungs, a substance in the form of a gas, vapor, fume, mist, or dust.

Inherent Risk

The hazard of a substance or process due to its particular characteristics, which cannot be changed, e.g., the flammable properties of gasoline.

Inherently Safe

An essential character of a process, system, or equipment that makes it without or very low in hazard or risk. For example, the use of non-combustible liquids instead of combustible liquids for an operation (e.g., transformer insulating oil) makes a process inherently safer by removing the hazard of fire should the liquid contact an ignition source.

Inhibitor

An agent that arrests or slows chemical action or material that is used to prevent or retard rust or corrosion.

Injury

Physical harm or damage to a person resulting in the marring of appearance, personal discomfort, infection, and/or bodily hurt or impairment. The definition of this word is frequently determined by the government agency or other organization using it.

Injury and Illness Prevention Program (IIPP)

A health and safety program that employers develop and implement to avoid injuries and illness in the workplace.

Injury (Occupational)

Any acute hurt, harm, or impairment to a worker that arises out of, or in the course of, employment and is due to an external cause.

Injury or Illness, OSHA

Per Occupational Safety and Health Administration (OSHA) regulations, 29 CFR 1904.46 for record-keeping purposes, an injury or illness is an abnormal condition or disorder. Injuries include cases such as, but not limited to, a cut, fracture, sprain, or amputation. Illnesses include both acute and chronic illnesses, such as, but not limited to, a skin disease, respiratory disorder, or poisoning.

In-Running Nip Point

A rotating mechanism that can seize and wind up loose clothing, belts, hair, body parts, etc. It exists when two or more shafts or rolls rotate parallel to one another in opposite directions. It also can occur between a rotating shaft and a fixed surface. See also **Pinch Point**.

Inspection

An examination, observation, measurement, or test undertaken to assess structures, systems, and components and materials, as well as operational activities, technical processes, organizational processes, procedures, and personnel competence. See also **Safety Inspection**.

Institute for Safety and Health Management (ISHM)

The Institute for Safety and Health Management is an organization founded to promote the establishment of professional safety management standards. The organization exists to promote the establishment of standards for the profession and recognizes safety and risk management professionals who, through demonstrated professional experience and the passing of a comprehensive exam, have met the ISHM's requirements for mastering the safety management body of knowledge. It issues a credential of Certified Safety and Health Manager (CSHM) to individuals who demonstrate this knowledge and experience and have passed their examination.

Insurable Interest

Refers to the relationship of the party of whom the insurance coverage is written to the peril being insured. The party must either own or have a financial stake in the subject that may suffer the damage or loss, which would involve out-of-pocket expense to that party if no insurance indemnity were available.

Insurance

The making of a legal and enforceable contract between one party (insurer or underwriter) with another (insured) whereby in consideration of a sum of money (premium), the insurer agrees to pay an agreed amount of money to the insured if and when the latter may suffer some loss or may be injured by some event (designated contingencies), the occurrence of which is described in the contract of insurance, which is usually a policy. See also **Deductible, Insurance**.

Insurance Institute for Property Loss Reduction

An association of insurance companies formed to reduce fatalities, injuries, and the loss of property resulting from all types of natural hazards in the United States. It focuses on improving construction and building techniques to minimize the damage from natural hazards.

Insurance Institute of America (IIA)

An organization that develops, publishes, and administers a variety of continuing education programs (e.g., accounting, finance, general insurance, claims, underwriting, loss control management, premium auditing, etc.) for professionals in the insurance business and related fields, including professional certifications to individuals in the property and liability insurance business such as the Associate in Risk Management (ARM) designation.

Insurance Services Office (ISO), Inc.

A nonprofit association of insurance companies that provides services for participating companies. It provides statistical, actuarial, and underwriting information for numerous affiliated insurance companies and more than a dozen lines of insurance. It maintains one of the largest private databases in the world for insurance premiums and losses paid. Included in these lines is liability, automobile, boiler and machinery, homeowners, farm, and commercial fire insurance. ISO is a voluntary, nonprofit, unincorporated association of insurers. Previous to 1971, the functions performed by ISO were undertaken by various insurance organizations in different states. ISO gathers data that are used to establish rates for fire protection policies for residential and commercial properties. ISO developed the municipal grading schedule, which is commonly used to establish a basis of insurance rates for municipalities. It is an evaluation of the fire protection features of cities and towns based on seven factors: climatic conditions, water supply, fire department, fire service communications, fire safety control, building codes, and a survey report.

Insurance Survey

A physical appraisal of a property to determine its value, current condition to accepted protection standards and practices, and risks. Based on the survey, a rating of the property is made to comparable installations to determine the insurance requirements.

Insured

The person who has purchased a policy of insurance and is protected by it.

Insured Costs

Incident losses, which are covered by workers' compensation, medical, or other insurance programs. They comprise the insured element of the total incident cost.

Intensity Level, Sound

The sound-energy flux density level. In decibels of sound, intensity level equals 10 times the logarithm to the base 10 of the ratio of the intensity of this sound to the reference intensity. See also **Decibel (dB)**.

Interlock

A device that interacts with another device or mechanism to govern succeeding operations. The device is usually an instrument, switch, or a control: (1) an instrument that will not allow one part of a process to function unless another part is functioning, (2) a device such as a switch that prevents a piece of equipment from operating when a hazard exists, or (3) to arrange the control of machines or devices so that their operation is interdependent in order to ensure their proper coordination. For example, an interlock on an elevator door will prevent the car from moving unless the door is properly closed.

Interlocking Guard

A form of guard that has a movable part connected with the machinery so that the parts of the machine causing a hazard cannot be set in motion until the guard is in place. Before the guard can be opened sufficiently to allow access to the work area, the power is turned off and the motion is braked to prevent potential injuries to workers. It may also be called Interlocked Barrier Guard. See also **Barrier Guard**; **Interlock; Machine Guarding**.

International Building Code (IBC)

A model building code developed by the International Code Council (ICC). It has been adopted throughout most of the United States. Most of the International Building Code deals with fire prevention, in that the International Building Code handles fire prevention in regard to construction and design of buildings while the International Fire Code addresses fire prevention basically for ongoing operations.

International Code Council (ICC)

Established in 1994 as a non-profit organization dedicated to developing a single set of comprehensive and coordinated national model codes and standards used to construct residential and commercial buildings, including homes and schools. Its mission is to provide the highest quality codes, standards, products, and services for all concerned with the safety and performance of the built environment. It is composed of representatives of the Building Officials and Code Administrators International, Inc. (BOCA), International Conference of Building Officials (ICBO), and Southern Building Code Congress International, Inc. (SBCCI) organizations. See also **International Fire Code® (IFC)**.

International Fire Code® (IFC)

Regulations governing the safeguarding of life and property from all types of fire and explosions hazards, handling or use of hazardous materials, and the use and occupancy of buildings and premises. Topics include general precautions against fire, emergency planning and preparedness, fire department access, fire hydrants, automatic sprinkler systems, fire alarm systems, hazardous materials storage and use, and fire safety requirements for new and existing buildings and premises. It was first published in 2000 and is provided by the International Code Council (ICC), which consists of representatives of the Building Officials and Code Administrators International, Inc. (BOCA), International Conference of Building Officials (ICBO), and Southern Building Code Congress International (SBCCI). International Fire Code® is a registered trademark of the International Code Council.

International Safety Equipment Association (ISEA)

A trade association in the United States for companies that manufacture safety and personal protective equipment. Its member companies are manufacturers of protective clothing and equipment used in factories, construction sites, hospitals and clinics, farms, schools, laboratories, emergency response, and in the home. These companies make products for head, eye and face, respiratory, hearing, hand, and fall protection; high visibility apparel and headwear; environmental monitoring instruments; emergency eyewash and shower equipment; first aid kits; protective apparel; ergonomic protective equipment; respiratory protective escape devices; and personal hydration systems. The ISEA was organized in 1933. The association provides a forum through which its members can work to promote the standardization of safety equipment, represent the industry's interests before government bodies, and interpret government actions to the industry, collect and disseminate information about the industry, maintain links to other organizations in the safety industry worldwide, and promote the proper use of personal protective equipment.

International Safety Rating System (ISRS)

A safety evaluation tool to measure an organization's effectiveness in safety management by independently auditing the level of safety applied to 20 categories. The 20 categories include leadership, leadership training, planned inspections and maintenance, critical task analysis and procedures, incident investigation, task observation, emergency preparedness, rules and work permits, incident analysis, knowledge and skills training, personal protective equipment, occupational health and hygiene, system evaluation, engineering and change management, individual communication, group communication, safety promotion, hiring and placement, materials and services management, and off-the-job safety. The organization is given a score based on a scale from 0 to 10. The ISRS is a proprietary product owned and maintained by Det Norske Veritas (DNV), a leading measurement and assessment consultant for health, safety, environmental, and quality management systems. It was originally developed in 1976 by Frank Bird, a safety management consultant. It has found wide use as a safety benchmark and comparison tool in the industry.

International Society for Fall Protection (ISFP)

A non-profit organization that works with worldwide safety professionals to improve access to fall protection information. Any company or individual interested in fall protection may be a member. The ISFP was organized in 1988, and since then it has been working to reduce fall-related injuries and fatalities by promoting research and facilitating communication among industry professionals. It organizes international symposia and conferences; cooperates with various governmental, public education, and R&D agencies; and maintains a Web site for its purposes.

Intrinsically Safe

A circuit or device in which any spark or thermal effect in normal or abnormal conditions is incapable of causing ignition of a mixture of flammable or combustible material in air under prescribed hazardous atmosphere conditions (i.e., combustible vapors) due to insufficient ignition energy. No single device or wiring is intrinsically safe by itself (except for battery-operated self-contained apparatus such as portable pagers, transceivers, gas detectors, etc., which are specifically designed as intrinsically safe self-contained devices) but is intrinsically safe only when employed in a properly designed, intrinsically safe system. Intrinsically safe devices or equipment is used where it would be impractical to install or use explosionproof equipment due to its cost. The intrinsically safe designation is limited to devices and wiring in which the output or consumption of energy is small, the amount of supply voltage, resistance, capacitance, inductance, the manner in which a circuit is broken, its material, shape of contacts, and type of gas or vapor to be encountered. Intrinsically safe devices in the United States are commonly tested to Underwriters Laboratories (UL) 913, Standard for Safety, Intrinsically Safe Apparatus and Associated Apparatus for Use in Class I, II, and III, Division 1, Hazardous (Classified) Locations and installed per ANSI/ISA-RP 12.6, Recommended Practice for the Installation of Intrinsically Safe Systems for Hazardous Classified Locations.

Ionizing Radiation

A radiation source (alpha particles, beta articles, gamma rays, X-rays, neutrons, and cosmic rays) that may be considered a health hazard to individuals if the exposure is sufficiently strong or of duration considered detrimental. Ionizing radiation sources may be found in a wide range of occupational settings, including health care facilities, research institutions, nuclear reactors and their support facilities, nuclear weapon production facilities, and various other similar manufacturing settings. See also **Radiation**.

Irritant

A chemical that is not corrosive but causes a reversible inflammatory effect on living tissue by chemical action at the site of contact. A chemical is considered a skin irritant if, when tested on the intact skin of albino rabbits by the methods of Consumer Products Safety Commission (CPSC) 16 CFR Part 1500.41 for an exposure of four or more hours or by other appropriate techniques, it results in an empirical score of 5 or more. A chemical is classified as an eye irritant if so determined under the procedure listed in CPSC 16 CFR Part 1500.42 or other approved techniques. Some examples of irritants are chlorine, nitric acid, and various pesticides.

ISEA (ANSI) 107, High Visibility Garment Standard

A voluntary industry consensus standard, developed by the International Safety Equipment Association (ISEA) in 1999, that specifies the requirements for personal protective equipment (PPE) that is capable of visually signaling the user's presence. For a garment to be labeled American National Standards Institute (ANSI) Class 1, 2, 3, or E it is required to first be ANSI/ISEA 107 certified. This certification requires many tests by an accredited laboratory to make sure that the garment lives up to the requirements of the standard. Some tests/requirements include:

Minimum widths of retroflective or combined-performance materials
Spacing between multiple bands
Distance from bottom edge of garment
Placement of materials on sleeves
Gaps to enable fastening
Contiguous 360-degree visibility
Placement of material on legs
Ergonomics
Luminance factor of background material
Luminance of retroflective and combined performance material

Colorfastness of materials to perspiration, laundering, and other exposure

Dimensional changes in materials

Mechanical properties of materials: bursting strength, tear resistance, water repellency, etc.

Measurement of color

Performance after varied temperature exposures

Proper product labeling including care and usage instructions

Isocyanate Exposure

A health hazard from working with isocyanate. Isocyanates are compounds contained in the isocyanate group. They react with compounds containing alcohol groups to produce polyurethane polymers, which are components of polyurethane foams, thermoplastic elastomers, spandex fibers, and polyurethane paints. Isocyanates are the raw materials that make up all polyurethane products. Work that may involve exposure to isocyanates include painting, foam-blowing, and the manufacture of many polyurethane products, such as chemicals, polyurethane foam, insulation materials, surface coatings, car seats, furniture, foam mattresses, under-carpet padding, packaging materials, shoes, laminated fabrics, polyurethane rubber, and adhesives, and during the thermal degradation of polyurethane products. Health effects of isocyanate exposure include irritation of skin and mucous membranes, chest tightness, and difficulty breathing. Isocyanates include compounds classified as potential human carcinogens and are known to cause cancer in animals. The main effects of hazardous exposures are occupational asthma and other lung problems, as well as irritation of the eyes, nose, throat, and skin.

Isolate

Blocking a flow path to prevent flow through a system.

Isolating Device

A particular valve, switch, or circuit breaker that isolates stored or supplied energy to a work area.

Isolation

A loss control measure against identified risks by segregating the identified hazard to a specific (remote) location to protect the surrounding area from its effects and vice versa. Examples include placement of a chemical plant or process in a remote location and enclosure of an individual in an acoustic booth or enclosure to protect against noise exposure.

J

Job Hazard Analysis (JHA)
See **Job Safety Analysis**.

Job Safety Analysis (JSA)
A safety management risk assessment technique that is used to define and control the hazards associated with a process, job, or procedure. The Job Safety Analysis ensures that the hazards involved in each step of a task are reduced to as low as reasonably practical (ALARP). The assessment starts with a summary of the entire job process. The job is broken into smaller steps and listed in a tabular form. The hazards for each step are then identified and listed. This is repeated for each step in the process and a method of safe work is identified. It may be also called a Job Hazard Analysis (JHA).
See also **As Low As Reasonably Practical (ALARP)**.

Job Safety Training
Training associated with or emphasizing the safety aspects of a job and the hazards of tasks and their interrelationships within a job.

J

K

Key Event (of an Accident)
That event (or events) in a series of events in an incident that determine the exact time, place, type, and extent of consequences of the incident.

Key Performance Indicator (KPI), Safety
Measurement statistics that may be utilized to evaluate the safety performance of an organization. KPIs are usually categorized as leading or lagging. Lagging indicators are common incident statistics, such as injuries, fatalities, motor vehicle accidents (MVAs), and fires and are considered lagging due to the fact that they materialize after the incidents. Leading indicators are usually the number of management safety inspections, the number of safety meetings, levels of safety training, etc., and are considered proactive safety activities to prevent an incident. See also **Lagging Indicator**; **Leading Indicator**; **Safety Dashboard**.

Kickback
A term used to describe the reaction of a piece of material (usually wood) when being cut by a circular saw, which is rotating in the opposite direction from the material as it is fed into the saw. Since the saw blade is rotating at a high rate of speed the force that can be exerted on the material is substantial and can cause serious injury to the operator.

Kill Switch
See **Deadman Switch**.

Knit Wrist Cuffs
Knit wrist cuffs are designed to hold a glove in place on the hand and to prevent debris from entering the glove.

L

Labeled

Equipment or materials to which is attached a specified label, symbol, or other identifying mark of an inspection agency, nationally recognized testing laboratory, or organization acceptable to the authority having jurisdiction and concerned with product evaluation that maintains periodic inspection of the production of labeled equipment or materials. The provision of a label by the manufacturer on a fire protection product indicates compliance with appropriate standards or performance in a specified manner for fire safety. See also **Approved**; **Classified**; **Factory Mutual (FM)**; **Listed**; **Underwriters Laboratories (UL)**.

Laboratory Safety

Addressed by the Occupational Safety and Health Administration (OSHA) in specific standards for general industry, such as OSHA Standards 29 CFR 1910.1000, Air Contaminants; 29 CFR 1910.1000, Table Z-1, Limits for Air Contaminants; 1910.1450, Occupational Exposure to Hazardous Chemicals in Laboratories; 29 CFR 1910.1450, App A, National Research Council Recommendations Concerning Chemical Hygiene in Laboratories (Non-Mandatory); and 29 CFR 1910.1450, App B, References (Non-Mandatory). Laboratory safety is also addressed by NFPA 45, the National Fire Protection Association (NFPA) Standard on Fire Protection for Laboratories, and the International Codes by the International Code Council (ICC).

Laceration

A term applied to cleanly cut incised wounds as well as jagged, irregular, blunt breaks, or tears through the skin. Severity extends from small cuts that can be taped together to severe wounds with damage to underlying structures. The term is also applied to wounds of the mucous membranes and the surface of the eyes.

Lacrimater

A material that causes excessive tearing of an eye due to direct contact or vapors.

Ladder Safety

Falls are the major concern when working with ladders. Factors that contribute to ladder falls are ladder slip (top or bottom), over-reaching, slipping on rungs or steps, defective equipment, and improper ladder selection for the task. Therefore, safety factors to be considered when using portable ladders include placement, securing or tying down, climbing style, angle of inclination, three-point contact, and tasks to be performed. The Occupational Safety and Health Administration (OSHA) sets minimum requirements for the use of ladders in business and industry. Additionally, many states set their own code and regulations under the Occupational Safety and Health Act, which supersede the national OSHA standards. The adequacy of ladders and the work practices followed by employees using them are regulated by OSHA in four sections: Portable Wood 29 CFR 1910.25, Portable Metal 29 CFR 1910.26, Fixed Ladders 29 CFR 1910.27, and ladders used in Construction Industry 29 CFR 1910.1053. These sections specify the standards to which all portable ladders must be manufactured, care and placement of ladders in the workplace, and the safe use of ladders on the job. Ladders are required to meet certain American Ladder Institute (ALI) safety codes based on the material and type of ladder. For Wood Ladders,

TABLE L.1
ANSI Duty Rating for Ladders

Type	Supporting Capability	Use
1AA	170 kg (375 lbs.)	Extra heavy duty industrial
1A	136 kg (300 lbs.)	Extra heavy duty industrial
1	114 kg (250 lbs.)	Heavy duty industrial
11	102 kg (225 lbs.)	Medium duty commercial
111	91 kg (200 lbs.)	Light duty household

ALI (ANSI) A14.1 applies; Metal Ladders, ALI (ANSI) A14.2; Fixed Ladders, ALI (ANSI) A14.3; Job Made Ladders ALI (ANSI) A14.4; Fiberglass Ladders ALI (ANSI) A14.5; and Mobile Ladders ALI (ANSI) A14.7. ANSI codes also have an established Duty Rating. This rating identifies which portable ladder is intended for the conditions under which the ladder can be safely used, highlighted in Table L.1.
See also **American Ladder Institute (ALI)**; **Cage, Ladder Safety**.

Ladder Safety Device

Any device, other than a cage or well, designed to eliminate or reduce the possibility of falls while using a ladder and which may incorporate such features as life belts, friction brakes, and sliding attachments. See also **Cage, Ladder Safety**.

Lagging Indicator

A measurement statistic that may be utilized to evaluate the safety performance of an organization by management. Lagging indicators are commonly incident statistics, such as injuries, fatalities, motor vehicle accidents (MVAs), fires, etc., and are considered lagging due to the fact that they materialize after the incidents occur. See also **Key Performance Indicator (KPI), Safety**; **Leading Indicator**; **Safety Dashboard**.

Laminated Glass

A type of safety glass that holds together when shattered. If it breaks, it is held in place by an interlayer, typically of polyvinyl butyral (PVB), between its two or more layers of glass. The interlayer keeps the layers of glass bonded even when broken, and its high strength prevents the glass from breaking up into large, sharp pieces. This produces a characteristic "spider web" cracking pattern when the impact is not enough to completely pierce the glass. Laminated glass is normally used when there is a possibility of human impact or where the glass could fall if shattered. Skylights and automobile windshields typically use laminated glass. In geographical areas requiring hurricane-resistant construction, laminated glass is often used in exterior commercial applications. See also **Safety Glass**, **Tempered Glass**, **Window Film, Safety and Security**.

Lanyard

A flexible line with a positive means to lock end connections closed (i.e., locking-type snap hooks with a self-closing, self-locking keeper) that is part of a personal fall arrest system, which is used to secure the wearer of a full body harness to a lifeline or a point of anchorage to prevent harm to the individual should he or she inadvertently fall from a height. A specially designed shock-absorbing lanyard with a built-in shock absorber is also available that elongates during a fall, so that fall arresting forces are significantly reduced when compared to a traditional web or rope lanyard. See also **Lifeline**; **Personal Fall Arrest System**.

Laser Burn

A tissue injury caused by a beam of coherent light. Laser beams are particularly hazardous to the eye, where they may cause burns of the cornea, lens, or retina with consequent effects on visual acuity. A laser burn of the retinal fovea may destroy central vision sufficiently to produce legal blindness.

Laser Hazard

LASER is an acronym that stands for Light Amplification by Stimulated Emission of Radiation. The laser produces an intense, highly directional beam of light. The most common cause of laser-induced tissue damage is thermal in nature, where the tissue proteins are denatured due to the temperature rise following absorption of laser energy. The human body is vulnerable to the output of certain lasers, and under certain circumstances, exposure can result in damage to the eye and skin.

Laser Safety Glasses

Glasses constructed to provide protection from the intense light of lasers and have filtered lenses to suit different lasers. They are required to meet American National Standards Institute (ANSI) Z87 and be clearly marked with the manufacturer's name. See also **Safety Glasses**.

Latent Failure

A failure in a component as a result of a hidden defect.

Latent Fault

A fault that is present but hidden from regular means of detection. Typically these faults can only be identified as part of an incident or a detailed proof test.

Law of Unintended Consequences

A common humorous expression that implies that any purposeful action will produce some unintended, unanticipated, and usually unwanted consequences. Stated in other words, each cause has more than one effect, and these effects will invariably include at least one unforeseen side effect. The unintended side effect can be more significant than the intended effect. Like Murphy's Law, it is a humorous expression rather than an actual law of nature. May also be called Law of Unforeseen Consequences. See also **Murphy's Law**.

Layers of Protection Analysis (LOPA)

A method of analyzing the likelihood (frequency) of a harmful outcome event based on an initiating event frequency and on the probability of failure of a series of independent layers of protection capable of preventing the harmful outcome. LOPA is a recognized technique for selecting the appropriate Safety Integrity Level (SIL) of a safety instrumented system (SIS) per the requirements of ANSI/ISA-84.00.01 (see Figure L.1). See also **Safety Integrity Level (SIL)**; **Safety Instrumented System (SIS)**.

Leading Indicator

Measurement statistics that may be utilized to evaluate the safety performance of an organization by management. Leading indicators are usually the number of management safety inspections, number of safety meetings, levels of safety training, etc., and are considered leading due to the fact that they materialize normally before an incident occurs and therefore are leading. They are commonly considered proactive safety activities to prevent an incident. See also **Key Performance Indicator (KPI), Safety**; **Lagging Indicator**; **Safety Dashboard**.

Less Than Adequate (LTA)

A deficiency not meeting requirements. Commonly used in the safety profession to describe safety features that are not in accordance with regulations or accepted practices for industry.

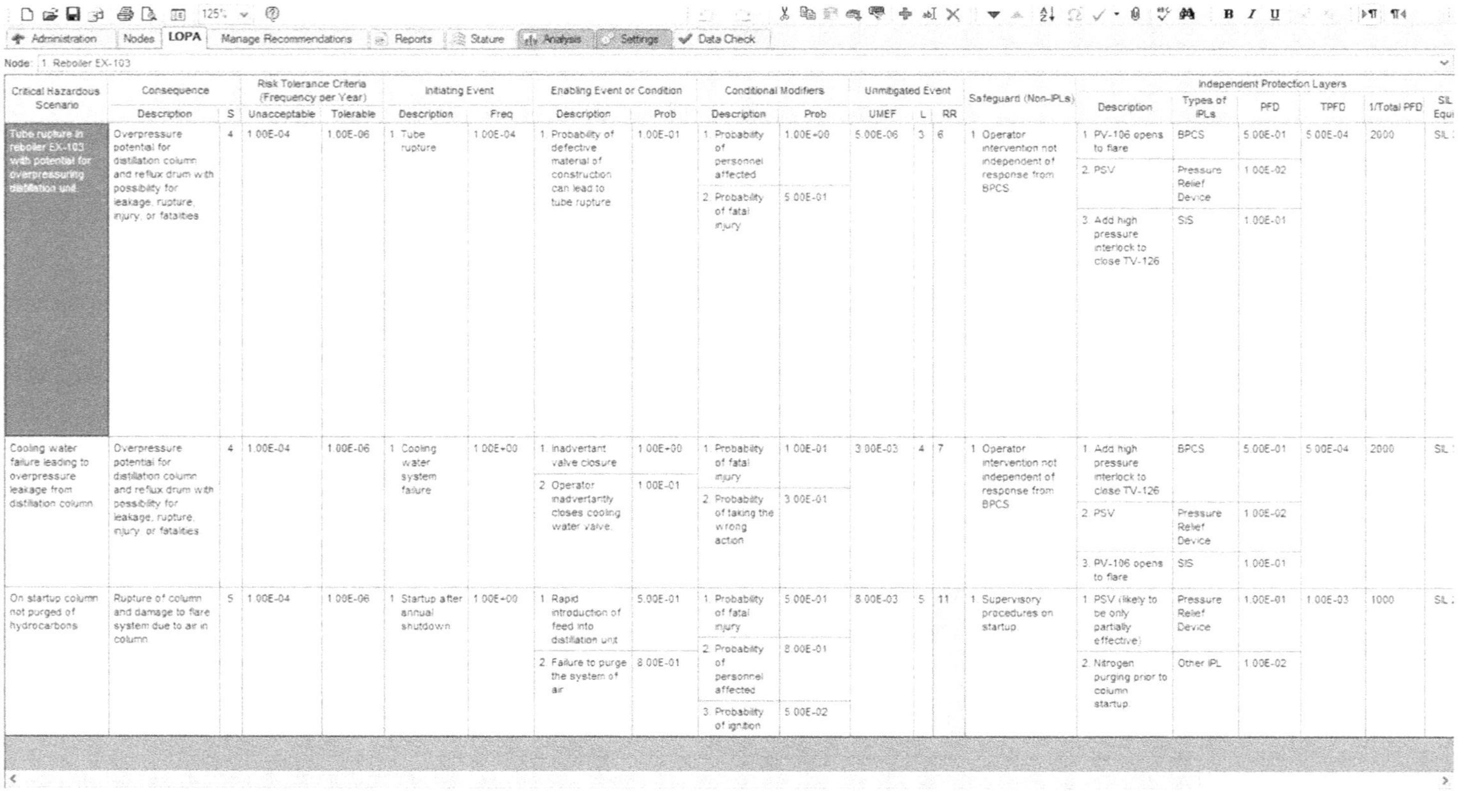

Node: 1. Reboiler EX-103

Critical Hazardous Scenario	Consequence Description	S	Risk Tolerance Criteria (Frequency per Year) Unacceptable	Tolerable	Initiating Event Description	Freq	Enabling Event or Condition Description	Prob	Conditional Modifiers Description	Prob	Unmitigated Event UMEF	L	RR	Safeguard (Non-IPLs)	Independent Protection Layers Description	Types of IPLs	PFD	TPFD	1/Total PFD	SIL Equi
Tube rupture in reboiler EX-103 with potential for overpressuring distillation unit	Overpressure potential for distillation column and reflux drum with possibility for leakage, rupture, injury, or fatalities	4	1.00E-04	1.00E-06	1. Tube rupture	1.00E-04	1. Probability of defective material of construction can lead to tube rupture	1.00E-01	1. Probability of personnel affected	1.00E+00	5.00E-06	3	6	1. Operator intervention not independent of response from BPCS	1. PV-106 opens to flare	BPCS	5.00E-01	5.00E-04	2000	SIL
									2. Probability of fatal injury	5.00E-01					2. PSV	Pressure Relief Device	1.00E-02			
															3. Add high pressure interlock to close TV-126	SIS	1.00E-01			
Cooling water failure leading to overpressure leakage from distillation column	Overpressure potential for distillation column and reflux drum with possibility for leakage, rupture, injury, or fatalities	4	1.00E-04	1.00E-06	1. Cooling water system failure	1.00E+00	1. Inadvertant valve closure	1.00E+00	1. Probability of fatal injury	1.00E-01	3.00E-03	4	7	1. Operator intervention not independent of response from BPCS	1. Add high pressure interlock to close TV-126	BPCS	5.00E-01	5.00E-04	2000	SIL
							2. Operator inadvertantly closes cooling water valve.	1.00E-01	2. Probability of taking the wrong action	3.00E-01					2. PSV	Pressure Relief Device	1.00E-02			
															3. PV-106 opens to flare	SIS	1.00E-01			
On startup column not purged of hydrocarbons	Rupture of column and damage to flare system due to air in column	5	1.00E-04	1.00E-06	1. Startup after annual shutdown	1.00E+00	1. Rapid introduction of feed into distillation unit	5.00E-01	1. Probability of fatal injury	5.00E-01	8.00E-03	5	11	1. Supervisory procedures on startup	1. PSV (likely to be only partially effective)	Pressure Relief Device	1.00E-01	1.00E-03	1000	SIL
							2. Failure to purge the system of air	8.00E-01	2. Probability of personnel affected	8.00E-01					2. Nitrogen purging prior to column startup.	Other IPL	1.00E-02			
									3. Probability of ignition	5.00E-02										

FIGURE L.1 Example of Layers of Protection Analysis (LOPA) Worksheet. (Worksheet Example courtesy of Dyadem.)

Lethal Concentration (LC)

The concentration of a substance that is sufficient to kill a test animal. See also **Lethal Dose (LD)**.

Lethal Concentration 50

A lethal concentration that kills 50 percent of test animals within a specific time.

Lethal Dose (LD)

An indication of the lethality of a given substance or type of radiation. Because resistance varies from one individual to another, the lethal dose represents a dose (usually recorded as dose per kilogram of subject body weight) at which for a given percentage of subjects a fatality will occur. The most commonly used lethality indicator is the LD50, a dose at which 50 percent of subjects will die.

Lethal Dose 50

The dose required to produce the death of 50 percent of the exposed population within a specific time.

Level, Noise

The logarithm of the ratio of the measured quantity to a reference quantity of the same kind. The base of the logarithms, the reference quantity, and kind of level must be specified.

Level A Clothing

Protective clothing that should be worn when the highest level of respiratory, skin, and eye protection is needed.

Level B Clothing

Protective clothing that should be worn when the highest level of respiratory protection is needed, but a lesser level of skin protection.

Level C Clothing

Protective clothing that should be worn when the criteria for using air-purifying respirators are met.

Level D Clothing

Protective clothing that should be worn only as a work uniform and not on any site with respiratory or skin hazards.

Liability

Responsibility to another person for negligence causing harm.

Liability Insurance

Insurance that obligates the insurance company to pay any liability for which the insured may be covered and, also at the expense of the company, to defend any damage suits brought against it to enforce such liability, thus protecting it from liability and expense of litigation growing out of claims in which it is involved. Insurance that agrees to reimburse the policyholder for sums that may be required to pay others as the result of negligence.

Lien

A right or claim for payment against a workers' compensation case.

Life Jacket

See **Personal Flotation Device (PFD)**.

Life Line

A rope attached to a life ring or ring life buoy that is thrown to an individual in the water to assist in the efforts to return the individual to a safe location. The life ring or buoy can be pulled by rescuers via a rope. (Note: **Life Line** is a different term than **Lifeline**, which refers to a line [i.e., rope] utilized for water life saving activities and is defined elsewhere in this chapter). See also **Personal Flotation Device (PFD)**.

L

Life Preserver

See **Personal Flotation Device (PFD)**.

Life Ring

See **Personal Flotation Device (PFD)**.

Life Safety

The preservation of life from an incident or its associated hazards.

Life Safety Code (LSC)

A fire code (NFPA 101) developed by the National Fire Protection Association (NFPA) for the preservation of life from a fire event or its associated hazards. It is primarily concerned with exit facilities and arrangements and protection against fire events. As a result of the large life loss of the Triangle Shirtwaist Fire in New York in 1911, the NFPA was challenged to provide life safety measures for factories and loft buildings, control of smoking in hazardous areas, improved exits, and provision of fire drills. NFPA 101, Life Safety Code, is the standard that is used to delineate the appropriate fire exit requirements.

Life Safety Systems

The embodiment of measures taken or prescribed to prevent the endangerment of individuals threatened by a fire event or its associated hazards. In buildings, life safety systems primarily consist of exit facilities, emergency lighting, fire sprinklers, and standpipe systems. In high-rise buildings, stairways serve as vertical emergency exits and also all elevators are automatically shut down to prevent the possibility of people becoming trapped in them. Emergency generator systems are provided to permit the operation of one elevator at a time to rescue people trapped in them by a power failure. Generators also serve other vital building functions such as emergency lighting and fire pumps. Fire-suppression systems often include sprinklers, but, if none are required by building codes, a separate piping system is provided with electric pumps to maintain pressure and to bring water to fire-hose cabinets throughout the building. There are also exterior connections at street level for portable fire-truck pumps. The fire hoses are so placed that every room is accessible; the hoses are intended primarily for professional fire fighters but may also be used by the building occupants.

Lifeboat

A small boat kept on a ship or fixed offshore structures for use in an emergency, typically one of a number on deck, suspended from davits, dropped directly to the sea from an angle of storage (see Figure L.2). A rescue lifeboat is a vessel that is used to attend a ship or vessel in distress, or its survivors, to rescue crewmen and passengers.

Lifeguard

An individual responsible for overseeing the safety of the users of a body of water and its environs, such as a swimming pool, a water park, or a beach. Lifeguards are qualified strong swimmers, trained and certified in water rescue, in using a variety of aids and equipment depending on the requirements of their particular venue, and in first aid. In some areas, lifeguards may form part of the provided emergency services and respond to incidents, and in some communities the lifeguard service also carries out mountain rescues, or may function as the primary Emergency Medical Service (EMS) provider.

Lifeline

A rope designed and utilized to provide fall protection that must be secured above the point of operation to an anchorage or structural member. It connects to an anchorage at one end to hang vertically (vertical lifeline) or to an anchorage at both ends to stretch horizontally (horizontal lifeline) and serves as a method to connect other components of a personal fall arrest system to the anchorage. It must also be capable

FIGURE L.2 Lifeboat.

of supporting a minimum specified dead weight. They may also be called static lines, drop lines, safety lines, rat lines, or scare lines. (Note: **Lifeline** is a different term than **Life Line**, which refers to a line [i.e., rope] utilized as a part of a fall arrest system and is defined elsewhere in this chapter). See also **Fall Arrest System**.

Light Duty

A temporary change in one's job assignment to accommodate work restrictions due to injury or health concerns.

Lightning Protection System

A system designed to protect structures and buildings from destructive lightning strikes by draining lightning surge-induced charges harmlessly to ground. A lightning protection system usually consists of strike termination devices, electrical conductors, ground terminals, and surge suppression devices.

Likelihood

The expected frequency (or probability) of an event's occurrence. The frequency of an event is often expressed in events per year or events per million hours. It is one of the two components used to define a risk (consequence and likelihood). See Table L.2. See also **Probability**.

Limit Switch

A switch fitted to elevators, traveling cranes, etc., in order to cut off the power supply if, for example, the elevator compartment or equipment travels beyond a certain specified limit.

LIPE

An acronym for Life safety, Incident stabilization, Property conservation, and protection of the Environment. Commonly referred to during hazardous material incident planning and response.

Listed

Equipment, materials, or services included in a list published by a qualified or approved testing agency, inspection agency, or other organization whose listing states that the equipment, material, or service meets the appropriate designated standards or has been tested or evaluated and found suitable for use for a specified purpose. Listing for electrical devices typically means testing by Underwriters Laboratories (UL) or Factory Mutual (FM) laboratories for applications according to

TABLE L.2
Typical Likelihood Rating Descriptions

Rating	Likelihood (Probability)
1	Frequency: 0.0 to 1×10^{-6} (never to 1 in 1,000,000 years)
	Scenario: Should not occur in the life of the process and there is no historical industry experience to suggest it will occur
	Layers of Protection: Four or more independent, highly reliable safeguards are in place; failure of three safeguards would not initiate an unwanted event
2	Frequency: 1×10^{-6} to 1×10^{-4} (1 in 1,000,000 years to 1 in 10,000 years)
	Scenario: Similar events are unlikely to occur, but have historically occurred in this type of process somewhere within the industry
	Layers of Protection: Three independent, highly reliable safeguards are in place; failure of two safeguards would not initiate an unwanted event
3	Frequency: 1×10^{-4} to 1×10^{-3} (1 in 10,000 years to 1 in 1000 years)
	Scenario: This particular scenario is likely to occur somewhere in the industry during the life of this general type of process
	Layers of Protection: Two independent, highly reliable safeguards are in place; failure of one safeguard would not initiate an unwanted event
4	Frequency: 1×10^{-3} to 1×10^{-2} (1 in 1000 years to 1 in 100 years)
	Scenario: This particular scenario will almost certainly occur somewhere in the industry during the life of this specific type of process (but not necessarily at this location)
	Layers of Protection: Single layer of safeguard and operator interface are in place to prevent unwanted events
5	Frequency: 1.0 to 1×10^{-2} (always to 1 in 100 years)
	Scenario: This particular scenario has occurred somewhere in the industry in this specific process or is likely to occur at this location during the life of this facility
	Layers of Protection: Procedures or operator interface relied upon to prevent unwanted events

the National Electrical Code. See also **Approved**; **Classified**; **Factory Mutual (FM)**; **Underwriters Laboratories (UL)**.

Litigated Claim

A workers' compensation claim in which an attorney is involved, most likely on behalf of the injured worker.

Load Limit

The upper weight limit capable of safe support by a vehicle or floor. The designer of floors to meet load limit requirements should consult American National Standards Institute (ANSI)/ASCE Standard 7-95, Minimum Design Loads for Buildings and Other Structures.

Load Weight or Allowable Load

Refers to highest weight of load to be safely carried by a vehicle.

Local Emergency Planning Committee (LEPC)

A committee appointed by the state emergency response commission, as required by Superfund Amendment and Reauthorization Act (SARA) Title III, to formulate a comprehensive emergency plan for its jurisdiction. They are composed of representatives of the following groups and organizations: elected officials, public safety and health personnel, local environment and transportation agencies, community groups and media, and representatives of affected facilities. LEPCs are responsible for the development and maintenance of emergency plans for response to chemical emergencies, such as pipeline incidents. See also **Incident Action Plan**; **SARA Title III**.

Local Exhaust Ventilation

A ventilation system that captures and removes the contaminants at the point they are being produced before they escape into the workroom air. They typically incorporate a hood, enclosure, or inlet to collect the agent, ductwork, a filter or air cleaning device, a fan or other air moving device, and discharge ductwork to the exhaust point. Local exhaust is required for moderate to highly toxic air contaminants. American National Standards Institute (ANSI) Z9.2, Fundamentals Governing the Design and Operation of Local Exhaust Ventilation Systems, is the standard for local systems.

Lockdown

Similar to the shelter-in-place procedure, a lockdown is utilized when a situation occurs that may be a hazard to health or is life-threatening in a school or similar environment. It is intended to limit access and hazards by controlling and managing staff and occupants in order to increase safety and reduce possible victimization. In a lockdown, the building will have restricted access until the "All Clear" is given, or until individuals are directed by emergency personnel or staff. A lockdown may be called by administrative officials, law enforcement agencies, or other emergency responders. A lockdown may be called for a variety of reasons, including weapons, intruders, police activity in or around the facility, contamination or hazardous materials, or terrorist events. See also **Shelter-in-Place**.

Lockout

The placement of a lockout device on an energy isolating device in accordance with an established procedure ensuring that the energy isolation device and the equipment being controlled, e.g., electrical, hydraulic, mechanical, compressed air, or any other source that might cause unexpected movement to be disengaged or blocked; electrical sources must be de-energized and locked or positively sealed in the off position and cannot be operated until the lockout device is removed. It is a program or procedure that prevents injury by eliminating unintentional operation or release of energy within machinery or processes during set-up, start-up, cleaning and clearing jams, or maintenance repairs. It may also be called "lock off." See also **Lockout/Tagout (LOTO)**.

Lockout Device

A positive means such as a lock, either key or combination type, to hold an energy isolating device in the safe (e.g., not operating) position and prevent the energizing of a machine or equipment. Included are blank flanges and bolted slip blinds.

Lockout Hasp

A locking mechanism that allows several individuals to place their individual locks on a single device, so that the specific device cannot be operated unless all the individual locks are removed.

Lockout/Tagout (LOTO)

An Occupational Safety and Health Administration (OSHA) standard, 29 CFR 1910.47, and a basic safety concept of specific practices and procedures used to safeguard employees from the unexpected energization or startup of machinery and equipment, or the release of hazardous energy during service or maintenance activities. This requires that a designated individual turns off and disconnects the machinery or equipment from its energy source(s) before performing service or maintenance and that the authorized employee(s) either lock or tag the energy-isolating device(s) to prevent the release of hazardous energy and take steps to verify that the energy has been isolated effectively. Lockout and Tagout (LOTO) is addressed in specific federal

L

standards for general industry, marine terminals, longshoring, and the construction industry. OSHA estimates that compliance with lockout and tagout requirements prevents about 120 fatalities and 50,000 injuries each year. Workers injured on the job from exposure to hazardous energy lose an average of 24 workdays for recuperation. In a study conducted by the United Auto Workers (UAW), 20 percent of the fatalities (83 of 414) that occurred among their members between 1973 and 1995 were attributed to inadequate hazardous energy control procedures, specifically lockout and tagout procedures. See also **Electrical Safety**.

Logging Safety

The prevention of injuries and fatalities in the logging industry. Logging has consistently been one of the most hazardous industries in the United States. The use of chain saws and logging machines poses hazards wherever they are used. As loggers use their tools and equipment, they deal with massive weights and the irresistible momentum of falling, rolling, and sliding trees and logs (Figure L.3). The hazards are more acute when dangerous environmental conditions are factored in, such as uneven, unstable, or rough terrain; inclement weather including rain, snow, lightning, winds, and extreme cold; and/or remote and isolated work sites where health care facilities are not immediately accessible. In 2007, the U.S. logging industry employed 101,000 workers and accounted for 88 fatalities, resulting in a fatality rate of 87.1 deaths per 100,000 workers for that year. This rate is over 23 times higher than the overall fatality rate in the United States in 2007 (3.7 deaths per 100,000). This high risk for fatal work injuries highlights the need for rigorous research and intervention

L

FIGURE L.3   Logging industry safety.

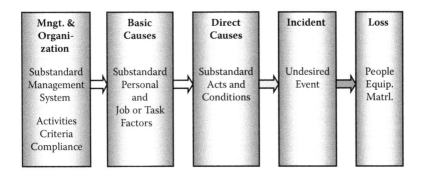

Mngt. & Organi- zation	Basic Causes	Direct Causes	Incident	Loss
Substandard Management System	Substandard Personal and Job or Task Factors	Substandard Acts and Conditions	Undesired Event	People Equip. Matrl.
Activities Criteria Compliance				

FIGURE L.4   Loss-causation model.

programs to improve the safety for this industry. A new set of national occupational safety and health goals for the logging industry have been developed as part of the National Occupational Research Agenda process and are undergoing public review and comment.

Loss

For the insurance industry it is defined as the amount the insurer is required to pay because of a judgment, by virtue of the terms of the insurance contract. Business operations usually have loss defined as the entire property damage and may also include business interruption impacts.

Loss-Causation Model

An incident evaluation tool to investigate causes leading to incidents and as a framework for incident investigation. It models the flow of actions that lead to an incident or loss (Figure L.4). It was originally developed by Herbert W. Heinrich and later modified by Frank Bird. See also **Domino Theory**.

Loss Control

A program whose objective is to minimize incident-based losses. Total loss control is based on studies of near misses or non-injury or damage incidents, and on analysis of both direct and indirect incident causes (root causes). Both injuries and property damages are included in the analysis. Loss control activities include fire prevention, education, inspection, overhaul, and salvage.

Loss Control Management

The application of safety management techniques and skills through a program of activities directed at risk avoidance, loss prevention, and loss reduction aimed at minimizing loss from the risks of the business.

Loss History

An organization's history of losses (claims). Insurance companies view loss history as an indication of an organization's propensity for losses in the future. See also **Loss Trend**.

Loss Prevention

A program to identify and correct potential incident concerns before they result in an actual event resulting in a loss.

Loss Prevention Engineer (LPE)

An individual who by education, training, and experience is familiar with the nature and characteristics of industrial hazards. This individual typically assists in identifying hazards and develops appropriate controls for these hazards that, when effectively implemented, prevent occupational injury, illness, and property damage.

This person should have knowledge of industrial hygiene and toxicology, design of engineering hazard controls, fire protection, ergonomics, system and process safety, safety and health program management, incident investigation and analysis, product safety, construction safety, education and training methods, measurement of safety performance, human behavior, environmental safety and health, and safety, health, and environmental laws, regulations, and standards. See also **Safety Professional**; **Safety Representative**.

Loss Prevention Engineering

See **Safety Engineering**.

Loss Prevention Foundation

A nonprofit organization, created in 2006 to serve individuals primarily in the retail loss prevention sector. Its goal is to advance the loss prevention profession by providing relevant, convenient, and challenging educational resources. They support this through two certification programs: LPQualified (LPQ) and LPCertified (LPC), on-line educational resources, and their loss prevention professional membership program. The certification programs are designed to give individuals the opportunity to gain valuable knowledge about retail, business, leadership, and loss prevention.

Loss Ratio, Insurance

A fraction calculated by dividing the amount of the losses by the amount of the premiums. It is expressed as a percentage of the premiums. Various bases are used in calculating the loss ratio, e.g., earned premium loss ratio, written premium loss ratio, etc.

Loss Reduction

Activities undertaken to reduce the magnitude or severity of injuries, illnesses, property damages, and business operation interruptions from incidents.

Loss Reserve

An estimate of the amount an insurer expects to pay for losses incurred but not yet due for payment.

Loss Trends

Projections of future incident impacts based on analysis of historical loss patterns. They are not entirely accurate since they cannot predict the actual loss that will occur. See also **Loss History**.

Lost Time Incident (LTI)

See **Lost Time Injury or Illness (LTI)**.

Lost Time Injury or Illness

A work injury or illness that results in death or disability and in which the injured person is unable to report for duty on the next regularly scheduled shift. See also **Injury**.

Lost Workdays

The number of workdays (consecutive or not) beyond the day of injury or onset of illness that an employee was away from work or limited to restricted work activity because of an occupational injury or illness. See also **Days of Disability**.

Lost Workday Cases

Cases that involve days away from work or days of restricted work activity, or both.

Loudness

The intensity attribute of an auditory sensation, in terms of which sounds may be ordered on a scale extending from soft to loud. Loudness depends primarily upon the sound pressure of the stimulus, but it also depends upon the frequency and wave form of the stimulus.

Loudness Level

A subjective method for rating loudness in which a 1000-Hz tone is varied in intensity until it is judged by listeners to be equally as loud as a given sound sample. The loudness level in phones is taken as the sound pressure level, in decibels, of the 1000-Hz tone.

Low Energy Circuit

Electrical circuits that do not contain sufficient energy to produce incendiary sparks. Typically signal and communication circuits are considered low energy circuits.

Lower Confidence Limit (LCL)

A statistical procedure used in analyzing sampling data to estimate the probability that the true value of the sampled quantity is lower than that obtained.

Lower Explosive Limit (LEL)

The minimum concentration of combustible gas or vapor in air below which propagation of flame does not occur on contact with an ignition source. The lower limits of flammability of a gas or vapor at ordinary ambient temperature expressed in percent of the gas or vapor in air by volume. See also **Explosive Limits**.

Lower Flammable Limit (LFL)

Synonymous with lower explosive limit (LEL). See also **Lower Explosive Limit (LEL)**; **Upper Flammable Limit (UFL)**.

Lumen

Under the common system, it was defined as the flux on one square foot of a sphere, one foot in radius, with a light source of one candle at the center that radiates uniformly in all directions. Under SI, luminous flux is measured in lumen units (symbol: lm) and has as its formula cd × sr, which are the SI Base Units of candela and steradian for a solid angle. Thus, the lumen is the luminous flux emitted in a solid angle of one lumen uniformly distributed in a solid angle of one steradian by a point source having a uniform intensity of one candela. See also **LUX**.

Luminescent

Emitting light not due to high temperatures, usually caused by excitation by rays of a shorter wavelength.

LUX

The lux (symbol: lx) is the SI unit of illuminance and luminous emittance. The luminance produced by a luminous flux of one lumen uniformly distributed over a surface of one square meter. See also **Footcandle (fc)**.

M

Machine Guarding
The installation of equipment or devices on machines to eliminate hazards to individuals created by operation of the machines. These usually consist of gates, covers, shields, housings, or deflectors. Machine guarding hazards are addressed in specific Occupational Safety and Health Administration (OSHA) standards for the general industry, marine terminals, longshoring, and the construction and agriculture industries. Machine guards are normally grouped into four categories: fixed barrier machine guards, interlocking machine guards (electrical, mechanical, or hydraulic), adjustable machine guards, and self-adjustable machine guards. Fixed barrier machine guards are machine guards that are permanent parts of the machine. They generally provide the maximum amount of protection and require the least amount of maintenance. Interlocking machine guards shut down the machine in use when the guard is opened or removed. These machine guards can also provide a maximum amount of protection, but can require more maintenance than fixed barrier machine guards. Adjustable machine guards are positioned or arranged to work with the size of the material the operator is working with. Adjustable machine guards are easy to disengage by the operator and generally require more maintenance than other guards. Self-adjustable machine guards allow the material being used in the machine to enter the hazard area and places a barrier between the operator and the material as it is processed. Once the material has finished going through the machine, the guard snaps completely back into place. Like adjustable machine guards, self-adjustable machine guards may require frequent maintenance, and often do not provide maximum protection. All machine guards must prevent fingers, arms, and any part of an individual from coming into contact with a hazard zone; they must be designed and installed correctly; and the machine guards must fit the machine it is used on perfectly. See also **Adjustable Guard**; **Barrier Guard**; **Fixed Guard**; **Guard, Interlocking Barrier**; **Interlocking Guard**; **Point of Operation Guarding**.

Machine Safety
The prevention of injuries or fatalities as a result of contact with machinery or equipment. National Institute for Occupational Safety and Health (NIOSH) data from 1980 to 1998 indicates that occupational injury from machinery was ranked third after motor vehicle and homicide as cause of death in the United States. Fatalities from incidents related to machines were 13 percent of the total. The leading industries for these incidents included agriculture, mining, manufacturing, and construction. The injuries included struck by or against an object, caught in or compressed by equipment, caught in or crushed in collapsing materials, and being struck by material ejected from machinery.

Machinery Safety System
An integrated total system, including the pertinent elements of the mechanical actions, the controls, the safeguarding, and any required supplemental safeguarding, and their interfaces with the operator, and the environment, designed, constructed, and arranged to operate together as a unit, such that a single failure or single operating error will not cause injury to personnel due to point of operation hazards.

Major Disaster

As defined by the Stafford Disaster Relief and Emergency Assistance Act (42 U.S.C. 5122), a major disaster is any natural catastrophe (including any hurricane, tornado, storm, high water, wind-driven water, tidal wave, tsunami, earthquake, volcanic eruption, landslide, mudslide, snowstorm, or drought), or, regardless of cause, any fire, flood, or explosion, in any part of the United States, which in the determination of the U.S. President causes damage of sufficient severity and magnitude to warrant major disaster assistance under this Act to supplement the efforts and available resources of States, tribes, local governments, and disaster relief organizations in alleviating the damage, loss, hardship, or suffering caused thereby.

Major Injury

An injury where there is loss of time to the injured person and a medical expense.

Makeup Air

Clean tempered outdoor air supplied to a workspace to replace the air removed by an exhaust ventilation system or by an industrial process that has removed air from the environment.

Malpractice

A type of negligence in which a professional, under a duty to act, fails to follow generally accepted professional standards, and that breach of duty is the cause of injury to a plaintiff who suffers damages. It is committed by a professional or the professional's subordinates or agents on behalf of a client or patient and causes damages to the client or patient.

Management of Change (MOC)

A method to evaluate, authorize, implement, communicate, and document changes to process technology, chemicals, equipment, procedures, facilities, buildings, or organizations as to their potential for hazards, potential consequential loss, the magnitude of the potential risk, and the impact on facility operation. Control by means of elimination or mitigation of the hazards and/or the consequences should then be implemented to minimize potential risk. See Figure M.1 for a flowchart of a typical MOC procedure.

Management Oversight Risk Tree (MORT)

The management oversight risk tree (MORT) was developed in the early 1970s for the U.S. Energy Research and Development Administration as a safety analysis method that would be compatible with complex, goal-oriented management systems. The MORT chart is a logic/decision diagram that arranges safety program elements in an orderly and logical manner, referred to in a "tree" fashion. Its analysis is carried out by means of a fault tree, where the top event is "Damage, destruction, other costs, lost production or reduced credibility of the enterprise in the eyes of society." The tree gives an overview of the causes of the top event from management oversights and omissions or from assumed risks or both. The MORT tree has more than 1500 possible basic events compressed to 100 generic events, which have been identified in the fields of accident prevention, administration, and management. MORT is used in the analysis or investigation of incidents and events, and evaluation of safety programs. Research literature cites its usefulness where it was revealed that where "normal" investigations revealed an average of 18 problems (and recommendations), "complementary" investigations with MORT analysis revealed an additional 20 contributions per case. See also **Fault Tree Analysis (FTA)**.

Manual on Uniform Traffic Control Devices (MUTCD)

Defines the standards used nationwide to install and maintain traffic control devices on all public streets, highways, bikeways, and private roads open to public traffic. It is published by the Federal Highway Administration under 23 CFR Part 655, Part F.

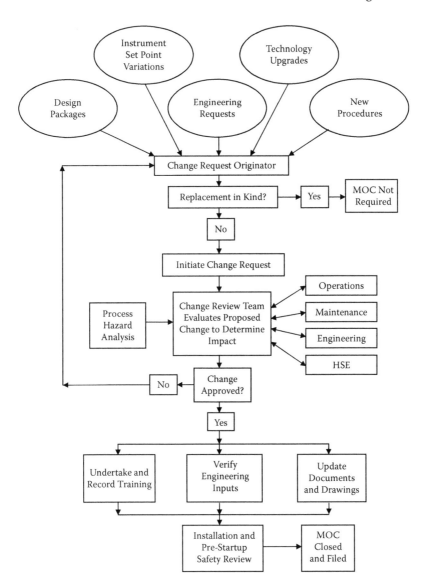

FIGURE M.1   MOC process flowchart.

Manual Pull Station (MPS), Fire Alarm
A switch provided on a fire alarm system that is manually activated to indicate a fire event (see Figure M.2). The switch is configured to conspicuously identify it as a fire alarm device and is usually fitted with a tamper device (break glass, rod, or cover) to discourage or prevent false activation. It sends a signal to a central monitoring station for notification of location and activation of alarms.

Margin of Safety
A border, edge, or limit beyond which a particular behavior, condition, or situation becomes hazardous or unsafe.

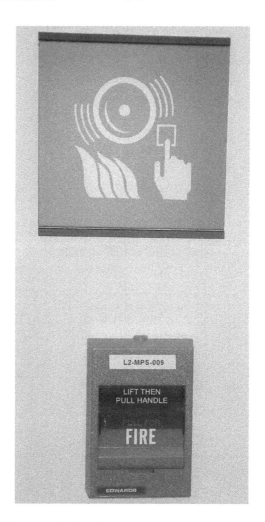

FIGURE M.2   Fire alarm manual pull station.

Maritime Safety Standards

   Generally refers to Occupational Safety and Health Administration (OSHA) standards and requirements that are defined in 29 CFR Part 1915 (Occupational Safety and Health Standards for Shipyard Employment), Part 1917 (Marine Terminals), and Part 1918 (Safety and Health Regulations for Longshoring). See also **General Industry Safety Standards**; **Safety of Life at Sea, SOLAS Regulations**.

Markov Analysis

   A fault propagation method used to analyze failure rate or probability for safety instrumented functions. A diagram is constructed to represent the system under consideration including the logical relationships between its components. In Markov analysis there are a group of circles, each of which represents a system state. The different states are connected with transitions, which are shown as arrows and indicate paths to move from one state to another. The transitions are quantified using either failure rates when the transition is from an acceptable state to a failed state or repair rates when the transition is from a failed state back to an acceptable state. As with other models, there are several solution methods to obtain results. For safety instrumented system applications, the method using steady state equations are not appropriate; instead, numeric discrete time solutions are preferred.

Mask
A protective covering for the face or head, such as a wire screen, a metal shield, a respirator, or a gas mask.

Masking (Noise)
The stimulation of one ear of a subject by controlled noise to prevent hearing with that ear the tone or signal given to the other ear. This procedure is used where there is at least a 15 to 20 dB difference in the two ears.

Material Safety Data Sheet (MSDS)
An information sheet on properties of a material that meet certain combustible, toxic, or other hazardous threshold properties that are specified by the Environmental Protection Agency regulation Title III of the Superfund Amendment and Reauthorization Act (SARA). They also contain emergency and first aid procedures for the specified material. MSDSs are required to be provided and maintained by organizations that have hazardous materials. They are also required to provide copies to the local fire department for the purposes of fire fighter protection and pre-planning. See also **Hazard Communication (HAZCOM); Hazardous Material; SARA Title III**.

Maximum Foreseeable Loss (MFL)
A term used in the insurance industry that is normally defined as the worst case scenario under which an estimate is made of the maximum dollar amount that can be lost if a catastrophe occurs such as a hurricane or firestorm. See also **Probable Maximum Loss (PML)**.

Maximum Occupancy Allowances
The allowable maximum number of individuals that are allowed to occupy a structure. Maximum occupancy postings are primarily provided at structures to ensure that the available exits for the facility are adequate and panic will not ensue during an emergency evacuation. The maximum occupancy postings are based on measurable standards for exit door provisions. These provisions are most commonly cited in the National Fire Protection Association (NFPA) 101, Life Safety Code (LSC). See also **Overcrowding**.

Maximum Permissible Concentration (MPC)
The concentrations set by the National Committee on Radiation Protection (NCRP), which are the recommended maximum average concentrations of radionuclides to which an employee may be exposed, assuming that the employee works 8 hours a day, 5 days a week, 50 weeks a year.

Maximum Permissible Dose (MPD)
A dose of ionizing radiation not expected to cause appreciable bodily injury to an individual at any time during their life.

Maximum Use Concentration (MUC)
The product of the protection factor of the respiratory protection equipment and the permissible exposure limit (PEL).

Mean Time Between Failure (MTBF)
The average time between successive failures, estimated by the total measured operating time of a population of items divided by the total number of failures within the population during the measured time period. Alternatively the MTBF of a repairable item can be also calculated by the ratio of total operating time to the total number of failures.

Mean Time to Fail Spurious

The mean time until a failure of the system causes a spurious process trip.

Mean Time to Failure (MTTF)

The average amount of time until a system fails or its "expected" failure time. Please note that the MTTF can be assumed to be the inverse of failure rate (lambda) for a series of components, all of which have a constant failure rate for the useful life period of the components.

Means of Escape

A way out of a building or structure. It may not comply with the formal definition of means of egress, but does provide an alternative avenue to evacuate a facility. See also **Evacuation; Exit**.

Medical Surveillance

The systematic approach to monitoring health changes in workers to identify and determine which effects may be work related.

Medical Treatment

As defined by Occupational Safety and Health Administration (OSHA) reporting requirements, medical treatment (other than first aid) administered by a physician or by registered professional personnel under the standing orders of a physician. See also **First Aid**.

Medicare

A social insurance program administered by the U.S. government, providing health insurance coverage to eligible individuals who have been legal residents of the United States for 5 years and are aged 65 and over or are disabled and have been receiving either Social Security benefits or Railroad Retirement Board disability benefits for at least 24 months from the date of first disability payment.

M

Metal Fume Fever

A flu-like condition that is caused as a result of inhaling heated metal fumes.

Methanometer

An instrument to measure the percentage of methane in the air in underground coal mines and designed to alert miners to the presence of potentially dangerous concentrations of this gas, which could result in an explosion and/or fire. Methane (CH_4) is an odorless, tasteless, colorless, highly flammable, and lighter than air gas found in coal mining operations by the decomposition of coal and other carbonaceous materials. Methane's low density (approximately 50 percent of the density of air) causes it to concentrate in the higher parts of an underground mine if ventilation is insufficient to properly mix it with the mine air. Mine officials typically carry methanometers to evaluate gas levels in work areas as well as to inspect other areas of the mine. Usually if methane is measured at 1.25 percent then work will cease and equipment will be shut down, and if the methane level is measured at 2.50 percent then all personnel will be withdrawn to ensure their safety from risk of fire or explosion. Methane levels are regularly measured using handheld methanometers in close proximity to the actual immediate mining locations. Readings are also taken across intake airways to keep the methane level below 0.25 percent.

Mine Safety and Health Administration (MSHA)

A federal enforcement agency responsible for the health and safety of the nation's miners. The Mine Safety and Health Administration (MSHA), part of the U.S. Labor Department, helps to reduce deaths, injuries, and illnesses in the mining industry through regulation, inspection, and educational activities and programs. The agency develops and enforces safety and health rules applying to all U.S. mines, assists mine operators who have special compliance problems, and provides technical, educational, and other types of assistance. MSHA works cooperatively with industry, labor, and other federal and state agencies toward improving safety and health conditions for all miners. The MSHA's responsibilities are defined in the Federal Mine Safety and Health Act of 1977. The MSHA replaced the Mining Enforcement and Safety Administration in 1978, which were previously responsible for enforcing the health and safety regulations for mines in the United States.

Minor Injury
An injury where no lost time or major medical costs are involved. See also **Major Injury**.

Mistake
A human action that produces an unintended result.

Mists
Suspended liquid droplets generated by condensation from the gaseous to the liquid state or by breaking up a liquid into a dispersed state, such as by splashing, foaming, or atomizing. Mist is formed when a finely divided liquid is suspended in the atmosphere.

Misuse Mode and Effects Analysis (MMEA)
Assesses the likelihood of occurrence of potential misuse modes and their effect on safety before and after corrective actions. See also **Failure Mode and Effects Analysis (FMEA/FMECA)**.

Mitigation
The activities designed to reduce or eliminate risks to persons or property or to lessen the actual or potential effects or consequences of an incident. Mitigation measures may be implemented prior to, during, or after an incident. These measures are often gathered by lessons learned or recommendations due to earlier incidents. Mitigation involves ongoing actions to reduce exposure to, probability of, or potential loss from hazards. Measures may include zoning and building code requirements, engineering solutions, and analysis of hazard-related data. Mitigation can also include efforts to educate governments, businesses, and the public on measures they can take to reduce loss and injury.

Modified Work
A change in an employee's working conditions in order to accommodate work restrictions.

Moral (Ethical) Safety
The ability to maintain a set of standards, beliefs, and operating principles that are consistent, that guide behavior, and that are grounded in a respect for life. See also **Behavior-Based Safety (BBS)**; **Safety Culture**.

Morphological Approach
A structured analysis of an incident directed by insights from historical case studies of incidents, but not as rigorous as a hazard analysis. See also **Deductive Approach**; **Inductive Approach**.

MORT Chart or Tree
See **Management Oversight Risk Tree (MORT)**.

Motor Vehicle Accident (MVA)
An incident in which one vehicle hits a stationary object or another vehicle.

Motor Vehicle Nontraffic Accident
Any motor vehicle incident that occurs entirely in any place other than traffic.

Motor Vehicle Safety
The risk of roadway incidents associated with on-the-job operation of motor vehicles affects millions of U.S. workers. Data from the Bureau of Labor Statistics (BLS) indicate that in 2007, nearly 3.9 million workers in the United States were classified as motor vehicle operators. Over 40 percent (1.6 million) of these motor vehicle

operators were employed as heavy truck (including tractor-trailer) drivers. Other workers who use motor vehicles in performing their jobs are spread across numerous other occupations. These include workers who operate vehicles owned or leased by their employer, and those who drive personal vehicles for work purposes. Motor vehicle–related incidents are consistently the leading cause of work-related fatalities in the United States. Of approximately 5700 fatalities annually reported by the Bureau of Labor Statistics, 35 percent are associated with motor vehicles.

Between 2002 and 2007, on average:

1371 workers died each year from crashes on public highways.
330 workers died each year in crashes that occurred off the highway or on industrial premises.
363 pedestrian workers died each year as a result of being struck by a motor vehicle.

Motorcycle Safety

Per data from the National Highway Traffic Safety Administration (NHTSA), per mile traveled, motorcyclists in 2003 were about 32 times more likely than motor vehicle passengers to die in a motor vehicle accident and about six times more likely to be injured. Motorcyclist accident prevention efforts are targeted at improving skills and licensing, decreasing alcohol and drug impairment, and increased awareness by other drivers for road use/sharing with motorcyclists. Injury prevention has been aimed at the use of protective gear, including helmets, which have shown to be 37 percent effective in preventing fatal injuries to motorcyclists.

Multi-gas Detector

An air monitoring device that measures oxygen levels, explosive (flammable) levels, and one or two toxic gases such as carbon monoxide or hydrogen sulfide.

Murphy's Law

A popular engineering axiom or saying that if anything can go wrong it will. It is used as either a purely sarcastic musing that things always go wrong, or, less frequently, a reflection of the mathematical idea that, given a sufficiently long time, an event that is possible will almost surely take place. Although, in this case, emphasis is put on the possible unfortunate occurrences. See also **Law of Unintended Consequences**.

Murphy's Law Corollary

If anything can go wrong, it will in the worst possible aspect.

Musculoskeletal Disorder (MSD)

An injury or illness of soft tissues of the fingers, upper arms, shoulders and neck, lower back, or legs that is primarily caused or exacerbated by workplace risk factors such as sustained and repeated exertions or awkward postures and manipulations. Administrative controls for MSD hazards include employee rotation, job task enlargement, alternative tasks, and employer-authorized changes in work pace. Work practice controls for MSD hazards include use of neutral postures to perform tasks (straight wrists, lifting close to the body), use of two-person lift teams, and observance of micro breaks. See also **Repetitive Strain Injury (RSI)**.

Musculoskeletal Disorder (MSD) Hazard

The presence of risk factors in a job that occur at a magnitude, duration, or frequency that is reasonably likely to cause MSDs that can result in work restrictions or medical treatment beyond first aid.

Musculoskeletal Disorder (MSD) Incident

Per Occupational Safety and Health Administration (OSHA) regulation 29 CFR 1910.900, it is defined as an MSD that is work related and requires medical treatment beyond first aid, or MSD signs or MSD symptoms that last for 7 or more consecutive days after the employee reports them.

Musculoskeletal Disorder (MSD) Signs

Objective physical findings that an individual may be developing an MSD. Examples of MSD signs are decreased range of motion, deformity, decreased grip strength, and loss of function.

Musculoskeletal Disorder (MSD) Symptoms

Physical indications that an individual may be developing an MSD. MSD symptoms do not include discomfort. Examples of MSD symptoms include pain, numbness, tingling, burning, cramping, and stiffness.

Musculoskeletal Injuries (MSIs)

A work-related musculoskeletal disorder is an injury to the muscles, tendons, and/ or nerves of the upper body either caused or aggravated by work. Other names used to describe work-related musculoskeletal disorders include repetitive motion injuries, repetitive strain injuries, cumulative trauma disorders, soft tissue disorders, and overuse syndromes. Work-related musculoskeletal injuries (MSIs) affecting the upper body and limbs are now recognized as one of the leading causes of worker pain and disability.

Mutual-Aid Agreement

A formal agreement between agencies or jurisdictions to assist one another upon request for an incident by furnishing personnel, equipment, or expertise.

M

N

Narcosis
A physical condition in which the effects of a toxic substance exposure cause a reduction or dulling in consciousness, similar to an anesthetic effect.

National Electrical Code (NEC)
A code for the design and installation of electrical wiring and devices to safeguard life and property from fires. The original national electrical code was developed in 1897. It is currently issued as National Fire Protection Association (NFPA) 70, National Electrical Code (NEC).

National Electrical Manufacturers Association (NEMA)
A trade association for electrical equipment manufacturers, historically responsible for developing U.S. standards and now contributing to global standards and harmonization. It is the author of the NEMA 250 standard, an alpha-numeric coding system defining the level of ingress protection for electrical enclosures.

National Fire Protection Association (NFPA)
A nonprofit organization based in the United States (Quincy, Massachusetts), organized in 1986, and the leading authority on fire safety. It is dedicated to protecting lives and property from the hazards of fire. Its stated objective is to promote the science and improve the methods of fire protection, to obtain and circulate information on this subject, and to secure cooperation in the matter of common interest. It is responsible for the development and publication of the National Fire Codes® (through consensus committees), related handbooks, and numerous fire service training and public education materials.

National Floor Safety Institute (NFSI)
A nonprofit organization founded in 1997, whose mission is to aid in the prevention of slips and trips-and-falls through education, training, research, and standards development. It aims to assist businesses to resolve its slips, trips, and fall hazards. It is led by a 15-member Board of Directors representing product manufacturers, insurance underwriters, trade associations, and independent consultants. Its primary standard development includes the American National Standards Institute (ANSI) B101 committee for the measurement of the coefficient of friction of various walkway surfaces to determine how slippery their surfaces are and therefore may contribute to an incident. It recently has introduced a listing of products for rating walkway surfaces' resistance to slips and also can undertake testing of existing surfaces for slip resistance ratings. See also **Slips, Trips, and Falls**.

National Green Cross for Safety Medal
An award presented each year by the National Safety Council (NSC) since 1999. It recognizes organizations and their leaders for outstanding achievements in safety and health, community service, and responsible citizenship. To be considered for the medal, an organization and its leadership must demonstrate a superior record in advancing safety and health practices consistent with the mission of the National Safety Council.

National Highway Traffic Safety Administration (NHTSA)

The NHTSA was established by the Highway Safety Act of 1970 to carry out safety programs previously administered by the National Highway Safety Bureau. Specifically, the agency directs the highway safety and consumer programs established by the National Traffic and Motor Vehicle Safety Act of 1966, the Highway Safety Act of 1966, the 1972 Motor Vehicle Information and Cost Savings Act, and succeeding amendments to these laws.

National Incident Management System (NIMS)

As issued by the Department of Homeland Security in 2004, a system to provide a consistent nationwide approach for federal, state, local, and tribal governments; the private sector; and nongovernmental organizations to work together to prepare, prevent, respond, and recover from incidents occurring in the United States. The NIMS also incorporates a core set of concepts, principles, and terminology. This includes multi-agency coordination systems; training; identification and management of resources; qualification and certification; and the collection, tracking, and reporting of incident information and incident resources.

National Institute for Occupational Safety and Health (NIOSH)

The NIOSH was established to help ensure safe and healthful working conditions by providing research, information, education, and training in the field of occupational safety and health. It also makes recommendations for the prevention of work-related injury and illnesses; it has no authority to make regulations, except in the area of respirators. NIOSH provides national and world leadership to prevent work-related illness, injury, disability, and death by gathering information, conducting scientific research, and translating the knowledge gained into products and services. NIOSH is part of the Centers for Disease Control and Prevention (CDC) within the U.S. Department of Health and Human Services (HHS).

National Response Plan

A plan to integrate federal domestic prevention, preparedness, response, and recovery plans into one all-discipline, all-hazards plan.

National Safety Council (NSC)

A voluntary nongovernmental organization. Promotes incident reduction by providing a forum for the exchange of safety and health ideas, techniques, and experiences and the discussion of incident prevention methods. The organization offers background courses at its Safety Training Institute and home study courses for supervisors. It is also Secretariat to six American National Standards Institute (ANSI)-accredited standards committees, producing nearly 30 safety-related standards. In 1912, the First Cooperative Safety Congress met in Milwaukee, Wisconsin. It was comprised of a small group of industrial leaders from the American Midwest that were concerned for American workers' safety. The outcome of their gathering was a decision to form a permanent body devoted to the promotion of safety in U.S. industry. In Chicago on October 13, 1913, the National Council for Industrial Safety was formed. This later developed into the National Safety Council, which adopted the current name in 1914.

National Safety Education Center (NSEC)

A partnership between Northern Illinois University (NIU), the Construction Safety Council (CSC), and the National Safety Council (NSC). The NSEC provides Occupational Safety and Health Administration (OSHA)-approved training courses and also offers certificates, cards, Continuing Education Units (CEUs), and Certified Safety Professional (CSP) points for successful completion of courses.

National Transportation Safety Board (NTSB)

Established in 1967, the National Transportation Safety Board is an independent federal agency charged by Congress with investigating significant incidents. It derives its authority from Title 49 of the United States Code, Chapter 11. The NTSB investigates every civil aviation accident in the United States and significant incidents in the other modes of transportation (pipeline, aviation, railroad, highway, and marine).

The Safety Board determines the probable cause of:

All U.S. civil aviation accidents and certain public-use aircraft accidents
Selected highway accidents
Railroad accidents involving passenger trains or any train accident that results in at least one fatality or major property damage
Major marine accidents and any marine accident involving a public and a nonpublic vessel
Pipeline accidents involving a fatality or substantial property damage
Releases of hazardous materials in all forms of transportation
Selected transportation accidents that involve problems of a recurring nature

The NTSB is responsible for maintaining the government's database of civil aviation incidents, conducts special investigations and safety studies, and issues safety recommendations aimed at preventing future incidents. Safety Board investigators are on call 24 hours a day, 365 days a year. The NTSB has issued more than 12,000 recommendations in all transportation modes to more than 2200 recipients. Since 1990, the NTSB has highlighted some important issues on a Most Wanted list of safety improvements.

Nationally Recognized Testing Laboratory (NRTL)

An organization that is recognized by the Occupational Safety and Health Administration (OSHA) in accordance with 29 CFR 1910, which tests for safety, and lists or labels or accepts equipment or materials, and which meets OSHA-specific testing and examination requirements.

Natural Gas Pipeline Safety Act

A federal regulation enacted in 1968 that provided for federal government authority over interstate pipelines transporting hazardous liquids and natural gas. The Office of Pipeline Safety (OPS) was formed under the Department of Transportation (DOT) to set minimum safety standards for design, construction, inspection, testing, operation, and maintenance, as well as to perform inspections and enforce regulations. See also **Office of Pipeline Safety (OPS)**; **Pipeline and Hazardous Materials Safety Administration (PHMSA)**; **Pipeline Safety Regulations**.

Natural Ventilation

A type of general ventilation that depends on natural instead of mechanical means for air movement. Natural ventilation can depend on the wind or the difference in temperature from one area to another to move air through a building. Therefore, it is unpredictable and unreliable. See also **Local Exhaust Ventilation**.

Nature of Injury

The type or classification of the hurt, harm, or impairment received or inflicted.

Nature of Injury or Illness

Names the principal physical characteristic of a disabling condition, such as sprain/strain, cut/laceration, or carpal tunnel syndrome. Utilized by the Bureau of Labor Statistics for injury descriptions.

Near Accident
See **Near Miss**.

Near Miss
Terminology used to describe an event that had the potential to result in an incident, but where no such event occurred. Sometimes is referred to as a Near Accident. These may include events where injury or property damage could have occurred but did not; events where a major safety system failed to perform as designed, e.g., fire pump auto start malfunction; or events where potential environmental damage could result. The term is actually considered a technical misnomer by some, since the technical accuracy would to refer to such an event as a "near hit" or "near occurrence," i.e., an incident that nearly occurred, rather than an incident that nearly missed.

Near Miss Report (NMR)
Detailed information that has investigated a Near Miss, analyzed its occurrence, identified preventive measures, and is used for communicating corrective actions and preventive measures to reduce injuries and losses in the future within an organization.

Negative Pressure Respirator
A respirator in which the air pressure inside the facepiece is negative during inhalation with respect to the ambient air pressure outside the respirator. See also **Positive Pressure Respirator**.

Negligence
The lack of reasonable conduct or care, characterized by accidental behavior or thoughtlessness, that a prudent person would ordinarily exhibit. There need not be a legal duty.

N

Actionable negligence: The breach or non-performance of a legal duty, through neglect or carelessness, which results in damage or injury to another.

Comparative negligence: Where negligence by both the plaintiff and the defendant is concurrent and contributes to the injury. The plaintiff's damages are diminished proportionately, provided fault is less than the defendant's, and even by exercising ordinary care the plaintiff could not have avoided the consequences of the defendant's negligence.

Contributory negligence: Conduct by the injured person who should have known involved an unreasonable risk. Inattentiveness or carelessness when using an article known to be defective or hazardous, or disregard of warnings or instructions issued by manufacturers and sellers usually constitutes contributory negligence.

Degrees of negligence: "Ordinary" negligence is based upon the fact that one ought to have known the results of unsafe acts, while "gross" negligence rests on the assumption that one knew the results of acts but was recklessly or wantonly indifferent to the results. All negligence below that called "gross" or "ordinary" by the courts is "slight" negligence.

Neurotoxins
Neurotoxins are agents that can cause toxic effects on the nervous system that may produce emotional or behavior abnormalities.

Neutral Position
The body position that minimizes stresses on the body. Typically the neutral posture will be near the mid-range of any joint's range of motion. Used in an ergonomic evaluation.

Neutralization
Control strategy for hazardous substances (e.g., strong acids, caustics, oxidizers, etc.) whereby a neutralizing compound is added to a highly dangerous compound, e.g., an acid to an alkali, to reduce its immediate danger.

NFPA 70E, Standard for Electrical Safety in the Workplace
A National Fire Protection Association (NFPA) standard that addresses electrical safety issues including work practices, maintenance, special equipment requirements, installation and demolition of electrical conductors, electrical equipment, signaling and communications conductors and equipment, and raceways.

NFPA 101, Life Safety Code
See **Life Safety Code (LSC)**.

NFPA 170, Standard for Fire Safety and Emergency Symbols
A National Fire Protection Association (NFPA) standard containing the uniform symbols used for fire safety, emergency, and associated hazards. Its use is suggested for the general public, fire service, architectural and engineering drawings, insurance diagrams, firefighting operations, and pre-incident planning sketches.

NFPA 301, Code for Safety to Life from Fire on Merchant Vessels
A National Fire Protection Association (NFPA) fire code that provides the minimum requirements, based on NFPA 101, Life Safety Code, with due regard to function for various types of merchant vessels, for the design, operation, and maintenance of merchant vessels for safety to life from fire and similar emergencies.

NFPA 550, Guide to Fire Safety Concepts Tree
See **Fire Safety Concepts Tree**.

NFPA 610, Guide to Emergency and Safety Operations at Motorsports Venues
A National Fire Protection Association (NFPA) guideline for racetrack owners, operators, promoters, first responders, insurers, and sanctioning agencies for implementing (planning and training) safety and emergency response strategies.

NFPA 704, Standard System for the Identification of the Hazards of Materials for Emergency Response
A National Fire Protection Association (NFPA) standard for the system to determine the degree of health, flammability, and instability hazards of chemicals. The system also provides for the recognition of unusual water reactivity and oxidizers. The NFPA 704 ratings are displayed in markings that are sometimes referred to as the "NFPA Hazard Diamond." The four divisions are typically color-coded, with blue indicating level of health hazard, red indicating flammability, yellow (chemical) reactivity, and white containing special codes for unique hazards. Each of health, flammability, and reactivity divisions is rated on a scale from 0 (no hazard; normal substance) to 4 (severe risk) (see Table N.1). See also **DOT Hazard Class**; **Hazardous Materials Identification System (HMIS®)**; **NFPA Hazard Rating**.

NFPA Hazard Diamond
See **NFPA Hazard Rating**.

NFPA Hazard Rating
Classification of a chemical by a four-color diamond, subdivided into four more diamonds representing health, flammability, reactivity, and specific hazard level by a numbered hazard rating from 0 to 4. The left side is for health hazards, which are indicated by a black number with a blue background; top is for fire hazards, indicated by a black number with a red background; the right side is for reactivity, indicated by a black number with a yellow background; and the bottom is for specific hazards, indicated by a black acronym lettering or symbol with a white background (see Figure N.1). It is used for the rapid identification of hazardous

TABLE N.1
Description of Ratings Used on NFPA 704 Diamond

Health (Blue)

4   Very short exposure could cause death or major residual injury (e.g., hydrogen cyanide, phosphine)

3   Short exposure could cause serious temporary or moderate residual injury (e.g., chlorine gas)

2   Intense or continued but not chronic exposure could cause temporary incapacitation or possible residual injury (e.g., ethyl ether)

1   Exposure would cause irritation with only minor residual injury (e.g., acetone)

0   Poses no health hazard, no precautions necessary (e.g., lanolin)

Flammability (Red)

4   Will rapidly or completely vaporize at normal atmospheric pressure and temperature, or is readily dispersed in air and will burn readily (e.g., propane). Flash point below 23°C (73°F)

3   Liquids and solids that can be ignited under almost all ambient temperature conditions (e.g., gasoline). Flash point between 23°C (73°F) and 38°C (100°F)

2   Must be moderately heated or exposed to relatively high ambient temperature before ignition can occur (e.g., diesel fuel). Flash point between 38°C (100°F) and 93°C (200°F)

1   Must be heated before ignition can occur (e.g., soybean oil). Flash point over 93°C (200°F)

0   Will not burn (e.g., water)

Instability/Reactivity (Yellow)

4   Readily capable of detonation or explosive decomposition at normal temperatures and pressures (e.g., nitroglycerine, RDX)

3   Capable of detonation or explosive decomposition but requires a strong initiating source; must be heated under confinement before initiation; reacts explosively with water; or will detonate if severely shocked (e.g., ammonium nitrate)

2   Undergoes violent chemical change at elevated temperatures and pressures; reacts violently with water; or may form explosive mixtures with water (e.g., phosphorus, potassium, sodium)

1   Normally stable, but can become unstable at elevated temperatures and pressures (e.g., propene)

0   Normally stable, even under fire exposure conditions, and is not reactive with water (e.g., helium)

Special (White)

The white "special notice" area can contain several symbols. The following symbols are defined by the NFPA 704 standard:

W̶   Reacts with water in an unusual or dangerous manner (e.g., caesium, sodium)

OXY   Oxidizer (e.g., potassium perchlorate, ammonium nitrate, hydrogen peroxide)

SA   Simple asphyxiant gas (includes nitrogen, helium, neon, argon, krypton, and xenon)

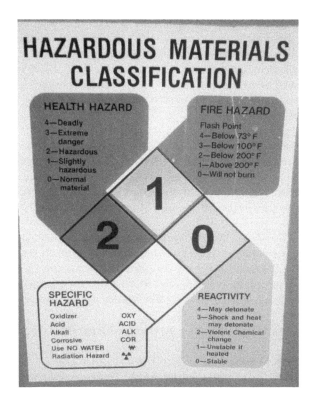

FIGURE N.1   NFPA hazard rating. (Photo courtesy of D. P. Nolan.)

materials characteristics during an incident by emergency responders. See also **Hazardous Materials Identification System (HMIS®)**; **NFPA 704, Standard System for the Identification of the Hazards of Materials for Emergency Response**.

NFPA National Fire Protection Codes

Approximately 300 consensus codes and standards intended to minimize the possibility and effects of fire and other similar and related risks, which are published by the National Fire Protection Association (NFPA). The codes themselves have no enforcement authority. Only when they are adopted by an organization that has jurisdictional authority do they become enforceable or applicable by law. NFPA Technical Committees and Code Making Panels serve as the principal consensus bodies responsible for developing and updating all NFPA codes and standards. Membership in the committees is based on technical expertise, professional standing, commitment to public safety, and the ability to bring to the table the point of view of a category of interested people or groups. The Technical Committee makeup consists of a balance of affected interests, with no more than one-third of the Committee from the same interest category. To take any action a consensus must be reached by the committee.

Nip Point

See **Pinch Point**.

No Observable Adverse Effect Level (NOAEL)

In toxicology, the concentration of a substance at which exposure produces no evidence of injury or impairment. The Environmental Protection Agency (EPA) defines

NOAEL as an exposure level at which there are no statistically or biologically significant increases in the frequency or severity of adverse effects between the exposed population and its appropriate control; some effects may be produced at this level, but they are not considered as adverse, or as precursors to adverse effects. In an experiment with several NOAELs, the regulatory focus is primarily on the highest one, leading to the common usage of the term NOAEL as the highest exposure without adverse effects.

Node

As identified in a safety review (e.g., process hazard analysis [PHA], What-If Analysis, Hazard and Operability Study [HAZOP]), a defined part (section or subsystem or item of equipment) of a process that has a design intention that is specific and distinct from the design intention of other process parts, which allows the study team to analyze the specific equipment or system in an organized fashion.

Noise

Any undesired sounds usually resulting in an objectionable or irritating sensation. Legally, noise is exposure to sounds exceeding an average of 90 dB of noise for eight hours per day. A TWA (time-weighted average) of 90 dB equals the current maximum legal noise exposure (in the United States) for an individual without existing hearing damage. The exposure limit for someone with existing hearing damage is 85 dB.

Noise, Continuous

Noise that exists with negligibly small fluctuations in level within the period of observation.

Noise, Damage Risk Criteria

The suggested baseline of noise tolerance, which, if not exceeded, should result in no hearing loss due to noise. A damage risk criteria statement should include a specification of such factors as time of exposure, noise level, frequency, amount of hearing loss considered significant, percentage of the population to be protected, and method of measuring the noise.

Noise, Impact

Noise that is generated from striking an object.

Noise, Impulse

A noise of short duration (typically less than one second), especially of high intensity, abrupt onset, and rapid decay, and often rapidly changing spectral composition. Impulse noises are characteristically associated with such sources as explosions, impacts, firearm discharges, sonic booms, and many industrial processes.

Noise, Intermittent

Noise levels that are interrupted by intervals of relatively low sound levels.

Noise Attenuation

The reduction in sound pressure level incident upon the ear when hearing protection is worn.

Noise Control

Engineering measures aimed at the reduction of noise at the source; precluding the propagation, amplification, and reverberation of noise; and isolating workers. The term is meaningful only when noise control components and the points of observation are fully specified.

Noise Criterion Curve

An established noise level that is intended to permit adequate speech communications for a given area and is also used to minimize human annoyance.

TABLE N.2
Noise Protection Levels

Insufficient protection	Greater than 85 dB
Acceptable protection	80 to 84 dB
Optimal protection	75 to 79 dB
Acceptable protection	70 to 74 dB
Overprotection	69 dB and less

Noise Dosimeter

An instrument that integrates a function of sound pressure over a period of time in such a manner that it directly indicates a noise dose. See also **Dosimeter**.

Noise Exposure

A cumulative acoustic stimulation that reaches the ear of a person over a specified period of time such as a work shift, a day, a working life, or a lifetime.

Noise Hazard

An acoustic stimulation of the ear that is likely to produce noise-induced permanent hearing loss in some of the exposed population.

Noise-Induced Hearing Loss (NIHL)

See **Hearing Loss, Noise-Induced (NIHL)**.

Noise Level

The ambient sound pressure level that exists in a given location. Sometimes referred to as background noise level or reference sound level, noise level is measured using a frequency-weighting filter, the most commonly utilized being the A-weighting scale; thus, resulting measurements are denoted dB(A), or decibels on the A-weighting scale.

Noise Overprotection

The National Institute of Occupational Safety and Health (NIOSH) estimates that 90 percent of noise-exposed workers in the United States are exposed to 95 dB of TWA (time-weighted average), meaning they need only 10 dB of protection or attenuation. Workers who are overprotected from the noise exposure may feel they may not hear communications properly and might not be safe. There are no U.S. standards for overprotection for noise exposure; however, European guidance, EN 458, states that workers' attenuated noise levels should be no lower than 75–84 dB for ideal communication in noise (Table N.2).

Noise Reduction

A decrease of the sound pressure level at a specified observation point, which is attributable to a designated structure. Noise reduction is also used to designate the differences in sound pressure levels existing at two different locations at a single time, when the designated structures are in position.

Noise Reduction Rating (NRR)

The NRR is a single-number rating method that attempts to describe a hearing protector based on how much the overall noise level is reduced by the hearing protector. The NRR is required by law to be shown on the label of each hearing protector sold in the United States. The NRR is specified by 40 CFR, Part 211, Product Noise Labeling, Subpart B.

Hearing Protection Devices. It (the NRR) is independent of the noise spectrum in which it is applied. The values of sound attenuation used for calculation of the NRR are determined in accordance with American National Standards Institute (ANSI) S3.19-1974, "American National Standard for the Measurement of Real-Ear Hearing Protector Attenuation and Physical Attenuation of Earmuffs." When estimating A-weighted noise exposures, it is important to remember to first subtract 7 dB from the NRR and then subtract the remainder from the A-weighted noise level. The NRR theoretically provides an estimate of the protection that should be met or exceeded by 98 percent of the wearers of a given device. In practice, this does not prove to be the case, so a variety of methods for "de-rating" the NRR have been discussed.

Non-auditory Effects of Noise

The effects of loud, continuous noise on stress, fatigue, health, work efficiency, and performance.

Noncausal Association

A statistical association between the occurrence of a factor and a disease in which the factor is not a cause of the disease.

Noncombustible

A material or substance that will not burn readily or quickly. Noncombustible implies a lower degree of fire resistance than fire resistive. See also **Fire Resistive**.

Noncombustible (Material)

A material that does not ignite, burn, support combustion, or release flammable vapors when subjected to fire or heat. Noncombustible materials will also not aid combustion or add appreciable heat to an ambient fire. Materials are tested for non-combustibility in accordance with American Society for Testing Materials (ASTM) E 136, Standard Test Method for Behavior of Materials in a Vertical Tube Furnace at 750°C (1382°F). This test places the sample material to be tested in a furnace chamber for three trials. The chamber is a tube made of refractory material with a diameter of 7.5 cm (2.95 inches) and a height of 17 cm (5.9 inches). The tube is heated by electrical heating coils. Before testing the material, the temperature in the tube is stabilized uniformly at 750°C (1382°F) for a period of at least 10 minutes (except for the ends). A material sample of 4 by 5 cm (3.6 by 1.9 inches) is placed in the center of the tube and thermocouple readings are taken, one from the center of the sample and one from a specific location in the furnace. Materials are classified as noncombustible if during the three separate tests the test does not (a) cause the temperature readings of the furnace thermocouple to rise by 50°C (122°F) or more above the initial furnace temperature, (b) cause the temperature readings of the specimen thermocouple to rise 50°C (122°F) or more above the initial furnace temperature, or (c) flame for 10 seconds or more. It should be remembered that all materials may burn if exposed to the appropriate conditions (e.g., high temperatures, pressures, etc.). In practical terms, noncombustible materials do not contribute to the incipient stage of a fire nor in general to its temperature or duration.

Nondestructive Testing

A test to determine the characteristics or properties of a material or substance that does not involve its destruction or deterioration (e.g., x-ray examination, ultra-high frequency sound, etc.). See also **Dye-Penetrant Testing**.

Non-disabling Injury

An occupational injury that does not result in death, permanent total disability, permanent partial disability, or temporary total disability.

Non-fatal Injury Accident

An incident in which at least one person is injured and no injury terminates fatally.

Nonflammable

A material or substance that will not burn readily or quickly.

Non-injury Accident
Any incident in which there is no personal injury or from which no personal injury results.

Non-ionizing Radiation
Described as a series of energy waves composed of oscillating electric and magnetic fields traveling at the speed of light. Non-ionizing radiation includes the spectrum of ultraviolet (UV), visible light, infrared (IR), microwave (MW), radio frequency (RF), and extremely low frequency (ELF). Lasers commonly operate in the UV, visible, and IR frequencies. Non-ionizing radiation is found in a wide range of occupational settings and can pose a considerable health risk to potentially exposed workers if not properly controlled. See also **Radiation**.

Not Otherwise Classified
A general category of items such as might appear in an incident causal classification system to permit the grouping of relatively infrequent dissimilar items.

Notice Label or Sign
As defined by American National Standards Institute (ANSI) Z535.4, a sign or label that is preferred to address practices not related to personal injury. The Safety Alert Symbol (i.e., "!" enclosed in a triangle) is not to be used with a Notice label or sign. When a Notice or Safety Instruction label or sign is needed for equipment, color definitions are provided in ANSI Z535.4 and ANSI Z535.2 Environmental and Facility Safety Sign standards (i.e., blue/white and green/white, respectively). See also **Danger Label or Sign**; **Notice Label or Sign**; **Safety Alert Symbol**; **Safety Sign**; **Warning Label or Sign**.

Notice of Probable Violation
A warning provided by the Occupational Safety and Health Administration (OSHA) on the possible citation of a facility for noncompliance with federal safety regulations. See also **OSHA Violation(s)**.

Nuclear Safety
The achievement of proper operating conditions, prevention of incidents, or mitigation of incident consequences, resulting in protection of workers, the public, and the environment from undue radiation hazards. This covers nuclear power plants as well as all other nuclear facilities, the transportation of nuclear materials, and the use and storage of nuclear materials for medical, power, industry, and military uses. In addition, there are safety issues involved in products created with radioactive materials. The Office of Nuclear and Facility Safety establishes and maintains the Department of Energy (DOE) requirements for nuclear criticality safety. The DOE's detailed requirements for criticality safety are contained in Section 4.3 of the DOE Order 420.1, Facility Safety. Criticality safety requirements are based on the documented safety analysis required by 10 CFR 830, Subpart B.

Nuisance Dust
Generally innocuous dust not recognized as the direct cause of a serious pathological condition. Nuisance particulate is a term used historically by the American Conference of Governmental Industrial Hygienists (ACGIH) to describe airborne materials (solids and liquids) that have little harmful effect on the lungs and do not produce significant disease or harmful effects when exposures are kept under reasonable control. Nuisance particulates may also be called nuisance dusts. High levels of nuisance particulates in the air may reduce visibility and can get into the eyes, ears, and nose. Removal of this material by washing or rubbing may cause irritation.

Nuisance Trip
See **Safe Failure**.

O

Occupancy
The use of a building or other structure. The contents of a building or other structure.

Occupational Disabling Injury (ODI)
An injury that occurs in the workplace and results in a disabling injury. It may be permanent total, permanent partial, or temporary.

Occupational Disease
A disease arising out of, and in the course of, employment, i.e., resulting from exposure to harmful chemical, biological, or physical agents.

Occupational Exposure Limit (OEL)
The eight-hour time-weighted average concentration specified for workroom air selected from the 1986–1987 Threshold Limit Values and Biological Exposure Indices as adopted by the American Conference of Governmental Industrial Hygienists; the Recommended Standards for Occupational Exposure set forth in the July 1985 summary of the National Institute for Occupational Safety and Health Recommendations for Occupational Health Standards; or the 1986 Workplace Environmental Exposure Levels set forth by the American Industrial Hygiene Association.

Occupational Hazard
Conditions that exist in a workplace environment that increase the probability of death, injury, or illness to a worker.

Occupational Health
The field of general preventative medicine that is concerned with the prevention and/or treatment of illness induced by factors in the workplace environment. The major disciplines involved are occupational medicine, occupational health nursing, epidemiology, toxicology, industrial hygiene, and health physics.

Occupational Illness
Any abnormal physical condition or disorder, other than one resulting from an occupational injury, caused by exposure to environmental factors associated with employment. It includes acute and chronic illness or disease, which may be caused by inhalation, absorption, ingestion, or direct contact. Because occupational illnesses are rarely attributable to a specific incident they should be reported in the year in which the illness was first diagnosed and reported to the employer.

Occupational Incident
An incident involving an injury to an employee.

Occupational Injury
An acute injury arising out of, and in the course of, employment resulting from exposure to traumatizing physical or chemical agents in the workplace. Examples include amputations, fractures, eye loss, lacerations, and traumatic deaths. The Occupational Safety and Health Administration (OSHA) states that the industries with the highest risk for occupational injuries include construction, mining, manufacturing, and wholesale.

Occupational Injury and Illness Classification System (OIICS)
The OIICS was developed by the Bureau of Labor Statistics (BLS) to provide a standardized coding system for characterizing work-related injuries and illnesses. The

classification system is used to code the case characteristics of injuries, illnesses, and fatalities in the Survey of Occupational Injuries and Illnesses (SOII) and the Census of Fatal Occupational Injuries (CFOI) programs. These programs contain the rules of selection, code descriptions, code titles, and indices for the following code structures: Nature of Injury or Illness, Part of Body Affected, Source of Injury or Illness, Event or Exposure, and Secondary Source of Injury or Illness. The OIICS was originally released in December 1992. The OIICS has five subcomponent structures (Table O.1). Two structures are used to describe the characteristics of the injury or illness (i.e., "Nature" and "Part of Body Affected") and three structures are used to describe the incident circumstances (i.e., "Source of Injury or Illness," "Secondary Source," and "Event or Exposure"). The BLS uses the OIICS to characterize occupational fatalities collected through their Census of Fatal Occupational Injuries (CFOI) and nonfatal injuries and illnesses collected through their annual survey of employers, the Survey of Occupational Injuries and Illnesses (SOII). In 1995, The American National Standards Institute (ANSI) adopted the OIICS as the basis for the ANSI Z16.2, Standard for Information Management for Occupational Safety and Health. The National Institute for Occupational Safety and Health (NIOSH) uses the OIICS to characterize nonfatal occupational injuries and illnesses treated in emergency departments.

TABLE O.1
OIICS Division Titles

Nature of Injury or Illness	Part of Body Affected	Source and Secondary Source	Event or Exposure
0 Traumatic Injuries and Disorders	0 Head	0 Chemicals and Chemical Products	0 Contact with Objects and Equipment
1 Systemic Diseases or Disorders	1 Neck including Throat	1 Containers	1 Falls
2 Infectious and Parasitic Diseases	2 Trunk	2 Furniture and Fixtures	2 Bodily Reaction and Exertion
3 Neoplasms, Tumors, and Cancer	3 Upper Extremities	3 Machinery	3 Exposure to Harmful Substances or Environments
4 Symptoms, Signs, and Ill-Defined Conditions	4 Lower Extremities	4 Parts and Materials	4 Transportation Incidents
5 Other Conditions or Disorders	5 Body Systems	5 Persons, Plants, Animals, and Minerals	5 Fires and Explosions
8 Multiple Diseases, Conditions, or Disorders	8 Multiple Body Parts	6 Structures	6 Assaults and Violent Acts
9999 Nonclassifiable	9999 Nonclassifiable	7 Tools, Instruments, and Equipment	9 Other Events and Exposures
		8 Vehicles	9999 Nonclassifiable
		9 Other Sources	
		9999 Nonclassifiable	

Occupational Injury and Illness Records, OSHA
A recording of each reportable occupational injury (including fatality) and illness required by every employer covered by the National System for Uniform Recording and Reporting of Occupational Injury and Illness.

Occupational Injury or Illness, Reportable, OSHA
Any disability or permanent impairment to an employee that results from any exposure in the work environment that (1) results in death, (2) prevents the employee from performing their normal assignment during the next regular or subsequent work day or shift, or (3) does not cause death or loss of time, but (a) results in transfer to another job or termination of employment, (b) requires medical treatment other than first aid, (c) results in loss of consciousness, (d) is diagnosed as an occupational illness, or (e) results in restriction of work or motion.

Occupational Safety
The prevention of personnel and environmental incidents in work-related environments or situations. Effective occupational safety efforts involve the control and elimination of recognized workplace hazards to attain an acceptable level of risk and promote the wellness of workers. Optimal occupational safety results from a continuous proactive process of anticipating, identifying, designing, implementing, and evaluating risk-reduction practices.

Occupational Safety and Health Act of 1970
The Occupational Safety and Health Act of 1970 (OSH Act) assigns the Occupational Safety and Health Administration (OSHA) two regulatory functions: setting standards and conducting inspections to ensure that employers are providing safe and healthful workplaces. The act states, "To assure safe and healthful working conditions for working men and women; by authorizing enforcement of the standards developed under the Act; by assisting and encouraging the States in their efforts to assure safe and healthful working conditions; by providing for research, information, education, and training in the field of occupational safety and health; and for other purposes." The Act was formulated by Congress since it found that personal injuries and illnesses arising out of work situations impose a substantial burden upon, and are a hindrance to, interstate commerce in terms of lost production, wage loss, medical expenses, and disability compensation payments. Congress declared it to be its purpose and policy, through the exercise of its powers to regulate commerce among the States and with foreign nations and to provide for the general welfare, to assure so far as possible every working man and woman in the Nation safe and healthful working conditions and to preserve our human resources. Prior to the adoption of the federal Occupational Safety and Health (OSH) Act, workplace safety and health issues had been almost entirely the responsibility of each U.S. state. In recognition of significant state-level involvement in safety and health, Section 18 of the OSH Act allows states and jurisdictions the option of developing and enforcing their own OSH programs. OSHA standards may require that employers adopt certain practices, means, methods, or processes reasonably necessary and appropriate to protect workers on the job. Employers must become familiar with the standards applicable to their establishments and eliminate hazards. Compliance with standards may include ensuring that employees have and use personal protective equipment when required for safety or health. Employees must comply with all rules and regulations that apply to their own actions and conduct. Even in areas where OSHA has not set forth a standard addressing a specific hazard, employers are responsible for complying with the OSH Act's "general duty" clause. The general duty clause [Section 5(a)(1)] states that each employer "shall furnish a place of

employment which is free from recognized hazards that are causing or are likely to cause death or serious physical harm to his employees" and covers all employers and their employees in the 50 states, the District of Columbia, Puerto Rico, and other U.S. territories. Coverage is provided either directly by the federal Occupational Safety and Health Administration (OSHA) or by an OSHA-approved state job safety and health plan. Employees of the U.S. Postal Service are also covered. The Act defines an employer as any "person engaged in a business affecting commerce that has employees, but does not include the U.S. or any state or political subdivision of a State." Therefore, the Act applies to employers and employees in such varied fields as manufacturing, construction, longshoring, agriculture, law and medicine, charity and disaster relief, organized labor, and private education. See also **Occupational Safety and Health Administration (OSHA)**.

Occupational Safety and Health Administration (OSHA)
A federal agency in the Department of Labor (DOL) responsible for standard-setting, regulating, and enforcing workplace codes, rules, and laws. Established in 1970, OSHA's role is to ensure safe and healthful working conditions for working individuals by authorizing enforcement of the standards developed under the Occupational Safety and Health Act, by assisting and encouraging the States in their efforts to ensure safe and healthful working conditions, and by providing for research, information, education, and training in the field of occupational safety and health. OSHA inspects about 40,000 facilities per year, while the 26 state-operated OSHA organizations inspect another 60,000. OSHA can issue citations, which can result in financial penalties up to $7,000 for non-serious violations, but can rise to $70,000 for repeat offenders. It has been estimated that since OSHA began, occupational deaths have been reduced by 62 percent and injuries have declined by 42 percent.

Occupational Safety and Health Codes and Standards
Rules of procedure designed to secure uniformity and protection of life and property having the force of law in certain jurisdictions. Examples include Occupational Safety and Health Administration (OSHA) Standards (29 CFR 1910) and the National Fire Protection Association (NFPA) Codes. See also **General Industry Safety Standards**.

Occupational Safety and Health Review Commission (OSHRC)
An independent federal agency created established to review Occupational Safety and Health Administration (OSHA) citations or penalties resulting from OSHA inspections of work places that are contested by employers, employees, or their representatives. The OSHRC was established through the OSH Act. The Review Commission functions as an administrative court, with established procedures for conducting hearings, receiving evidence, and rendering decisions by its Administrative Law Judges (ALJs). OSHRC's Rules of Procedure provide for two levels of adjudication. The first level is before an Administrative Law Judge. The second level is review of the ALJs' decisions by the agency's Commissioners in Washington, if one of the parties petitions for review. If review of the ALJ decisions by the Commissioners is not directed, the petitioning party may request review by an appropriate U.S. Circuit Court of Appeals. Any person who is adversely affected or aggrieved may also appeal the decision of the OSHRC Commissioners to an appropriate U.S. Court of Appeals. Review by a Court of Appeals must be sought within 60 days after the Commission's final decision is issued.

Odor Threshold
The odor threshold is the lowest concentration of a chemical in air that is detectable by smell. The odor threshold should only be regarded as an estimate. This is because

odor thresholds are commonly determined under controlled laboratory conditions using people trained in odor recognition. As well, in the workplace, the ability to detect the odor of a chemical varies from person to person and depends on conditions such as the presence of other odorous materials. Odor cannot be used as a warning of unsafe conditions since workers may become used to the smell (adaptation), or the chemical may numb the sense of smell, a process called olfactory fatigue. However, if the odor threshold for a chemical is well below its exposure limit, odor can be used to warn of a problem with your respirator.

Off-the-Job-Safety
Incident prevention activities or programs associated with non-job-related activities.

Office of Pipeline Safety (OPS)
A federal regulatory agency of the U.S. Department of the Transportation, Research, and Special Programs Administration. They are responsible for ensuring the safe, reliable, and environmentally sound operation of pipelines in the United Sates. They develop and implement pipeline safety regulations at the federal level and share regulatory responsibility with the states. The Pipeline and Hazardous Materials Safety Administration (PHMSA) is the primary federal regulatory agency responsible for ensuring that pipelines are safe, reliable, and environmentally sound. From the federal level, it oversees the development and implementation of regulations concerning pipeline construction, maintenance, operation, and emergency response of U.S. oil and natural gas pipeline facilities. See also **Pipeline and Hazardous Materials Safety Administration (PHMSA)**; **Pipeline Safety Regulations**.

Office of Railroad Safety
A federal regulatory agency of the U.S. Federal Railway Administration that promotes and regulates safety throughout the nation's railroad industry. The office executes its regulatory and inspection responsibilities through a diverse staff of railroad safety experts who cover hazardous materials, motive power and equipment, operating practices (including drug and alcohol, signal, and train control), and track structures. These experts include safety inspectors, program managers for highway-rail grade crossings and trespass prevention, bridge structure specialists, and industrial hygienists. It oversees and participates on the Railroad Safety Advisory Committee, which provides consensus recommendations from the industry on a range of regulatory issues. It also collects and analyzes rail-related incident data from the railroads and actively investigates serious events to determine their cause and compliance with existing laws and regulations, which are published for public review. See also **Railroad Crossings Safety**; **Railroad Safety**.

Office Safety
It is estimated that office workers sustain 76,000 fractures, dislocations, sprains, strains, and contusions each year. In office areas, trips and falls are the number one cause of injury. Office workers are also injured as a result of foreign substances in the eye, spilled hot liquids, burns from fire, and electric shock. The office may also contain hazards such as poor lighting, noise, poorly designed furniture and equipment, and machines that emit noxious gases and fumes. Even the nature of office work itself has produced a whole host of stress-related symptoms and musculoskeletal strains. For example, long hours at the video display terminal (VDT) can cause pains in the neck and back, eyestrain, and a general feeling of tension and irritability. Illness has increased among the office worker population. This may be attributed, in part, to the increased presence of environmental toxins within the office and to stress-producing factors associated with the automated office. Resulting illnesses may include respiratory problems, skin diseases, and stress-related conditions.

One-Call System

A one-call system is a system that allows excavators (individuals, commercial contractors, and governmental organizations) to make one telephone call to provide notification of their intent to dig to underground facility operators (i.e., buried pipelines). The one-call center will then notify all underground facility operator members of the intended excavation along with the date and location of the excavation. The facility operators or, in some cases, the one-call center can then locate the facilities before the excavation begins so that extra care can be taken to avoid damaging the facilities. All 50 states within the United States are covered by one-call systems. Most states have laws requiring the use of the one-call system at least 48 hours before beginning an excavation. Local one-call center numbers can be determined by calling, toll-free, 1-888-258-0808.

Organizational Error

A latent management system concern that may result in human error.

OSH Act

See **Occupational Safety and Health Act of 1970**.

OSHA Injuries and Illnesses Log (300)

Occupational Safety and Health Administration (OSHA)-required record keeping about injuries and illnesses caused by work-related activities that result in lost work time, fatalities, off-site treatment, and/or restricted work activity. They also have to classify the work-related injuries and illnesses and note the severity. Log information must be posted at the worksite (Figure O.1).

OSHA Inspection

An Occupational Safety and Health Administration (OSHA) worksite safety inspection. There are five types of site inspections that OSHA conducts. These are listed as follows in their order of importance, as determined by OSHA:

Imminent Danger: Imminent danger situations are given top priority. An imminent danger is any condition where there is reasonable certainty that a danger exists that can be expected to cause death or serious physical harm immediately or before the danger can be eliminated through normal enforcement procedures. When an imminent danger situation is found, the compliance officer will ask the employer to voluntarily abate the hazard and to remove endangered employees from exposure. Should the employer refuse, OSHA will apply to the nearest federal District Court for legal action to correct the situation.

Catastrophic and Fatal Accidents: Second priority is given to investigation of fatalities and catastrophes resulting in hospitalization of three or more employees.

Employee Complaints: Each employee has the right to request an OSHA inspection when the employee feels that he or she is in imminent danger from a hazard or when he or she feels that there is a violation of an OSHA standard that threatens physical harm. If the employee so requests, OSHA will withhold the employee's name from the employer.

Programmed High Hazard Inspections: OSHA establishes programs of inspection aimed at specific high hazard industries, occupations, or health hazards. Workplaces are selected for inspection on the basis of death, illness and injury rates, employee exposure to toxic substances, etc.

Re-inspections: Establishments cited for alleged serious violations may be re-inspected to determine whether the hazards have been corrected.

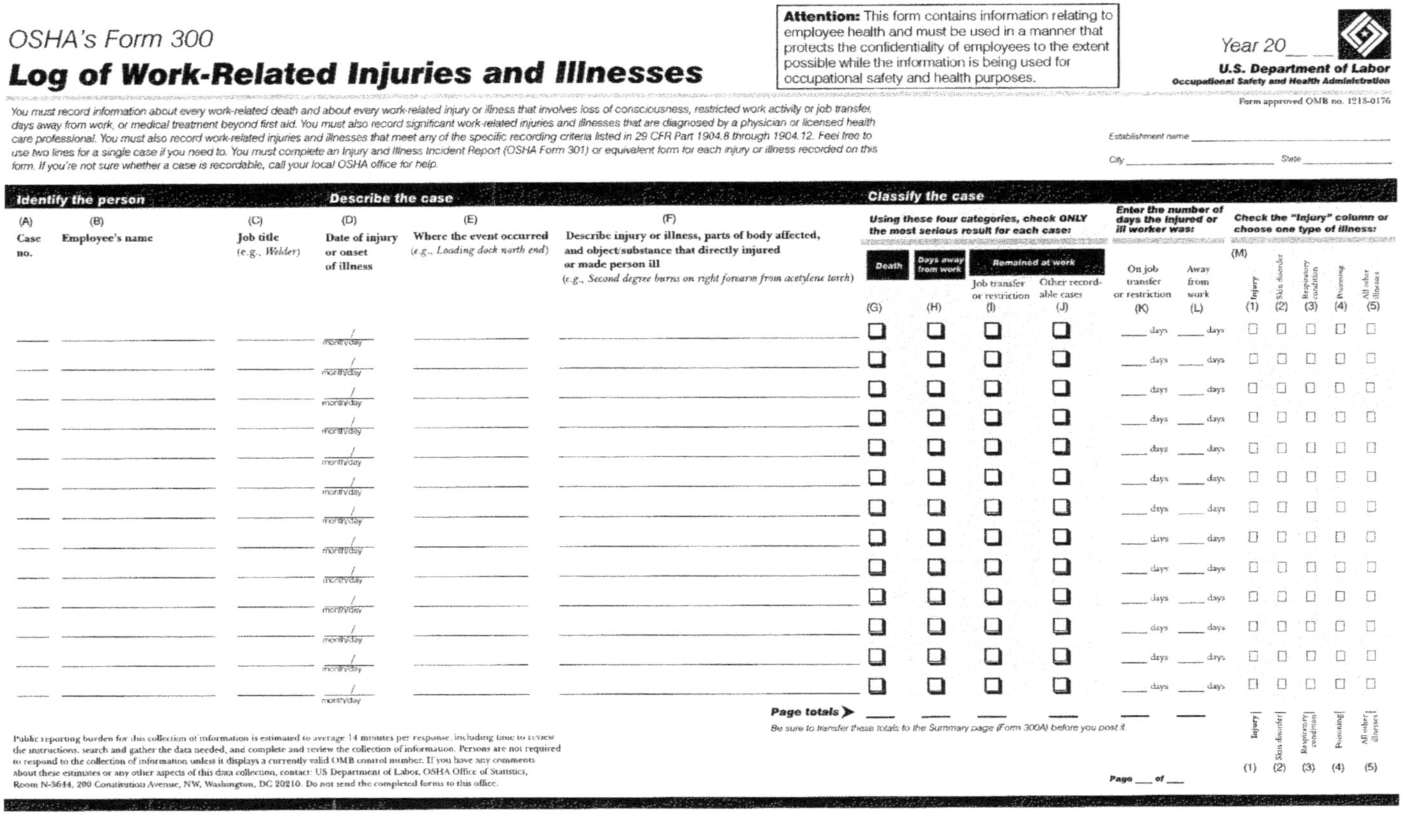

FIGURE O.1 Occupational Safety and Health Administration (OSHA) 300/300A Form. (Source: U.S. Department of Labor.)

OSHA Recordable Case

Work-related fatalities, injuries, or illnesses (other than first aid treatments) that involve medical treatment, loss of consciousness, restriction of work or motion, or transfer to another job, which must be recorded on the Occupational Safety and Health Administration (OSHA) 300 Log.

OSHA Reportable Event

An incident that causes any fatality or the hospitalization of five employees or more requires a notification report to the nearest Occupational Safety and Health Administration (OSHA) office.

OSHA Violation(s)

The types of citations and penalties that may be proposed by the Occupational Safety and Health Administration (OSHA) are from the OSH Act and are listed below. An organization has 15 working days from receipt of the citations and proposed penalties to either elect to comply with them, to request and participate in an informal conference with the OSHA area director, or to contest them before the independent Occupational Safety and Health Review Commission.

Other Than Serious Violation: A violation that has a direct relationship to job safety and health, but probably would not cause death or serious physical harm. The maximum proposed penalty for this type of violation is $7,000.

Serious Violation: A violation where there is substantial probability that death or serious physical harm could result, and that the employer knew, or should have known, of the hazard. The maximum proposed penalty for this type of violation is $7,000. Imminent danger situations are also cited and penalized as serious violations.

Willful Violation: A violation that the employer intentionally and knowingly commits. The employer either knows that the operation constitutes a violation, or is aware that a hazardous condition exists and made no reasonable effort to eliminate it. The penalty range for this type of violation is $5,000 to $70,000.

Repeated Violation: A violation of any standard, regulation, rule, or order where, upon re-inspection, another violation of the same previously cited section is found. Repeated violations can bring fines of up to $70,000.

Failure to Abate: Failure to correct any violations may bring civil penalties of up to $7,000 per day for every day the violation continues beyond the prescribed abatement date.

Other regulatory violations and penalties include:

Falsifying records, reports, or applications, which can bring a fine of $10,000 and/or six months in jail upon conviction.

Violations of posting requirements, which can bring civil penalties of up to $7,000.

Assaulting a compliance officer, or otherwise resisting, opposing, intimidating, or interfering with a compliance officer in the performance of his or her duties is a criminal offense, subject to a fine of not more than $5,000 and/or 3 years in jail.

Conviction of a willful violation that has resulted in the death of an employee can lead to individual fines of up to $250,000 and/or 6 months in jail and corporate fines of up to $500,000.

Overcrowding

The existence of more people in a building, structure, or portion thereof than have been authorized or posted by the local fire code official, or when the fire code official determines that a threat exists to the safety of the occupants due to persons sitting and/or standing in locations that may compromise, obstruct, or impede the use of aisles, passages, corridors, stairways, exits, or other components of the means of egress, i.e., that allows a safe passage from a facility. See also **Maximum Occupancy Allowances**.

Overload

Operation of equipment in excess of the normal full load rating. If it persists for a sufficient length of time it would cause damage or dangerous overheating. A fault, such as a short circuit or ground fault, is not an overload.

Overload, Electrical

Overcurrents that are large enough and persist long enough to cause damage or create a danger of fire. These currents in excess of rated ampacity produce effects in proportion to the degree and duration of the overcurrent. Electrical overloads will cause internal heating of the electrical conductor. Heating will occur in the entire length and cross section of the conductor from the power source to the load. If the overload is greater than five times the ampacity rating, the conductor may become hot enough to ignite fuels in contact with it as the insulation melts off.

Overrange Limit

The maximum input that can be applied to a device without causing damage or permanent change in performance.

Override Control

Generally, two control loops connected to a common final control element—one control loop being nominally in control with the second being switched in by some logic element when an abnormal condition occurs so that constant control is maintained. May also refer to a technique in which more than one controller manipulates a final control element. The technique is used when constraint control is important.

Oxidant

A chemical material that supports the combustion reaction process to combine with a fuel, e.g., oxygen, nitrous oxide, nitric oxide, chlorates, and chlorine.

Oxidizer

Any material that combines with a fuel source to support the combustion process. Free-burning fires need an oxygen level of 16 to 21 percent (12 percent for a smothering fire) to support combustion.

Oxygen Deficiency

Designates an atmosphere having less than the percentage of oxygen found in normal air. Normally, air contains approximately 21 percent oxygen at sea level. Oxygen deficiency occurs in atmospheres containing less than 19.5 percent oxygen by volume. When the oxygen concentration in air is reduced to approximately 16 percent, many individuals become dizzy, experience a buzzing in the ears, and have a rapid heartbeat. As the oxygen concentration falls below 16 percent, the brain sends commands to the breathing control center, causing the victim to breathe faster and deeper. As the oxygen level continues to decrease, full recovery is less certain. An atmosphere of only 4 to 6 percent oxygen will cause the victim to fall into a coma in less than 40 seconds. Oxygen must be administered within minutes to offer a chance of survival. Even when a victim is rescued and resuscitated, he or she risks cardiac arrest. As shown in Table O.2, the human body is adversely affected by lower concentrations. The Occupational Safety and Health Administration (OSHA)

O

TABLE O.2
Oxygen Deficiency Effects on the Human Body

Atmospheric Oxygen Concentration (%)	Possible Results
20.9	Normal
19.0	Some unnoticeable adverse physiological effects
16.0	Increased pulse and breathing rate, impaired thinking and attention, reduced coordination
14.0	Abnormal fatigue upon exertion, emotional upset, faulty coordination, poor judgment
12.5	Very poor judgment and coordination, impaired respiration that may cause permanent heart damage, nausea, and vomiting
Less than 10	Inability to move, loss of consciousness, convulsions, death

Source: Compressed Gas Association. Safety Bulletin, SB-2, Oxygen-Deficient Atmospheres, 2007.

requires employers to maintain workplace oxygen at levels between 19.5 and 23.5 percent. The Chemical Safety and Hazard Investigation Board (CSB) identified 85 nitrogen asphyxiation incidents that occurred in the workplace between 1992 and 2002 in the United States. In these incidents, fatalities occurred to 80 individuals and 50 were injured.

Oxygen-Enriched Atmosphere

An atmosphere containing more than 23.5 percent oxygen by volume or any atmosphere with a partial pressure of oxygen greater than 178 mm Hg.

Oxygen Monitor

A device that measures the percentage of oxygen present in the atmosphere at any point in time. In some instances it is linked to a visual or audible alarm when the level has decreased to a point of concern.

Oxygen Toxicity

A disorder associated with increased partial pressures of oxygen. There are two types of oxygen toxicity:

High pressure: Breathing 100 percent oxygen at pressures greater than 3 ATA (atmospheres absolute) may result in acute toxicity, producing convulsions.

Low pressure: Breathing 100 percent oxygen at 1 ATA for extended periods (24 hours or greater) may result in pulmonary dysfunction and pulmonary edema.

P

Pareto Chart

A type of chart that contains both bars and a line graph and displays the values in descending order as bars and the cumulative totals of each category, left to right, as a line graph. Its purpose is to highlight the most important among a (typically large) set of factors. In loss prevention evaluations it is used to display the number of failures of components by part number in descending order of failure rate or number of failures observed. Data may also be shown taking into account the total cost of each failure.

Particulate

A particle of solid or liquid matter.

Particulate Matter (PM)

A suspension of fine solid or liquid particles in air, such as dust, fog, fume, mist, smoke, or sprays. Particulate matter suspended in air is commonly known as an aerosol. See also **Respirable-Sized Particulates**.

Pathogen

A disease-causing agent.

Percent Impairment of Hearing

An estimate of an individual's ability to hear correctly. It is usually determined by conducting an audiogram test. The specific rule for calculating this quantity varies from state to state according to law.

Performance-Based Regulation

A code or standard that expresses requirements for a building, structure, or system in terms of functional objectives and performance requirements. This is opposed to prescriptive requirements that require specific features and are mandatory.

Perimeter Tape

See **Barrier Tape**.

Permanent Disability

A permanent impairment; includes any degree of impairment, from an amputation of a part of a finger or a permanent impairment of vision to seriously and permanently nonreversible, nonfatal injuries.

Permanent Disability Benefits

Payments provided when an employee's work injury permanently limits the kinds of work an employee can do.

Permanent Impairment

See **Permanent Disability**; **Permanent Total Disability**.

Permanent Partial Disability

Any injury other than death or permanent total disability that results in the loss, or complete loss of use, of any member or part of a member of the body, or any permanent impairment of functions of the body or part thereof, regardless of any pre-existing disability of the injured member or impaired body function. These cases are used in computing American National Standards Institute (ANSI) Standard Z16.2 injury rates whether or not time is lost. Also refers to a term used by the state workers' compensation office to describe an individual who is stable, but may have permanently lost some function due to a workplace injury. The determination normally has to be stated by a medical office, which is used to determine financial benefits to the individual, which also takes into account the individual's age, education, and ability to adapt to work. See also **Permanent Total Disability**.

Permanent Total Disability

Any injury other than death that permanently and totally incapacitates an employee from following any gainful occupation, or which results in the loss, or the complete loss of use, of any of the following in one accident: (a) both eyes; (b) one eye and one hand, or arm, or leg, or foot; or (c) any two of the following not on the same limb: hand, arm, foot, or leg. The definition can vary from state to state. See also **Permanent Partial Disability**.

Permissible Exposure Limit (PEL)

The permissible exposure, inhalation, or dermal exposure limit specified in 29 CFR Part 1910, Subparts G and Z (OSHA standards for Occupational Health and Environmental Control and Toxic and Hazardous Substances). This is the 8-hour time-weighted average (TWA) or ceiling concentration above which workers may not be exposed. The use of personal protective equipment (PPE) is advisable where there is a potential for exposure.

Permissible Noise Exposure

Noise exposure limits cited in Occupational Safety and Health Administration (OSHA) regulations, 29 CFR 1910.95 (b) (2) for worker hearing protection. The noise exposure limits are shown in Table P.1.

Permit-Required Confined Space

Permit-required confined space (permit space) means a confined space that has one or more of the following characteristics: contains or has a potential to contain a hazardous atmosphere; contains a material that has the potential for engulfing an entrant; has an internal configuration such that an entrant could be trapped or asphyxiated by inwardly converging walls or by a floor which slopes downward and tapers to a smaller cross-section; contains any other recognized serious safety or health hazard.

P

Permit to Work

A formal written or verbal approval to operate a planned procedure, which is designed to protect personnel working in hazardous areas or activities. See also **Work Permit**.

TABLE P.1
Permissible Noise Exposure Limits

Duration per Day, Hours	Sound Level dBA, Slow Response
8	90
6	92
4	95
3	97
2	100
1	102
1/2	105
1/4 or less	115

Personal Air Samples
Air samples taken with a pump that is directly attached to the worker with the collecting filter and cassette placed in the worker's breathing zone (required under Occupational Safety and Health Administration [OSHA] asbestos standards and Environmental Protection Agency [EPA] worker protection rule).

Personal Alert Safety Systems (PASS)
An individual protective device that emits an audible alarm to notify others and assists in locating a firefighter in danger. The personal alert safety system (PASS) device includes a motion detector that senses movement and automatically sounds an alarm signal if no movement is sensed for 30 seconds in case a firefighter is incapacitated and cannot activate the alarm. Requirements for PASS devices are specified in NFPA 1982, Standard on Personal Alert Safety Systems (PASS) for Fire Fighters.

Personal Factor
The mental or bodily characteristic that permitted or occasioned an act which contributed to an incidental occurrence.

Personal Fall Arrest System
A system used to arrest an individual's fall. It consists of a substantial anchorage, full body harness, and lanyard, and may include a deceleration device, lifeline, or suitable combinations of these. The components are sometimes referred to as ABCs, i.e., A for anchor point means, B for body harness, and C for connecting device (lanyards). See also **ASSE Z359.1, Safety Requirements for Personal Fall Arrest Systems, Subsystems, and Components; Fall Restraint System**.

Personal Flotation Device (PFD)
A generic term for personal devices used to help prevent drowning should an individual inadvertently fall in the water or need assistance while swimming. These include life buoys, life jackets, life rings, ring life buoys, work vests, and life vests. Life rings and ring life buoys are used to throw to individuals who may inadvertently fall overboard or otherwise need assistance while in the water, usually with a rope (life line) attached to retrieve, rescue, or assist the individual (see Figure P.1). Life jackets and vests are normally worn as a precautionary measure during routine operations. A personal flotation device is nomenclature used by the U.S. Coast Guard, while Personal Life Saving Appliances is nomenclature used by the International Maritime Organization in relation to Safety of Life at Sea, i.e., SOLAS regulations.

Personal Gas Monitor
A device carried by an individual to detect toxic gases or an oxygen-deficient atmosphere, which would cause harm to the health of an individual. They generally have an audible, visual, and vibrating alarm that activates at a preset level to warn the wearer of the immediate hazard of the surrounding atmosphere.

Personal Measurement
A measurement collected from an individual's immediate environment.

Personal Protection
The action of shielding the body against contact with known chemical or physical hazards in the environment.

Personal Protective Equipment (PPE)
Devices worn by the worker to protect against hazards in the environment. For example, respirators, gloves, and hearing protectors. The use of personal protective equipment is the least preferred method of protection from hazardous exposures. It can be unreliable and, if it fails, the person can be left completely unprotected. This is why engineering controls are preferred. Sometimes, personal protective equipment

FIGURE P.1 Life ring with lifeline provision. (Photo courtesy of NOAA.)

may be needed along with engineering controls. For example, a ventilation system (an engineering control) reduces the inhalation hazard of a chemical, while gloves and an apron (personal protective equipment) reduce skin contact. In addition, personal protective equipment can be an important means of protection when engineering controls are not practical, for example, during an emergency or other temporary conditions such as maintenance operations. Recommended personal protective equipment is often listed on Material Safety Data Sheets (MSDSs). The Occupational Safety and Health Administration (OSHA) requires the use of personal protective equipment to reduce employee exposure to hazards when engineering and administrative controls are not feasible or effective in reducing these exposures to acceptable levels. Employers are required to determine if Personal Protective Equipment (PPE) should be used to protect their workers. If PPE is to be used, a PPE program should be implemented. This program should address the hazards present; the selection, maintenance, and use of PPE; the training of employees; and monitoring of the program to ensure its ongoing effectiveness.

Personal Sampling

The process of measuring an individual's exposure to airborne contamination in the workplace. A sampling device is fitted to suspected individuals who may be exposed. Personal sampling devices include gas monitoring badges, impingers, and filtration devices.

Personnel Accountability

The ability to account for the location and welfare of individuals involved in an incident. See also **Evacuation**.

pH

The degree or acidity or basicity (alkalinity) of a solution rated from 0 to 14, with neutrality stated as 7. Roughly, pH can be divided into the following ranges:

0–2: Strongly acidic
3–5: Weakly acidic
6–8: Neutral
9–11: Weakly basic
12–14: Strongly basic

Physical Factors, Unsafe

Environmental factors conducive to an incident occurrence, which include physical or environmental hazards.

Physical Hazard

A workplace exposure to an individual that may result in physical injury. Table P.2 highlights the most common construction physical hazards cited by the Occupational Safety and Health Administration (OSHA) in 1991.

Physical Relief Device

Mechanical equipment that performs an action to relieve pressure when the normal operating range of temperature or pressure has been exceeded. Physical relief devices include pressure relief valves, thermal relief valves, rupture disks, rupture pins, and high temperature fusible plugs.

Physical Safety

The avoidance of unnecessary risks, and maintaining good health practices. Also includes freedom from substance abuse and other addictions; healthy, safe, relational sexual behaviors; and an absence of any kind of violence—physical, emotional, sexual, or verbal—including suicidality and self-destructive behavior. Physical safety

P

TABLE P.2
The 10 Most Frequently Cited OSHA Construction Standards

Rank	Description	OSHA Standard
1	Fall protection—guarding floors and platforms	1926.500 (d) (1)
2	PPE—head protection	1926.100 (a)
3	Electrical ground fault protection	1926.404 (b) (1) (i)
4	Electrical—grounding unavailable	1926.404 (f) (6)
5	Trench/excavation—protective systems	1926.625 (a) (1)
6	Scaffolding—guardrail specifications	1926.451 (d) (10)
7	PPE—appropriate for job	1926.28 (a)
8	Ladders/stairways—missing rails	1926.1052 (c) (1)
9	Fire protection—approved containers or tanks	1926.152 (a) (1)
10	General—housekeeping	1926.25 (a)

FIGURE P.2 Example of caution sign for pinch points.

is usually what people think of when describing the sense of being safe, since without it, other forms of safety are difficult to achieve. See also **Physical Hazard**.

Pinch Point

A pinch point is any point at which it is possible for a person or part of a person's body to be caught between moving parts of a machine, or between the moving and stationary parts of a machine, or between material and any part of the machine (Figure P.2). A pinch point does not have to cause injury to a limb or body part, although it might cause injury—it only has to trap or pinch the person to prevent them from escaping or removing the trapped part from the pinch point. Guards are typically provided to create a physical barrier to prevent anyone from reaching into, through, over, under, or around the guard to make contact with, or at, the pinch point. Most often pinch-point injuries are the result of workers who are not properly trained, don't realize the dangers of machinery, or take shortcuts to get the work done more quickly. Sometimes referred to as a Nip Point.

Pipeline and Hazardous Materials Safety Administration (PHMSA)

An agency within the U.S. Department of Transportation (DOT) that works to protect the public and the environment by ensuring the safe and secure movement of hazardous materials to industry and consumers by all transportation modes. Created in 2004, the PHMSA develops and enforces regulations for the safe, reliable, and environmentally sound operation of the 2.3 million mile pipeline transportation system and the nearly one million daily shipments of hazardous materials by land, sea, and air. The federal regulations it mainly focuses on include the Hazardous Material Regulations, 49 CFR, Parts 100 to 185, and Pipeline Safety Regulations, 49 CFR, Parts 190 to 199. The rules governing pipeline safety are included in Title 49 of the Code of Federal Regulations (CFR), Parts 190-199. Individual states may have additional or more stringent pipeline safety regulations. See also **Office of Pipeline Safety (OPS)**.

Pipeline Safety Regulations

The rules governing pipeline safety, which are stated in Title 49 of the Code of Federal Regulations (CFR), Parts 190-199. The main sections are as follows: Part 190 describes the procedures used by the Office of Pipeline Safety (OPS) in carrying out their regulatory duties. This part authorizes the OPS to inspect pipelines and describes the procedures by which the OPS can enforce the regulations. This part also describes the legal rights and options that the operating companies have in response to OPS enforcement actions. Part 191 describes requirements for operators of gas pipelines (including gas gathering, transmission, and distribution systems) for reporting of incidents, safety-related conditions, and annual summary data. Part 192 prescribes a wide variety of minimum safety requirements for gas pipelines. These regulations contain sections applicable to gas gathering, transmission, and distribution lines. Part 193 addresses safety standards for liquefied natural gas (LNG) facilities. Part 194 contains requirements for oil spill response plans. This part is intended to reduce the environmental impact of oil discharged from onshore oil pipelines. Part 195 prescribes the safety standards and reporting requirements for oil and carbon dioxide pipelines. As with the gas regulations, these regulations include detailed requirements on a broad spectrum of areas related to the safety and environmental protection of hazardous liquid pipelines. Part 195 also includes minimum requirements for operator qualification of individuals performing tasks required by the regulations. Parts 196-197 are currently reserved for future use. Part 198 prescribes regulations governing grants-in-aid for state pipeline safety compliance programs. Part 199 requires operators of gas and hazardous liquid pipelines to establish programs for preventing alcohol misuse and to test employees for the presence of alcohol and prohibited drugs and provides the procedures and conditions for this testing. See also **Office of Pipeline Safety (OPS)**; **Pipeline and Hazardous Materials Safety Administration (PHMSA)**.

Point of Operation Guarding
Machine guards, such as a hard guards or safety light screens, that are designed to protect personnel from hazardous machine motion when close to the machine's point of operation.

Point of Perception
The time and place at which the individual actually first perceived, that is, "saw, heard, smelled, or felt" the hazard, or the unusual or unexpected movement or condition that could be taken as a sign that the incident is about to occur.

Point of Possible Perception
The place and time at which the unusual or unexpected movement or condition could have been perceived by a normal person. This point always comes at or before the point of (actual) perception.

Poison
Any substance that when taken into the body is injurious to health.

Policy, Insurance
The contract issued by the insurance company to the insured.

Policy, Safety
A declaration by an organization of how it manages loss prevention at its facilities. It reflects the organization's culture, values, and commitment towards health and safety in its workplace.

Policy Period
The length of time a policy is in force, from the beginning or effective date to the expiration date.

Pollution
Synthetic contamination of soil, water, or the atmosphere beyond that which is natural.

Positive Isolation
To totally isolate a piece of equipment or plant from a live process. Simple isolation valves that are closed would not be considered positive isolation because they could

be subject to leakages. Double block and bleed, total blinding, or removal/disconnection of a pipe are considered positive isolation. An electrical device is not considered positively isolated unless the wiring is disconnected or its power feed (i.e., circuit breaker) is removed.

Positive Opening Safety Contacts

Contacts of safety switches that are forced open when actuated, without reliance upon spring action.

Positive Pressure Respirator

A respirator in which the pressure inside the respiratory inlet covering exceeds the ambient air pressure outside the respirator.

Post-incident Trauma

Psychological stress that may occur to individuals after they return from a response to a stressful emergency incident.

Postal Service Safety Enhancement Act

A federal regulation mandated in 1998 that makes the Occupational Safety and Health Act of 1970 applicable to the U.S. Postal Service in the same manner as any other employer.

Potential Exposure

Potential exposure is when the possibility exists that an employee could be exposed to a hazard because of work patterns, past circumstances, or anticipated work requirements, and it is reasonably predictable that employee exposure could occur at some time during the entry.

Potential Hazard

A situation, thing, or event having latent characteristics conducive to an incident occurrence.

Potential Incident

A condition (such as an unidentified hazard) or an event (such as a near miss) or sequence of events that does not have actual consequences, but that could, under slightly different circumstances, have unwanted consequences.

Potential Years of Life Lost

A measure of the relative impact of various diseases and injuries on a defined population. Potential years of life lost due to a particular cause is the sum of the years that the persons in a population would have lived had they experienced normal life expectancy.

Powered Air-Purifying Respirator

A positive pressure device that utilizes a blower and battery to force ambient air through cartridges and/or filters to the facepiece or hood.

ppb

Parts per billion. Used for measuring very low concentrations or contamination levels.

PPE Decontamination Shower

A safety shower specifically designed to introduce a detergent in the water flow to ensure a thorough decontamination of Personal Protective Equipment (PPE). See also **Damping Shower**; **Safety Shower**.

ppm

Parts per million. Used for measuring very low concentrations or contamination levels.

ppt

Parts per trillion. Used for measuring extremely low concentrations.

Precautions

Actions taken in advance to reduce the probability of an incident.

Pre-fire Plan

A description of the potential hazard and recommended actions to be taken by emergency and firefighting personnel during a fire or explosion event for a specific hazard, location, or facility based on previous inspections and surveys of identified hazards. Pre-fire plans note the structural features, physical layout, special hazards, installed protection systems, fire hydrant locations, water supplies, and similar features pertinent to firefighting operations. Pre-fire plans should be routinely updated or revised as changes occur in a facility or location.

Pre-incident Planning

A description of the potential hazards and recommended actions to be taken by public emergency response agencies or private industry for response to emergency incidents at a specific facility or location (Figure P.3). See also **Credible Scenario**.

Pre-startup Safety Review (PSSR)

An audit check performed prior to equipment operation to ensure that adequate Process Safety Management (PSM) activities have been performed. The check should verify that (1) construction and equipment are satisfactory, (2) procedures are available and adequate, (3) a Process Hazard Analysis (PHA) has been undertaken and recommendations resolved, and (4) the employees are trained. It is part of the requirements of the PSM program outlined in the Occupational Safety and Health Administration (OSHA) regulation 29 CFR 1910.119.

Preliminary Hazard Analysis (PHA)

An early or initial screening study for the identification, qualitative review, and ranking of process hazards, typically conducted during an initial evaluation of existing facilities or a project's conceptual design. Recommendations for the mitigation of identified hazards are provided. See also **Preliminary Risk Analysis**; **Process Hazard Analysis (PHA)**.

FIGURE P.3 Pre-incident planning.

Preliminary Hazard List (PHL)

A line item inventory of system hazards, with no evaluation of probability, severity, or risk.

Preliminary Risk Analysis

A qualitative investigative safety review technique that involves a disciplined analysis of the event sequences, which could transform a potential hazard into an incident. In this technique, the possible undesirable events are identified first and then analyzed separately. For each undesirable event or hazard, possible improvements or preventive measures are then formulated. The result from this methodology provides a basis for determining which categories of hazard should be looked into more closely and which analysis methods are most suitable. Such an analysis also proves valuable in the working environment to which activities lacking safety measures can be readily identified. With the aid of a frequency/consequence diagram, the identified hazards can then be ranked according to risk, allowing measures to be prioritized to prevent incidents. It is sometimes is referred to as a Preliminary Hazard Analysis (PHA).

Premium

The amount paid by an insured (your business) to an insurance company to obtain or maintain an insurance policy.

Pressure Demand Respirator

A positive pressure atmosphere-supplying respirator that admits breathing air to the facepiece when the positive pressure is reduced inside the facepiece by inhalation.

Pressure Relief Device

A mechanism that vents fluid from an internally pressurized system to counteract system overpressure; the mechanism may release all pressure and shut the system down (as does a rupture disc) or it may merely reduce the pressure in a controlled manner to return the system to a safe operating pressure (as does a spring-loaded safety valve).

Pressure Safety Valve

A valve that opens at a preset pressure to relieve excessive pressure within a vessel or piping. It may also be called a Pressure Relief Valve, Relief Valve, or Pop Valve.

P

Pressurization

A protection technique that enables the hazardous area electrical classification of an enclosed space to be reduced by purging the enclosure with clean air or an inert gas, and then by maintaining a minimum prescribed positive pressure to prevent the ingress of a flammable atmosphere. Depending upon the original area classification and the type of purge controller employed, the reduction may either be to a lower hazard classification or to a general-purpose classification. Special dilution provisions apply to enclosures containing apparatus with an internal source of released gas, such as a gas analyzer. Pressurized enclosures may be operated on the basis of "leakage compensation" to economize on purge media, or "continuous flow" where interior cooling or dilution of internally released gas is required.

Prevention (of Incident)

The science and art representing control of worker performance, machine performance, and environment to avoid, intervene, or eliminate the occurrence of an incident. Prevention connotes correction of conditions as well as elimination or isolation of hazards.

Prevention through Design (PtD)

A principle highlighted in the American National Standards Institute (ANSI) voluntary consensus standard Z10-2005, Paragraph 5.1.2, and is typically defined as addressing occupational safety and health needs in the design process to prevent or minimize the work-related hazards and risks associated with the construction, manufacture, use, maintenance, and disposal of facilities, materials, and equipment. The National Institute for Occupational Safety and Health (NIOSH) has partnered with the American Industrial Hygiene Association (AIHA), the American Society of Safety Engineers (ASSE), the Center to Protect Workers' Rights, the National Safety Council (NSC), the Occupational Safety and Health Administration, ORC Worldwide, the Regenstrief Center for Healthcare Engineering, and some commercial providers of workers' insurance for the development of a National Initiative on Prevention through Design.

Proactive Safety Measures

Programs or activities that are undertaken to prevent an incident. Sometimes referred to as leading safety indicators. These may include preparation and implementation of safety policies, standards and procedures, safety communication activities, risk assessments, management of change, audits and inspections, training, continuous safety improvement activities, etc. See also **Leading Indicator**.

Probability

The projected frequency of the occurrence of an event, usually based on statistical analysis of past similar events. Used for determining risk levels. May be qualitatively assessed or based on historical data for various types of hazard assessments. May also be sometimes referred to as Likelihood. See also **Likelihood; Risk**.

Probability of Failure on Demand (PFD)

A value that indicates the probability that a device or system will fail to respond to a demand in a specified interval of time. PFD equals 1 minus Safety Availability. See also **Safety Availability**.

Probable Maximum Loss (PML)

A term used in the insurance industry generally defined as the anticipated value of the largest loss that could result from the destruction and the loss of use of property, with the normal functioning of passive protective features (firewalls, a responsive fire department), and proper functioning of most (perhaps not all) active suppression systems (e.g., fire protection sprinklers). This loss estimate is usually smaller than the Maximum Foreseeable Loss, which assumes the failure of all active protective features. Underwriting decisions could be influenced by PML evaluations, and the amount of reinsurance ceded on a risk could be predicated on the PML valuation. See also **Maximum Foreseeable Loss (MFL)**.

Process Critical Equipment

Rotating equipment including turbines, electric driven pumps, compressors or generators handling combustible, flammable or toxic materials, and use drivers equal to or greater than 1000 HP. Process critical equipment also includes rotating equipment that is categorized as critical by a Process Hazard Analysis.

Process Hazard Analysis (PHA)

An organized formal review to identify and evaluate and hazards with industrial facilities and operations to enable their safe management. The review normally employs a qualitative technique to identify and access the importance of hazards as a result of identified consequences and risks. Conclusions and recommendations are provided for risks that are deemed at a level not acceptable to the organization. Quantitative methods may be also employed to embellish the understanding of the consequences and risks that have been identified.

Process Safety

A discipline that focuses on the prevention of fires, explosions, and unexpected chemical releases at facilities that employ processing of materials or chemicals. It

P

does not encompass the normal occupational safety concerns, such as slips, trips, and falls. Since 1984, there has not been another incident having as strong an industry-wide impact or receiving as much global attention as Bhopal (December 1984). According to the U.S. Occupational Safety and Health Administration, since 1992, on-site fatalities from process safety incidents have dropped by over 60 percent. Fatalities have been contained to the plant site, and when off-site releases have occurred, emergency procedures have kept injuries low. On the other hand, the U.S. Chemical Safety Board has identified 167 incidents in the last 25 years based on chemical reactivity alone, mostly at smaller manufacturers and companies whose main business was not chemistry. It has been stated that nearly all of these incidents could have been prevented had basic process safety guidelines and references been consulted.

Process Safety Management (PSM)

A comprehensive set of plans, policies, procedures, practices, and administrative, engineering, and operating controls designed to ensure that barriers to major incidents are in place, in use, and effective. Its emphasis is on the prevention of major incidents rather than specific worker health and safety issues. PSM focuses its safety activities on chemical-related systems, such as water treatment plants and chemical manufacturing plants, wherein there are large piping systems, storage, and blending and distributing activities.

Its main elements include:

Employee Participation
Process Safety Information
Operating Procedures
Training
Contractors
Pre-startup Safety Reviews
Mechanical Integrity
Hot Work Permits
Management of Change
Incident Investigation
Emergency Planning and Response
Compliance Audits

It is required by the Occupational Safety and Health Administration (OSHA) PSM regulation, 29 CFR 1910.119.

Process Safety Time

The period of time between a trip point being reached and a hazardous event occurring if no safety measures such as a shutdown are taken.

Product Liability

The liability a merchant or a manufacturer may incur as the result of some defect in the product sold or manufactured, or the liability a contractor might incur after job completion from improperly performed work. The latter part of product liability is called completed operations.

Product Safety

A scientific process to assess the probability that exposure to a product during any stage of its life cycle will lead to an unacceptable impact on human health or the environment. May also be referred to as product risk characterization or product risk assessment.

Professional Safety Academy (PSA)

An academy established in 1999 by the American Society of Safety Engineers (ASSE) to offer career support to ASSE members and the profession. Its program includes an annual Professional Development Conference and Exposition, workshops and seminars offered throughout the United States, and professional networking. See also **American Society of Safety Engineers (ASSE)**.

Program Protection Factor (PPF) Study

A study that estimates the protection provided by a respirator within a specific respirator program. Like the Effective Protection Factor (EPF), it is focused not only on the respirator's performance, but also on the effectiveness of the complete respirator program. PPFs are affected by all factors of the program, including respirator selection and maintenance, user training and motivation, work activities, and program administration. See also **Effective Protection Factor (EPF) Study**.

Proof Test

Testing of safety system components to detect any failures not detected by automatic on-line diagnostics, i.e., dangerous failures, diagnostic failures, and parametric failures, followed by repair of those failures to an equivalent as new state. Proof testing is a vital part of the safety life cycle and is critical to ensuring that a system achieves its required safety integrity level throughout the safety life cycle.

Property Conservation

A management philosophy to prevent losses or to minimize the frequency and severity of incidental losses. It is a combination of ensuring the necessary physical protection and the proper employee responses to preserve a company's human and physical assets. These assets can be threatened by perils such as fire, wind, collapse, flood, earthquake, explosion, theft, and/or equipment breakdown. Typically, terminology used in the insurance industry.

Property Damage (PD)

An incident wherein damage to or destruction of any property is the immediate and direct result. It does not include incidents resulting in loss of human life or personal injury.

Property Damage Liability Insurance

Insurance that provides for financial protection against liability for damages to the property of another, including loss of the use of the property, as distinguished from liability for bodily injury to another.

Property Insurance

Insurance that indemnifies an individual or organization with interest in physical property for its loss or the loss of its income-producing abilities.

Protection

Preservation from injury or harm.

Protection Factor

A quantitative measure of the fit of a particular respirator to a particular individual.

Protection Factor Study

A study that determines the protection provided by a respirator during use. This determination is generally accomplished by measuring the ratio of the concentration of an airborne contaminant (e.g., hazardous substance) outside the respirator (Cout) to the concentration inside the respirator (Cin) (i.e., Cout/Cin). Therefore, as the ratio between Cout and Cin increases, the protection factor increases, indicating an increase in the level of protection provided to employees by the respirator. There are four types of protection factor studies: Effective Protection Factor (EPF), Program Protection Factor (PPF), Workplace Protection Factor (WPF), and Simulated Workplace Protection Factor. See also **Effective Protection Factor**

(EPF) Study; **Program Protection Factor (PPF) Study**; **Simulated Workplace Protection Factor (SWPF) Study**, **Workplace Protection Factor (WPF) Study**.

Protection Layer

See **Independent Protection Layer (IPL)**.

Protective Clothing

An article of clothing furnished to an employee that is worn for personal safety and protection in the performance of work assignments in potentially hazardous areas or under hazardous conditions.

Protective Equipment

A device or item to be worn, used, or put in place for the safety or protection of an individual or the public at large, when performing work assignments or in entering hazardous areas or under hazardous conditions. Common examples include hearing protection, respirators, barricades, traffic cones, lights, safety lines, and flotation devices. See also **Personal Protective Equipment (PPE)**.

Protective Hand Cream

A cream designed to protect the hands and other parts of the skin from exposure to harmful substances.

Proximate Cause

The cause that directly produces the effect without the intervention of any other cause. The cause nearest to the effect in time or space.

Prudent

Cautious, careful, attentive, discrete, circumspect, and sensible as applied to action or conduct.

Psychogenic Deafness

Hearing loss originating or produced by the mental reaction of an individual to the individual's physical or social environment. Is it also called functional deafness or feigned deafness.

Psychological Safety

Refers to the ability to be safe with oneself, to rely on one's own ability to self-protect against any destructive impulses coming from within oneself or deriving from other people, and to keep oneself out of harm's way. It includes self-discipline, self-esteem, self-control, self-awareness, and self-respect. Psychological safety is the ability to be able to direct one's attention and focus, to know oneself, to feel effective in the world, to be able to exercise self-control and self-discipline, to have a sense of internal authority that is fair and non-abusive, and to be able to express one's sense of humor, creativity, and spirituality.

Public Fire Service

A global term that represents firefighters organized and supported by the local authorities or government. It may also be called Fire Department, fire brigade, fire and emergency services, and fire/rescue.

Purging

A method by which gases, vapors, or other air contaminants are displaced from a confined space. It is accomplished so that subsequent natural ventilation will not result in the reinstatement of an undesired atmosphere. Purging is commonly accomplished by the use of a protective or inert gas (e.g., nitrogen, carbon dioxide, etc.) at a sufficient flow and positive pressure to reduce the concentration of any flammable gas, vapor, or toxic materials initially present to an acceptable level. It may also be called de-gassing. See also **Inerting**.

Purging, Clearing
Replacement of a substance so rapidly so that there is minimum mixing, thus reducing the duration of any explosive mixture.

Purging, Dilution
Introduction of adequate quantities of an inert gas to an enclosure to ensure that explosive mixtures cannot form.

Purging, Displacement
Replacement of one substance with another without appreciable mixing. Displacement purges in pipelines are typically accomplished by separating the two substances with an inert fluid or a mechanical scraper.

Pyrophoric Material
The property of a material to ignite spontaneously in air under ordinary conditions, generally taken as a chemical with an auto-ignition temperature in air at or below 54.4°C (130°F). Certain very finely divided powders overall present so much available surface area to the air that they can spontaneously combust and are called pyrophoric substances. An alloy composed of iron and of rare earth metals, called misch metal, is pyrophoric. When scratched it gives off sparks capable of igniting flammable gases. It is used in cigarette lighters, miners' safety lamps, and automatic gas-lighting devices. Iron sulfide may form in hydrocarbon tanks for vessels that upon exposure to air will spontaneously combust. Ignition of pyrophoric materials can be prevented by keeping them damp until they can be safely disposed of (i.e., steaming out vessels).

Pyrotechnics
Explosive commodities that have extremely high flame speeds and heat releases. Pyrotechnics are primarily used in fireworks displays and munitions.

P

Q

Qualified Expert

An individual who, by virtue of certification by appropriate boards or societies, professional licenses, or academic qualifications and experience, is duly recognized as having expertise in a relevant field of specialization, e.g., fire safety, medical physics, radiation protection, occupational health, quality assurance, or any relevant engineering or safety specialty.

Qualitative Fitting Test (QLFT)

A pass or fail fit test in which the respirator wearer assesses the fit of the respirator by being subjected to a challenge agent that can be adequately detected by the senses.

Qualitative Risk Analysis

An evaluation of risk based on the observed hazards and protective systems that are in place as opposed to an evaluation that uses specific numerical techniques. See also **Danger Analysis**; **Quantitative Risk Analysis (QRA)**.

Quantitative Fitting Test (QNFT)

A fit test in which the assessment of the adequacy of the respirator fit is numerically measured by comparing respirator leakage with ambient concentrations of the challenge agent.

Quantitative Risk Analysis (QRA)

An evaluation of both the frequency and the consequences of potential hazardous events to make a logical decision on whether the installation of a particular safety measure can be justified on safety and loss control grounds. Frequency and consequences are usually combined to produce a measure of risk, which can be expressed as the average loss per year in terms of injury or damage arising from an incident. The risk calculations of different design alternatives can be compared to determine the safest and most economical options. Calculated risk may be compared to set criteria that have been accepted by society or required by laws. See also **Qualitative Risk Analysis**.

Q

R

Radiation

The emission and propagation of energy through space or a medium in the form of electromagnetic waves (gamma or x-rays) or particles (alpha and beta). Electromagnetic radiation may be ionizing or non-ionizing; all particulate radiation is ionizing. Certain human body parts are more specifically affected by exposure to different types of radiation sources. Several factors are involved in determining the potential health effects of exposure to radiation. These include the size of the dose (amount of energy deposited in the body), the ability of the radiation to harm human tissue, and which organs are affected (Table R.1). The most important exposure factor is the amount of the dose the amount of energy actually deposited in your body. The more energy absorbed by cells, the greater the biological damage. Health physicists refer to the amount of energy absorbed by the body as the radiation dose. The absorbed dose, the amount of energy absorbed per gram of body tissue, is usually measured in units called rads. Another unit of radiation is the rem, or roentgen equivalent in man. To convert rads to rems, the number of rads is multiplied by a number that reflects the potential for damage caused by a type of radiation. For beta, gamma, and x-ray radiation, this number is generally one. For some neutrons, protons, or alpha particles, the number is 20.

Radiation Control

Engineering measures, including devices, used to prevent worker exposure to harmful levels of ionizing and non-ionizing radiation.

Radiation Dose

The amount of radiation delivered to a specified area or the whole body. Dose rate is the dose delivered per unit of time. The term dose- or dosage- is also used generally to express the amount of energy or substance absorbed in a unit volume by an organ or individual.

TABLE R.1
Radiation Exposure Health Effects

Dose (rem)	Effects
5–20	Possible late effects; possible chromosomal damage
20–100	Temporary reduction in white blood cells
100–200	Mild radiation sickness within a few hours; vomiting, diarrhea, fatigue, reduction in resistance to infections
200–300	Serious radiation sickness effects in 100–200 rem and hemorrhage, exposure is a Lethal Dose (LD) to 10–35 percent of the population after 30 days LD 10–35/30
300–400	Serious radiation sickness, bone marrow and intestine destruction, LD 50–70/30
400–1000	Acute illness, early death, LD 60–95/30
1000–5000	Acute illness, early death, LD 100/10

Radiation Dosimetry

Measurement of the amount of radiation delivered to or absorbed at a specific place. A dose meter or dosimeter is an instrument that measures radiation dose. Personnel dosimetry is accomplished with such devices as the film badge, thermoluminescent dosimeter, or pocket ionization chamber. In this way continuous recording of cumulative radiation dose can be maintained.

Radiation Monitoring

Measurement of the amount of radiation present in a given area. Radiation exposures can be limited by monitoring and limiting access and stay time in areas in which high levels of radiation are present.

Radioactive Contamination

The deposition or presence of radioactive material in a place where it is not desired and may be harmful. In regard to personnel there are two types of radioactive contamination: external and internal. External radioactive contamination is the presence of radioactive material on the skin. Internal radioactive contamination is the presence of radioactive material within the body due to ingestion, inhalation, or absorption.

Railings

A railing is any one or a combination of those railings constructed in accordance with specific requirements typically consisting of a vertical barrier erected along exposed edges of floor openings, wall openings, ramps, platforms, and runways to prevent the falls of individuals. See also **Handrail**; **Stair Railing**; **Standard Railing**.

Railroad Crossings Safety

According to the Federal Railroad Administration, there were 2398 incidents at railroad crossings in 2008. Two hundred eighty-six of these incidents involved fatalities. There are approximately 150,000 miles of railroad track in the United States. Railroad tracks are private property, and a train always has the right-of-way. Trains cannot stop quickly as it takes almost a mile and a half for a train operating at a normal speed to stop, making these a collision hazard where vehicle and pedestrian crossings exist. Railroad crossings are marked with passive and active warning devices. Passive devices include signs and pavement markings in advance of the crossing. Active devices include gate arms, warning lights, and bells. There are approximately 139,862 public railroad crossings in the United States. More than 50,000 of those crossings are gated, while almost 90,000 are equipped only with warning lights. A gate does not appear to make a crossing safer, as incident statistics indicate approximately one-half of all crossing incidents occur where no gates are present.

Railroad Safety

The Federal Railway Administration promotes and regulates safety throughout the U.S. railroad industry. The office undertakes regulatory and inspection responsibilities through a staff of railroad safety experts who cover hazardous materials, motive power and equipment, operating practices (including drug and alcohol, signal, and train control), and track structures (Figure R.1).

Random Failure

A failure occurring at a random time that results from one or more degradation mechanisms. Random failures can be effectively predicted with statistics and are the basis for the probability of failure on demand-based calculations requirements for safety integrity level. See also **Systematic Failure**.

Random Noise

A sound or electrical wave whose instantaneous amplitudes occur as a function of time according to a normal (Gaussian) distribution curve. Random noise is an

FIGURE R.1   Railroad safety.

oscillation whose instantaneous magnitude is not specified for any given instant of time. The instantaneous magnitudes of a random noise are specified only by probability functions giving the fraction of the total time that the magnitude, or some sequence of the magnitudes, lies within a specific range.

Range of Motion

The limits of movement defined at a joint or landmark of the body. Stresses on the connective tissues at a joint increase as the joint moves towards the limit of its range of motion. Used in ergonomic evaluations.

Raynaud's Syndrome

Blood vessels of the hand are damaged from repeated exposure to vibration over a long period of time. The skin and muscles do not get the necessary oxygen from the blood and eventually die. Symptoms include intermittent numbness and tingling in the fingers; pale, ashen, and cold skin; and eventual loss of sensation and control in the hands and fingers. May also be called White Finger.

Recommended Exposure Limit (REL)

The National Institute of Occupational Safety and Health (NIOSH) recommended exposure limit. A short-term exposure limit (STEL) is designated by "ST" preceding the value; unless noted otherwise, the STEL is a 15-minute TWA exposure that should not be exceeded at any time during a workday. "TWA" indicates a time-weighted average concentration for up to a 10-hour workday during a 40-hour workweek. A ceiling REL is designated by "C" preceding the value; unless noted otherwise, the ceiling value should not be exceeded at any time. It is similar to the Permissible Exposure Limit (PEL) and Threshold Limit Value (TLV). See also **Ceiling Level or Limit**.

Recordable Hearing Loss

The amount of decline in hearing that requires recording on Occupational Safety and Health Administration (OSHA) health and safety records. OSHA requires reporting on its injury and illness log (Form 300), when work-related hearing loss occurs, which is a standard threshold shift, and when the average hearing thresholds are below 25 dB hearing loss (HL) or higher at STL frequencies in the ear (2000, 3000, and 4000 Hz on the audiogram). This means a hearing loss is recordable when a significant noise-induced shift in hearing occurs, and that shift is out of the normal range of hearing.

Recordable Injuries and Illnesses

Occupational Safety and Health Administration (OSHA) recordable cases, which include work-related injuries and illnesses that result in one or more of the following: death, loss of consciousness, days away from work, restricted work activity or job transfer, medical treatment (beyond first aid), significant work-related injuries or illnesses that are diagnosed by a physician or other licensed heath care professional (these include any work-related case involving cancer, chronic irreversible disease, a fracture or cracked bone, or a punctured eardrum); additional criteria include any needle-stick injury or cut from a sharp object that is contaminated with another person's blood or other potentially infectious material, any case requiring an employee to be medically removed under the requirements of an OSHA health standard, and tuberculosis infection as evidenced by a positive skin test or diagnosis by a physician or other licensed health care professional after exposure to a known case of active tuberculosis.

Recovery

Activities and programs designed to return operations at a site to pre-incident levels as quickly as possible. See also **Response**.

Recovery Plan

Directions for restoration of services and facilities; long-term care and treatment of affected persons; additional measures for social, political, environmental, and economic restoration; evaluation of the incident to identify lessons learned; post-incident reporting; and development of initiatives to mitigate the effects of future incidents.

Red Crescent

See **Red Cross**.

Red Cross

A symbol or icon of humanitarian organizations whose goals are to assist individuals in need and to prevent and relieve suffering. The proposal of the Red Cross as a symbol of humanitarian effort was made in 1863 by the International Committee for Relief to the Wounded (the original name of the International Committee of the Red Cross, which was later adopted in 1876). The first use of the Red Cross was undertaken during a battle in Denmark in 1864, primarily as a symbol of protection for medical personnel in the field, i.e., a white armlet with a red cross attached to it. The American Red Cross organization was founded by Clara Barton in 1881. Its activities include disaster relief; community services that help the needy; support and comfort for military members and their families; the collection, processing, and distribution of lifesaving blood and blood products; educational programs that promote health and safety; and international relief and development programs. The Red Crescent is also used as a similar symbol or icon. The color red is associated with life force and also emergency concern, and the cross is a symbol of faith. See also **Blue Cross**; **Green Cross**.

Registry of Toxic Effects of Chemical Substances (RTECS)

A listing of toxic data obtained from scientific literature maintained by the Department of Health and Human Services.

Regulatory Authority
An agency established by a provincial, federal, or territorial government that has the authority to make or enforce (or both) regulations regarding occupational health and safety.

Relative Risk Analysis
A hazardous material risk that is evaluated in comparison to another risk. The type of risk analysis used should be appropriate for the available data and to the exposure, frequency, and severity of potential loss.

Relative Risk Index Model, Pipeline
A relative risk index model is an analytical model or tool that is used to calculate a numerical score, representing the relative risk of a pipeline segment. This score is calculated based on variables that represent characteristics of the pipeline segment and the perceived importance of these characteristics to the risk of the segment.

Relief Valve
An automatic pressure-relieving device actuated by the pressure upstream of the valve and characterized by opening pop action with further increase in lift with an increase in pressure over popping pressure. See also **Pressure Relief Device**.

Repetitive Strain Injury (RSI)
According to the U.S. Department of Labor, Occupational Safety and Health Administration (OSHA), repetitive strain injuries are the most common and costly occupational health problem, affecting hundreds of thousands of American workers, and costing more than $20 billion a year in workers' compensation. According to the U.S. Bureau of Labor Statistics, nearly two-thirds of all occupational illnesses reported were caused by exposure to repeated trauma to workers' upper body (the wrist, elbow, or shoulder). One common example of such an injury is carpal tunnel syndrome.

Replacement Air
The volume of outdoor air that is delivered to a building in a controlled manner to assist in control of contaminants and to replace air being exhausted.

Replacement Cost
The cost of replacing property without a reduction for depreciation. By this method of value determination, damages for an insurance claim would be the amount needed to replace the property using new materials. Replacement cost refers to the amount that an entity would have to pay to replace an asset at the present time. This may not be the "market value" of the item. Replacement cost coverage is designed so the policyholder will not have to spend more money to get a similar new item and so that the insurance company does not pay for intangibles. May also be called replacement value.

R

Residual Risk
The risk remaining after preventive measures have been taken. Although preventive measures may have been implemented and effective, a residual risk will also be present if an operation or a facility continues to exist.

Resource Conservation and Recovery Act (RCRA)
A statute enacted in 1976 to regulate the treatment, storage, and disposal of hazardous chemical waste. It is administered by the Environmental Protection Agency (EPA).

Respirable-Sized Particulates
Particulates (solid or liquid matter such as dust, fog, fume, mist, smoke, or sprays) that are of a size that allows them to penetrate deep into the lungs upon inhalation.

Respirator
A protective device for the human respiratory system designed to protect the wearer from inhalation of harmful air contaminants. There are two types of respiratory protective devices: (a) air purifiers, which remove the contaminants from the air

by filtering or chemical absorption before inhalation, and (b) air suppliers, which provide clean air from an outside source or breathing air from a tank. Respirators should only be used as a "last line of defense" when engineering control systems are not feasible. Engineering control systems, such as adequate ventilation or scrubbing of contaminants, should be used to negate the need for respirators. The National Institute for Occupational Safety and Health (NIOSH) issues recommendations for respirator use. Industrial-type approvals are in accordance to the NIOSH federal respiratory regulations are stated in 42 CFR Part 84.

Respiratory Diseases

Disease conditions due to toxic agents in the respiratory tract, e.g., pneumonitis, bronchitis, pharyngitis, rhinitis, or acute congestion due to chemicals, dusts, gases, fumes, or infectious agents.

Respiratory Irritants

Irritants affecting the respiratory tract, e.g., dusts, vapors, gases.

Respiratory Protection Program

Management programs designed to ensure employee respiratory protection as required by Occupational Safety and Health Administration (OSHA) regulation 1910.134 and National Fire Protection Association (NFPA) Standard 1500.

Respiratory Protective Equipment

Protective devices for the human respiratory system designed to protect workers from over-exposure by inhalation of air contaminants (air-purifying and air-supplied) and oxygen deficiency (air-supplied). See also **Respirator**.

Response

The activities that address the short-term direct effects of an incident. It includes the immediate actions to save lives and protect property. See also **Recovery**.

Rest Allowances

Recovery time, including regularly scheduled work breaks, usually provided in jobs where heavy physical work or exposure to environmental extremes occurs. Rest allowances are built into the job standard so that productivity ratings recognize the need for additional recovery time in these jobs. See also **Work Recovery Cycles**.

Restraint Systems

An assembly of components and subsystems, including the necessary connectors, used to (a) stabilize and partially support the user at an elevated work location and allow free use of both hands. This type of restraint system is referred to as a work positioning system or, simply, a positioning system; (b) restrict the user's motion so as to prevent reaching a location where a fall hazard exists. This type of restraint system is referred to as a travel restriction system.

Restricted Duty Injury or Illness (RDI)

As defined by Occupational Safety and Health Administration (OSHA) regulations 29 CFR 1904.7(b) (4) (i) (A) and (B), an injury or illness that prevents an employee from performing one or more of the routine functions of his or her job, or from working the full workday that he or she would otherwise have been scheduled to work; or when a physician or other licensed health care professional recommends that the employee not perform one or more of the routine functions of his or her job, or not work the full workday that he or she would otherwise have been scheduled to work. For record-keeping purposes, an employee's routine functions are those work activities the employee regularly performs at least once per week.

Retail Loss Prevention

Field, or known in the past as retail, security.

Retrieval Line
A line or rope secured to an anchor point or lifting device outside a confined space and attached to a full body harness, chest harness, or wristlets worn by an individual entering the space, for the purpose of pulling the individual out of the confined space. See also **Confined Space**.

Right-of-Way
The right of one vehicle or pedestrian to proceed in a lawful manner in preference to another vehicle or pedestrian approaching under such circumstances of direction, speed, and proximity as to give rise to danger of collision unless one grants precedence to the other.

Risk
An assessment (i.e., life loss, property damage, business economic interruption, environmental impact, etc.) of both the probability and consequence of all hazards (e.g., explosion, fire, smoke exposure, toxic vapor releases, etc.) of an activity or condition, i.e., $R = f\{P, C\}$. In the insurance industry, "risk" refers to the person or thing insured. Risks can be reduced in four main ways: Avoidance, Reduction, Retention, and Transfer.

Risk Analysis
The science of risks and their probability and evaluation. In application, a procedure to identify and quantify risks by establishing potential failure modes, providing numerical estimates of the likelihood of an event in a specified time period, and estimating the magnitude of the consequences.

Risk Analysis Methodologies
Various evaluation techniques used in the safety profession to identify hazards, determine their effects, rank their risk, and determine if additional safeguards are required. They include Qualitative Methodologies such as Preliminary Risk/Hazard Analysis (PHA), Hazard and Operability studies (HAZOP), and Failure Mode and Effects Analysis (FMEA/FMECA); Tree-Based Techniques, such as Fault Tree Analysis (FTA), Event Tree Analysis (ETA), Cause-Consequence Analysis (CCA), Management Oversight Risk Tree (MORT), and Safety Management Organization Review Technique (SMORT, a simplified MORT analysis); and Techniques for Dynamic systems such as Go Method (used in electronics reliability), Digraph/Fault Graph, Markov Modeling, Dynamic Event Logic Analytical Methodology (DYLAM), and Dynamic Event Tree Analysis Method (DETAM).

Risk Analyst
An individual who analyzes historical loss data, prepares loss models, estimates potential losses, and investigates and applies emerging research methodologies for preparing loss predictions. Primarily used in the insurance industry.

Risk and Insurance Management Society, Inc.® (RIMS)
A national nonprofit organization that provides ongoing education, training, and information to those in the risk management field. It was founded in the 1950s and represents approximately 3500 industrial, service, nonprofit, charitable, and governmental entities. There are also over 10,000 risk management professional members. Risk and Insurance Management Society, Inc. is registered with the U.S. Patent and Trademark Office.

Risk Assessment
The amount or degree of potential hazard perceived by a given set of parameters and operating conditions. Risk assessment provides a means to identify and rank risks and to obtain information on their extent and nature. By performing a risk assessment suitable risk control measures can be identified and assessed for adequacy. In

addition, it is an effective means of prioritizing actions. Risk assessment involves identification of hazards, probability of occurrence, severity, and subsequent consequences. Risk assessment may be either qualitative or quantitative in nature. It may also be called a probabilistic risk assessment (PRA) where probabilities of occurrence are used extensively in the analysis. See also **Qualitative Risk Assessment**; **Quantitative Risk Assessment (QRA)**.

Risk Avoidance

The philosophy of risk management that an entity will not enter into an operation that poses an exposure of loss or will eliminate the exposure to a potential loss. An example is the replacement of a manual handling operation by a mechanical handling system. Primarily used in the risk management practices for corporate cost benefits in the analysis of insurance applications.

Risk-Based Inspection (RBI)

A frequency of inspection based on an evaluation of the hazard versus severity, i.e., the risk the facility represents, whereby high-risk facilities would be inspected more frequently than low-risk facilities. This focuses resources more beneficially to the appropriate potential high-risk facilities to improve safety.

Risk Control

The provision of suitable measures or elements to eliminate or control real or potential hazards. Risk control may be done through risk avoidance, risk retention, risk transfer, or risk reduction measures. See also **Risk Avoidance**; **Risk Management**; **Risk Reduction**; **Risk Retention**; **Risk Transfer**.

Risk Engineer

An individual qualified to identify potential hazards and consequences of the hazards, assess the probability of those hazards occurring, and determine appropriate hazard protection measures based on a cost-benefit approach and regulatory requirements.

Risk Factor

A characteristic or agent whose presence increases the probability of occurrence of a disease or injury.

Risk Graph

A qualitative and category-based method of safety integrity level (SIL) assignment. Risk graph analysis uses four parameters to make a SIL selection: consequence, occupancy, probability of avoiding the hazard, and demand rate. Each of these parameters is assigned a category and a SIL is associated with each combination of categories. In some cases, quantitative tools, such as Layers of Protection Analysis (LOPA), are used to assist the analyst in determining which category to use, but typically the assignment is done qualitatively. Using the selected categories, the analyst follows the resulting path that leads to the associated SIL assignment.

Risk Insurance

The provision of financial reimbursement through an agency for an economic loss for specific conditions and periods for the payment of premium (annual fee based on the hazard of the risk involved).

Risk Integral

A summation of risk as expressed by the product of consequence and frequency. The integral is summed over all of the potential unwanted events that can occur. If calculating the risk integral for loss of life, the consequence of concern and thus the units of the integral are fatalities. It is useful in combination with event trees to determine a total value of risk for a group of related incidents.

Risk Life and Limb

A cliché meaning taking a grave chance or embarking on a dangerous enterprise. At some time in the past the saying was "risk limb and member." In Thomas Burton's Diary (U.K. Member of Parliament during 1653–1659) for 1658, it is stated, "It is not enough to serve in those offices, unless they venture life and member."

Risk Management

The systematic application of policies, practices, and resources to the assessment and control of risk affecting human health and safety and the environment, which is also combined with economic, political, legal, and ethical considerations to make decisions. Hazard, risk, and cost-benefit analysis are used to support development of risk reduction options, program objectives, and prioritization of issues and resources.

Risk Management Plan (RMP)

Part of the U.S. requirement under the Occupational Safety and Health Administration (OSHA) guidelines for managing risk when dealing with large quantities of certain materials. The objective of the RMP is to prevent serious chemical incidents that could cause harm to the public and the environment and to reduce the potential impact of a release. An RMP consists of a hazard assessment of chemicals involved in a process at a facility, a prevention program to prevent a release, and an emergency response program of what to do in case of such a release. These programs are summarized in the RMP, which will then be made available to the public and state and local government agencies. Section 112(r) of the amended Clean Air Act (CAA) of 1990 mandates the RMP regulations. This section complements and supports the Emergency Planning and Community Right-to-Know Act (EPCRA) of 1986. It incorporates the Process Safety Management (PSM) standards of the 1970 Occupational Safety and Health Administration (OSHA) Act as defined in 29 CFR 1910.119. The RMP rules are codified in 40 CFR Part 68. Covered facilities were initially required to comply with the rule in 1999.

Risk Management Techniques

Various methods available to handle a hazard, which primarily include risk avoidance, risk reduction, risk control, and risk acceptance.

Risk Matrix

A safety assessment tool used to compare assessed risk, based on the probability and consequences of a potential incident (Table R.2).

R

R1: No further action or safety studies required. Individual personnel judgment required for operation to occur.

R2: Document process safety studies, hazards, and risk-reducing measures. Consider feasibility and the cost/benefit of additional risk-reducing measures. Supervision approval required for the operation.

R3: Document process safety studies, evaluate feasibility of additional risk reducing features, and implement if worker and off-site exposure can be reduced to a lower level. Operating group approval is required for the operation.

R4: Document process safety studies, hazards, and risk-reducing measures. Identify additional risk-reducing measures and implement if worker and off-site exposure can be reduced to a lower level. A quantitative risk analysis is required to assess hazards. Divisional management (company) approval is required for the operation.

R5: Additional process safety studies and risk-reducing measures are mandatory to achieve lower risk. Corporate (parent company) senior management approval is required for the operation.

TABLE R.2
Risk Matrix Example

LIKELIHOOD	L5	R3	R3	R4	R5	R5
	L4	R2	R3	R3	R4	R5
	L3	R2	R2	R3	R4	R4
	L2	R1	R2	R3	R3	R4
	L1	R1	R1	R2	R3	R4
		C1	C2	C3	C4	C5

CONSEQUENCE

In this particular risk-ranking matrix, the risk level is not inversely equal, i.e., C4 and L1 do not carry the same risk as L4 and C1. Generally, the risk is considered higher when the consequences are more severe rather than when frequency is greater.

Risk Rating

A rating of a risk from lowest to highest or vice versa. A rating is usually derived from a risk matrix assessing the risk from frequency or probability (high to low) against its consequences (minor to severe), whereby the function of the two provide a relative ranking of the risk compared to other risks, and the result will indicate the level of action that is required to resolve the risk. See also **Risk Matrix**.

Risk Reduction

The lowering of a loss exposure through the provision of risk avoidance, prevention techniques, loss control measures, or risk financing instruments.

Risk Retention

A risk management strategy where the risk is retained within an organization and any consequent loss is financed by the organization. There are considered to be two types: risk retention with knowledge and risk retention without knowledge. With knowledge a conscious decision is made to meet any resulting loss from within the organization's resources. Decisions on which risk should be retained should only be decided after all the risks have been identified, measured, and evaluated. Without knowledge are those that arise from a lack of knowledge of the existence of a risk or an omission to insure against that risk. Risks that have not been identified and evaluated are a form of risk retention.

Risk Threshold

A level of risk at which society, organizations, employees, or an individual are prepared to accept in terms of risk acceptance, e.g., an individual crossing a road in heavy traffic.

Risk Transfer

Shifting a risk by the means of a two-party contract, typically an insurance contract.

Rollover Protective Structure (ROPS)

In relation to a machine (e.g., farm tractor, material handling equipment, surface mining equipment, construction, forestry, etc.), a structure that protects every operator of the machine who is wearing a seat belt from being crushed if the machine rolls over. Required by various federal regulations for agriculture, mining, construction, etc. For farm tractors, the ROPS must meet American Society of Agricultural Engineers (ASAE) Standard S306.3-1974 titled "Protective Frame for Agricultural Tractors

Test Procedures and Performance Requirements," and Society of Automotive Engineers (SAE) Standard J334-1970, titled "Protective Frame Test Procedures and Performance Requirements." For other applications, SAE J1040, "Performance Criteria for Rollover Protective Structures (ROPS) for Construction, Earthmoving, Forestry, and Mining Machines." See also **Agricultural Safety**.

Root Cause

Considered the basic reason that something occurs and can be reasonably identified, management has the ability to remedy, and for which recommendations for prevention can be determined. This is opposed to causal factors, which are considered the immediate causes of an incident. Root causes are usually deficiencies in safety management systems, but can be any factor that would have prevented the incident if that factor had not occurred. By directing corrective measures at root causes, it is hoped that the likelihood of problem recurrence will be minimized. However, it is recognized that complete prevention of recurrence by a single intervention is not always possible. See also **Causal Factor (CF)**.

Root Cause Analysis (RCA)

A class of problem solving methods aimed at identifying the root causes of problems or events. RCA is often considered to be an iterative process, and is frequently viewed as a tool of continuous improvement. Root cause analysis is not a single, sharply defined methodology; there are many different tools, processes, and philosophies of RCA in existence. Safety-based RCA descends from the fields of incident investigation and occupational safety and health. Root causes tend to be viewed as failed or missing safety barriers, unrecognized risks or hazards, or inadequate safety engineering. See also **Causal Factor (CF)**; **Causal Factor Chart (CFC)**; **Taproot® Investigation**; **Tier Diagramming**.

Root Cause Map™ (RCM)

A safety analysis tool used in root cause identification. It is a decision diagram, which is basically divided into two sections. The left is typically used to identify and categorize causal factors associated with equipment failures (e.g., design input/output, design review verification, equipment records, calibration program, preventive maintenance program, inspection/testing program, administrative management systems). The right side is used to identify and categorize causal factors related to personnel errors (e.g., human factors engineering, procedures, training, immediate supervision, communications, personal performance). See also **Causal Factor (CF)**; **Root Cause Analysis (RCA)**.

Routes of Entry

The pathway by which material may gain access to the body including inhalation, ingestion, and skin or eye contact. May also be called Routes of Exposure.

Rupture Disc

A diaphragm designed to burst at a predetermined pressure differential. A simple rupture disc device that is composed of a non-reclosing pressure relief device that relieves excessive static inlet pressure via a rupture disc.

R

S

Sabotage

Deliberate acts of destruction or obstruction for political advantage, economic harm, or other disruptive action or impact.

Safe

Relatively free from danger, injury, or damage or from the risk of danger. See also **Physical Safety**; **Psychological Safety**.

Safe and Sound

A cliché meaning out of a dangerous predicament. If one is "safe," one is probably "sound," and vice versa. The saying has appeared since 1529.

Safe Failure

A failure that does not have the potential to put the safety instrumented system in a dangerous or fail-to-function state. The situation when a safety-related system or component fails to perform properly in such a way that it calls for the system to be shut down or the safety instrumented function to activate when there is no hazard present.

Safe Failure Fraction (SFF)

The fraction of the overall failure rate of a device that results in either a safe fault or a diagnosed (detected) unsafe fault. The safe failure fraction includes the detectable dangerous failures when those failures are annunciated and procedures for repair or shutdown are in place.

Safe Holding Distance

A method of safeguarding that protects the operator by requiring the operator to hold the workpiece at a distance from the hazard area such that the operator cannot reach the hazard portion of the machine cycle.

Safe Opening Safeguarding

A method of safeguarding that limits access to the hazard area by the size of openings, or by closing off access, when the workpiece is in place in the machine.

Safe Position of Controls Safeguarding

A method of safeguarding that requires the operator to be positioned at the machine controls at a distance from the hazard area such that the operator cannot reach the hazard area during the hazardous portion of the machine cycle.

Safe Refuge

A location free of risk from an incident of concern. An area designated as safe refuge has the capability to provide protection against the designated emergency, i.e., fire, explosion, toxic gases, etc. See also **Evacuation**; **Shelter-in-Place**.

Safe State

The predetermined safe position of the process equipment device under control, as determined by operational experience, a preliminary hazards analysis, or formal Hazard and Operability (HAZOP) study. Unless otherwise specified, the safe-state is "de-energized," i.e., without power, pneumatic, or hydraulic supply. It is also the state of the process after acting to remove the hazard resulting in no significant harm.

Safe Working Load (SWL)

Generally taken as the maximum load that an item of equipment (e.g., crane) may raise, lower, or suspend under particular service conditions and arrangements. Each individual type of equipment has its own specific safe working load limits.

Safe Workplace

One in which the likelihood of all identifiable undesired events are maintained at an acceptable level.

Safeguard

A precautionary measure or stipulation. Usually equipment or procedures designed to interfere with an incident propagation to prevent or reduce incident consequences.

Safety

Freedom from incidents that result in injury, damage or harm, i.e., an acceptable level of risk (low probability of harm) from an individual or society level of perspective.

Safety Alert Symbol

A symbol used on a safety sign that indicates a potential personal injury hazard as defined in American National Standards Institute (ANSI) Z535.4. It is composed of an equilateral triangle surrounding an exclamation mark. The color of the sign signifies its hazard level: red for danger, orange for warning, and yellow for caution. The safety alert symbol is not to be used to alert persons to property-damage-only incidents. See also **ANSI Z535.4, Product Safety Signs and Labels**; **Color Coding**; **Safety Sign**; **Signal Word**.

Safety and Health Achievement Recognition Program (SHARP)

An Occupational Safety and Health Administration (OSHA) cooperative voluntary program that recognizes small employers who operate an exemplary safety and health management system. Acceptance into SHARP by OSHA is an achievement of status that demonstrates the business as a peer model for worksite safety and health. Upon receiving SHARP recognition, the business worksite is exempt from OSHA programmed inspections during the period that the SHARP certification is valid. To achieve SHARP status a small employer has to:

Request a consultation visit that involves a complete hazard identification survey. Involve employees in the consultation process.

Correct all hazards identified by the consultant.

Implement and maintain a safety and health management system that, at a minimum, addresses OSHA's 1989 Safety and Health Program Days Away, Restricted, or Transferred Management Guidelines.

Lower the company's (DART) rate and Total Recordable Case (TRC) rate below the national average.

Agree to notify the state Consultation Project Office prior to making any changes in the working conditions or introducing new hazards into the workplace.

Safety and Health Information Bulletins (SHIBs)

A communication tool the Occupational Safety and Health Administration (OSHA) uses to inform internal staff and the public of significant occupational safety and health issues concerning hazard recognition, evaluation, and control in the workplace and at emergency response sites. SHIBs replaced the OSHA Hazard Information Bulletins (HIBs) and Technical Information Bulletins (TIBs), which provided similar information.

Safety at Sea Act

Federal regulations, i.e., Title 15, Subpart H, CFR 970.205, 970.800, 970.801, regarding assuring the safety of life and property at sea. It addresses vessels documented in the United States and used in activities authorized under an issued license to comply with conditions regarding the design, construction, alteration, repair, equipment, operation, manning, and maintenance relating to vessel and crew safety and the safety of life and property at sea. See also **Safety of Life at Sea, SOLAS Regulations**.

Safety Audit

See **Audit, Safety**.

Safety Availability

The fraction of time that a safety system is able to perform its designated function when the process is operating. The safety system is unavailable when it has failed dangerously or is in bypass. Safety availability is equal to 1 minus the "Probability of Failure on Demand" of the safety function. See also **Probability of Failure on Demand (PFD)**.

Safety Award

An item provided to a worker, group, or organization to recognize its safety performance in accordance with an organization's goals. The award helps promote prevention of workplace injuries and illness and motivate employees to improve their safety performance. See also **Safety Recognition Program**.

Safety Belt (Waist Belt)

A robust and secure belt worn by an individual (e.g., telephone line worker, window washers, construction worker, etc.) attached to a secure object (telephone pole, window sill, anchor point, etc.) via a safety lanyard, to prevent injury due to falling. They are intended for use where mobility can be limited, and where the combined effects of the anchorage point position and length of the lanyard limits the potential drop of the individual in case the individual falls. Also can refer to a seat or torso belt securing a passenger in an automobile or airplane to provide body protection during a collision, sudden stop, air turbulence, etc.

Safety Cage

See **Cage, Ladder Safety**; **Ladder Safety Device**.

Safety Can

A small, approved metallic can of not more than 18.9 liters (5 gallons) capacity equipped with a spring closing lid, which will safely relieve pressures when exposed to a fire (Figure S.1).

Safety Case

A formal examination of methods to be adopted to reduce the risk of an incident. It is often used in high potential risk applications, i.e., petroleum installations, nuclear facilities, etc.

Safety Chain (Towing)

Chains that are utilized to prevent a trailer from separating from its towing vehicle in event of the hitch's failure. The chains should be arranged so they are crossed in an "X" fashion below the ball mount, with enough slack that they do not restrict turning or allow the coupler to hit the ground. This prevents the trailer tongue from dropping to the ground if the connection becomes undone. Safety chains should be rated to equal or greater than twice the maximum gross trailer weight rating.

Safety Checklist

A checklist of safety precautions or of hazards that is used as reminder for inspections, work operations, or other similar reviews (Table S.1).

S

FIGURE S.1 Safety gas can.

Safety Committee
A team of employees formed to advise and promote the safety and health of the workplace. Members should include representative employees and management from all sections of an organization.

Safety Communication
The collective means by which safety information is disseminated to employees, including the classroom, departmental safety meetings, and written communications such as posters, newsletters, and postings of regulatory agency inspection findings.

Safety Competition
A safety promotion and motivation program to enhance an organization's safety performance by organizing a competition among various groups or departments for the best safety achievements (e.g., incident statistics, proactive safety activities). The winner of the competition is provided with a recognition or award for their efforts.

Safety Consequences
A failure has safety consequences if it causes a loss of function or other damage that could cause injury or result in a fatality.

Safety Coordinator
Individuals within organizations who assist in implementing the workplace safety program in their respective areas.

Safety Coupling
A friction coupling adjusted to slip at a predetermined torque to protect the rest of the system from overload.

Safety Crossing
See **Crossing Guard Safety**; **Railroad Crossings Safety**.

Safety Cuffs
An extended piece of protective shirt material attached by a seam at the wrist. Safety cuffs provide additional protection to the wrist area and slide on and off easily. Safety cuffs are typically made of more rigid material and remain firm even when exposed to perspiration. See also **Gauntlets**.

TABLE S.1
Excavation Safety Checklist

Project/Activity _____ Date _____

Designated Competent Individual _____

Weather: _____   Soil Type: _____   Type of Protective System: _____

Depth: _____   Width: _____   Length: _____

Yes	No	Item to Be Checked
____	___	Excavations and protective system inspected by designated competent person daily, before start of work
____	___	Designated competent person has authority to remove workers from excavation immediately
____	___	Surface encumbrances supported or removed
____	___	Employees protected from loose soil or rock
____	___	Employees provided with hard hats
____	___	Spoils, materials, and equipment set back a minimum of 2 feet from the edge of the excavation
____	___	Barriers provided at all remote excavations, wells, pits, etc.
____	___	Reflective vest or other highly visible PPE worn by employees exposed to vehicular traffic
____	___	Employees prohibited from working or walking under suspended loads
____	___	Employees prohibited from working on faces of sloped or benched excavations above other individuals
____	___	Warning system established and used when mobile equipment is operating near edge of excavation
____	___	Utility companies contacted and/or utilities located
____	___	Exact location of utilities marked near or at excavation
____	___	Underground installations protected, removed, or supported when excavation is open
____	___	Precautions taken to protect workers from water accumulation
____	___	Water removal equipment monitored by designated competent person
____	___	Surface water diverted or controlled
____	___	Inspection conducted after each rainstorm
____	___	Atmosphere tested when there is a possibility of oxygen deficiency or buildup of hazardous gases
____	___	Oxygen content between 19.5 and 21 percent
____	___	Ventilation provided to prevent flammable gas buildup
____	___	Testing undertaken to ensure atmosphere remains safe
____	___	Emergency response equipment readily available
____	___	Employees trained in use of PPE and emergency response equipment
____	___	Safety harness and lifeline used for deep excavations

S

Safety Culture

The collective individual and group values, attitudes, competencies, and patterns of behavior that determine the commitment to, and style and proficiency of an organization's health and safety programs.

Safety Cut-Out

An overload protective device in an electric circuit.

Safety Dashboard

Typically an intranet Web page of an organization, which features various safety statistics (e.g., leading and lagging indicators). These are arranged in a dial or graph format, in which real-time statistics (i.e., weekly or monthly reports) are compared to its stated safety targets or goals, providing a performance rating similar to a vehicle operating display, hence the reference to a vehicle dashboard.

Safety Device

Any kind of device, item, or system that is used in or on equipment and that controls or monitors any aspect of the safety of the equipment and includes a safety relief device.

Safety Distance

The minimum distance from each control-actuating device of a two-hand control system to the hazard point such that the operator cannot reach the hazard point with a hand or other body part before cessation of motion of the hazardous portion of the machine cycle. See also **Separation Distance (Safety Light Screen)**.

Safety Education

The transmission of knowledge, skills, attitudes, motivations, etc., concerning the safety requirements of operations, processes, environments, etc., to workers, supervisors, managers, and others. See also **Fire Training**.

Safety Engineer

See **Loss Prevention Engineer (LPE)**.

Safety Engineering

Safety engineering is concerned with the planning, development, improvement, coordination, and evaluation of the safety component of integrated systems of individuals, materials, equipment, and environments to achieve optimum safety effectiveness in terms of both protection of people and protection of property.

Safety Equal

A slogan commonly used in industry to highlight the fact that safety considerations are equal to production concerns. Its use has recently risen, since there have been incidents where production concerns were considered of higher value than plant shutdown, but eventually this led to a higher losses. Another concern was that the slogan Safety First was not being seriously applied due to production demands. See also **Safety First; Safety Pays**.

Safety Factor

See **Factor of Safety (FS); Margin of Safety**.

Safety First

A slogan commonly used in industry to highlight the importance of safety in its operations. See also **Safety Equal; Safety Pays**.

Safety Flowchart

A general flowchart that identifies events that may occur at a facility during an incident. The flowchart can identify possible avenues the event may lead to and the protection measures available to mitigate and protect the facility. It will also highlight deficiencies. The use of a flowchart helps with the understanding of events by personnel unfamiliar with industry risks and safety measures. It portrays a step-by-step scenario that is easy to follow and explain. Preparation of in-depth risk probability analysis can also use the flowchart as a basis of the Event Tree or Failure Mode and Effects Analysis. See also **Brainstorming**; **Hazard and Operability Study (HAZOP); What-If Analysis**.

Safety Function

A specific task or purpose that must be accomplished for safety.

Safety Fuse

A device used to initiate a pyrotechnic device. A safety fuse consists of a black powder core in a textile tube, covered with asphaltum or other waterproofing agent, and having an outer wrapper of tough textile or plastic. They are made in a standard diameter designed to be crimped into blasting caps. Safety fuses are manufactured with specified burn times per 30 cm, e.g., 60 seconds, so that a length of fuse 30 cm long will take 60 seconds to burn. Manufacturers warn that although every effort is made to insure uniform burn times, safety fuses are subject to variation depending on conditions and should be used with adequate safety margins.

Safety Glass

Impact-resistant and shatterproof glass used as eye protection and for automobile windows and large architectural windows and doors. Also includes heat-treated glass that breaks into granules instead of sharp-edged strands. American National Standards Institute (ANSI) Z26.1, Standard for Safety Glazing Materials for Glazing Motor Vehicles and Motor Vehicle Equipment Operating on Land Highways, safety standard is available on the construction of glass for vehicles and also ANSI Z97.1, Safety Glazing Materials Used in Buildings—Safety Performance Specifications and Methods of Test, for buildings. See also **Laminated Glass**; **Tempered Glass, Window Film, Safety and Security**.

Safety Glasses

Protective eyewear from impacts that may have side shields. They may also be called Safety Spectacles. They are required to meet American National Standards Institute (ANSI) Z87 and be clearly marked with the manufacturer's name. See also **ANSI Z87.1-2003, Standard for Occupational and Educational Eye and Face Protection Devices**.

Safety Gloves

Gloves used to protect the hands of individuals from physical injury (cuts, abrasions, temperature extremes, electrical shock, etc.), chemical hazards, biological hazards, and radioactive hazards.

Safety Handbook

Generally a publication that provides information on an organization's safety policies, responsibilities, rules, procedures, and guidelines for its employees.

Safety, Health, and Environment (SHE or HSE)

Common grouping of disciplines in business and industry due to the overlapping responsibilities and often rearranged as HSE. Sometimes also combined with quality assurance (QHSE).

Safety Helmet

Rigid headgear of varying materials designed to protect the head, not only from impact, but from flying particles and electric shock, or any combination of the three. Safety helmets should meet the requirements of American National Standards Institute (ANSI) Standard Z89.1, Protective Headware for Industrial Workers. See also **ANSI Z89.1, Protective Headware for Industrial Workers**.

Safety in Chemical Engineering Education (SACHE) Program

A university curriculum program established in 1992 as a cooperative arrangement between the Center for Chemical Process Safety and engineering schools to provide

S

teaching materials and programs that bring elements of process safety into undergraduate and graduate chemical and biochemical product and process classes. Its aim is to ultimately provide safer operating processes and practices. The organization also issues a process safety message to plant operators and other manufacturing personnel via a monthly one-page publication titled the Process Safety Beacon. It covers the breadth of process safety issues. Each issue presents a real-life incident and describes the lessons learned and practical means to prevent a similar incident in a process facility.

Safety Inspection

A systematic assessment of safety standards and policies of a plant, place of work, ongoing operations, or equipment to determine if looking for unsafe acts and conditions exists.

Safety Instrumented Function (SIF)

A set of equipment intended to reduce the risk due to a specific hazard (a safety loop). Its purpose is to (1) automatically take an industrial process to a safe state when specified conditions are violated; (2) permit a process to move forward in a safe manner when specified conditions allow (permissive functions); (3) take action to mitigate the consequences of an industrial hazard. It includes elements that detect when an incident is imminent, decide to take action, and then carry out the action needed to bring the process to a safe state. Its ability to detect, decide, and act is designated by the safety integrity level (SIL) of the function. See also **Safety Integrity Level (SIL)**.

Safety Instrumented System (SIS)

Implementation of one or more Safety Instrumented Functions. A SIS is composed of any combination of sensor(s), logic solver(s), and final element(s). A SIS usually has a number of safety functions with different safety integrity levels (SIL) so it is best to avoid describing it by a single SIL. See also **Layers of Protection Analysis (LOPA); Safety Integrity Level (SIL)**.

Safety Integrity Level (SIL)

A quantitative target for measuring the level of performance needed for safety function to achieve a tolerable risk for a process hazard. It is a measure of safety system performance, in terms of the probability of failure on demand. There are four discreet integrity levels, SIL 1–4. The higher the SIL level, the higher the associated safety level and the lower the probability that a system will fail to perform properly. Defining a target SIL level for a process should be based on the assessment of the likelihood that an incident will occur and the consequences of the incident. Table S.2 describes SIL for different modes of operation.

The level of overall availability for a system component is calculated as 1 minus the sum of the average probability of dangerous failure on demand. SIL-1: availability of 90–99 percent; SIL-2: availability of 99–99.9 percent; SIL-3: availability of 99.9–99.99 percent. See also **Layers of Protection Analysis (LOPA); Safety Instrumented Function (SIF); Safety Instrumented System (SIS)**.

Safety Interlock Switch

A switch used on guard doors that is used to detect if the door is opened while the machine is running, and uses a coded actuator to prevent intentional defeat. Safety interlock switches use positive opening contacts, which ensure that the closed switching contact is forced open when the guard is opened, without reliance upon spring action.

Safety Issues

Deviations from current safety standards or practices, or weaknesses in facility design or practices identified by plant events, with a potential impact on safety because of their impact on defense in depth, safety margins, or safety culture.

TABLE S.2
Typical Industry SIL Level Rates

Low-Demand Mode SIL

SIL PFDavg RRF

$4 \geq 10^{-5}$ to $< 10^{-4} >$ 10,000 to $\leq$ 100,000

$3 \geq 10^{-4}$ to $< 10^{-3} >$ 1000 to $\leq$ 10,000

$2 \geq 10^{-3}$ to $< 10^{-2} >$ 100 to $\leq$ 1000

$1 \geq 10^{-2}$ to $< 10^{-1} >$ 10 to $\leq$ 100

High-Demand or Continuous Mode SIL

SIL PFDavg per hour

$4 \geq 10^{-9}$ to $< 10^{-8}$

$3 \geq 10^{-8}$ to $< 10^{-7}$

$2 \geq 10^{-7}$ to $< 10^{-6}$

$1 \geq 10^{-6}$ to $< 10^{-5}$

Safety Jargon

Terminology used in the loss prevention profession that may be unfamiliar to those outside the discipline, e.g., ALARP, HAZOP, SIL Analysis, Layers of Protection Analysis, etc.

Safety KPIs

See **Key Performance Indicator (KPI), Safety; Lagging Indicator; Leading Indicator**.

Safety Lamp or Davy Lamp

A lamp used in areas of combustible gas that prevents ignition of the gas by the provision of a wire screen, which encloses the lamp flame. The wire screen absorbs the heat of the (oil lamp) light source before it can contact a gas, thereby preventing its ignition. Invented by the British chemist Sir Humphry Davy (1778–1829) in 1815 for use by coal miners where firedamp was present. George Stephenson (1781–1848), a British inventor and engineer, also independently invented a similar miner's safety lamp at about the same time but shared credit for this invention with Sir Humphry Davy.

Safety Lanyard

See **Lanyard; Lifeline**.

Safety Layers

Passive systems, automatically or manually initiated safety systems, or administrative controls that are provided to ensure that the required safety functions are achieved. Often expressed as (a) hardware, i.e., passive and active safety systems; (b) software, including personnel and procedures as well as computer software; (c) management control, particularly preventing defense in depth degradation (through quality assurance, preventive maintenance, surveillance testing, etc.) and reacting appropriately to experience feedback from degradations that do occur, e.g., determining root causes and taking corrective actions.

Safety Lifecycle

The procedures to first analyze the situation and document the safety requirements (Analysis Phases). Then, translate these requirements into a documented safety system design, using appropriate software and hardware subsystems and design methodology (Realization Phases). Next, evaluate the system against the required integrity and reliability specifications and modify it as needed. Finally, operate and maintain the system according to accepted procedures (Operation Phases), and document the results to ensure that performance standards are maintained throughout the system's life.

Safety Lock

A lock that can be opened only by its own key. Often used to lock out the electrical energized sources used in equipment or machinery operation.

Safety Management Organization Review Technique (SMORT)

Safety management organization review technique (SMORT) is a simplified modification of MORT developed in Scandinavia. This technique is structured by means of analysis levels with associated checklists, while MORT is based on a comprehensive tree structure. Owing to its structured analytical process, SMORT is classified as one of the tree-based methodologies. The SMORT analysis includes data collection based on the checklists and their associated questions, in addition to evaluation of results. The information can be collected from interviews, studies of documents, and investigations. This technique can be used to perform detailed investigation of incidents and near misses. It also served well as a method for safety audits and planning of safety measures. See also **Management Oversight Risk Tree (MORT)**.

Safety Management System (SMS)

Management of safety in order to promote a strong safety culture and achieve high standards of safety performance. Similar to process safety management attributes. See also **Process Safety Management (PSM)**.

Safety Manual

Document required for equipment certified in accordance with IEC 61508 that describes the conditions of use for that equipment in safety applications. It typically includes equipment usage requirements or restrictions, environmental limits, optional settings, failure rate data, useful life data, common cause beta estimate, and inspection and test procedures.

Safety Measure

Any action that might be taken, condition that might be applied, or procedure that might be followed to fulfill the basic requirements of safety.

Safety Meeting

A periodic meeting held by employers to communicate and evaluate safety and health issues brought up by employees.

Safety Monitoring

Periodic checks on the implementation of an organization's safety standards and procedures.

Safety Net System

A method to protect workers that are exposed to falls of more than 182.9 cm (6 ft), primarily used on construction sites. It consists of a net arranged underneath a fall hazard to catch an individual to prevent injury. Occupational Safety and Health Administration (OSHA) regulation 1926.402 (c) specifies the arrangements and testing required for safety net systems.

Safety Newsletter

Typically a monthly publication used to communicate safety information to individuals of an organization. It may summarize the leading and lagging safety indicators of the organization, provide articles of safety interest, and highlight recent standards or regulatory changes that are pertinent to the operation.

Safety of Life at Sea, SOLAS Regulations

The minimum international standard for the construction, equipment, and operation of ships, compatible with their safety. The SOLAS Convention is generally regarded worldwide as the most important of all international treaties concerning the safety of merchant ships. The first version was adopted in 1914, in response to the Titanic disaster. Since 1948, it is maintained and issued by the International Marine Organization. It applies only to ships engaged on international voyages or passenger vessels having berth or stateroom accommodations for 50 or more passengers. See also **Maritime Safety Standards**; **Safety at Sea Act**.

Safety Officer

See **Safety Representative**.

Safety Orientation

Familiarization and training provided to individuals to communicate potential safety hazards and expected safe behavior. It is usually required as a new hire employment process, reassignment of an employee, or visiting requirements to high hazard environments. Subjects covered may include safety and health policy statement, dress code, housekeeping requirements, communication of hazards, provision and use of personal protective equipment, emergency procedures (fires, spills, etc.), reporting procedures for unsafe acts or procedures, incident reporting procedures, near miss reports, incident investigation procedures, lockout/tagout procedures, machine guarding, electrical safety requirements, confined space entries, medical support, ergonomic safety, fire prevention, required safety training courses or certifications, and individual safety responsibilities and organizational safety support.

Safety Pays

A slogan used to justify the utilization of safety procedures, controls, or equipment. Its basis is that the potential cost of injuries and damages outweighs the cost of safety improvements, and therefore it is beneficial from a business perspective. See also **Cost-Benefit Analysis**.

Safety Perception Survey

A safety examination tool that measures the attitudes, perceptions, and motivation of employees towards safety that influence its safety culture and safety behaviors. It typically consists of a specifically constructed set of questions that are answered with a ranking scale, that are submitted to a sample of the workforce, the results of which are analyzed for possible safety improvements.

Safety Pin

A pin with a covered point. It is usually constructed as a loop-shaped pin that fastens into itself with its point under a protective cover to prevent opening or injury. Can also be referred to as a pin that prevents detonation for explosives, i.e., a pin when properly inserted prevents inadvertent or premature detonation, e.g., in a grenade.

Safety Policy

See **Policy, Safety**.

Safety Poster

A communication and motivation tool used to easily highlight by a graphic means a safety message, which can be placed at a highly observable location for individuals to see.

Safety Procedure

A set of instructions designed for the protection of personnel. Typical safety procedures include incident reporting and investigation, first aid or medical assistance arrangements, waste disposal, emergency response plan instructions, work permits, confined space entries, personal protection requirements, etc.

Safety Professional

An individual who, by virtue of specialized knowledge and skill, training, and educational accomplishments, identifies hazards and develops appropriate controls for these hazards, that when effectively implemented, prevent occupational injury, illness, and property damage. These individuals may also have been awarded or earned the status of Certified Safety Professional by the Board of Certified Safety Professionals. See also **Loss Prevention Engineer (LPE)**; **Safety Professional**; **Safety Representative**.

Safety Program

All of an organization's written safety policies, standards, and procedures applicable to its operations. This typically includes corporate safety policy, management and employee responsibilities, safety rules, incident management, training, records, audits and inspections, disciplinary policy, incentive or motivation, and safety communication and committees.

Safety Recognition Program

A program to promote on-the-job and off-the-job safety awareness by recognizing individuals and group safety achievements that achieve an organization's safety goals. See also **Safety Award**.

Safety Representative

An individual that is charged with monitoring and assessing safety hazards or unsafe conditions, and developing and advising measures to prevent an incident from occurring. They may also be called Safety Officers. See also **Loss Prevention Engineer (LPE)**.

Safety Requirements Specification

A specification containing all the requirements of the safety functions that have to be performed by the safety-related system. It includes both what the functions must do and also how well they must do it. It is often a contractual document between companies and is one of the most important documents in the safety life cycle process.

Safety Rule

A directive stating minimum safeguarding requirements, procedures, personal protective equipment, and behavior for work activities to avoid injury and damage.

Safety Sampling

A systematic sampling of particular dangerous activities, processes, or areas.

Safety Shoe

Personal protective equipment used to protect the foot from a variety of hazards. They are required to meet certain industry performance standards to be acceptable as protective equipment for individuals. See also **ASTM F 2412, Test Methods for Foot Protection, and ASTM F 2413, Standard Specification for Performance Requirements for Foot Protection**.

Safety Shower

An emergency system designed to provide immediate water application to an individual's face or body who has been in contact with hazardous chemicals, chemical compounds, or fire. They can be used in four basic ways:

Dilution: The water reduces the concentration of chemical on the skin to an acceptable level.

Cooling or warming: Water warms or cools the body due to chemical reaction exposure, which has caused a temperature hazard.

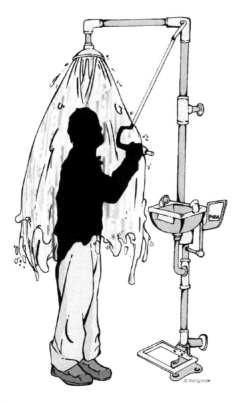

FIGURE S.2   Safety shower.

Irrigation: The water flushes the chemical away.
Extinguishment: The water can extinguish a clothing fire.

A safety shower provides cascading water over the entire body and face (Figure S.2). It is not meant for flushing the eyes with water, which should be provided by an eye wash facility. It is a requirement of Occupational Safety and Health Administration (OSHA) regulation 29 CFR 1910.151 (c) where individuals may be exposed to chemical hazards. See also **ANSI Z358.1, Standard for Emergency Eye Wash and Shower Equipment**; **Damping Shower**; **Eye Wash, Emergency**.

Safety Sign

A visual alerting device in the form of a sign, label, decal, placard, or other marking that advises the observer of the hazard(s) that can cause an incident and the level of hazard seriousness. It may also provide other directions to eliminate or reduce the hazard and may advise of the probable consequences of not avoiding the hazard. American National Standards Institute (ANSI) Z535.2 and ANSI Z535.3 provide guidance in the design, color, wording, and arrangement of safety signs (Figure S.3). See also **Critical Confusion**; **Safety Alert Symbol**; **Safety Symbol**.

Safety Standard

A document issued by a recognized authority that contains specifications for the avoidance of hazards. In the Occupational Safety and Health (OSH) Act two types of safety standards are recognized: an "established federal standard" and a "national

FIGURE S.3   Safety sign.

consensus standard." The term established federal standard means any operative occupational safety and health standard established by any agency of the United States and presently in effect, or contained in any Act of Congress in force on the date of enactment of the OSH Act. The term "national consensus standard" means any occupational safety and health standard or modification thereof which has been adopted and promulgated by a nationally recognized standards-producing organization under procedures whereby it can be determined by the Occupational Safety and Health Administration (OSHA) Secretary that persons interested and affected by the scope or provisions of the standard have reached substantial agreement on its adoption, was formulated in a manner which afforded an opportunity for diverse views to be considered, and has been designated as such a standard by the Secretary, after consultation with other appropriate federal agencies.

Safety Survey

See **Safety Inspection**.

Safety Symbol

A graphic representation that portrays a hazard or concern without the use of words. It may represent a hazard, hazardous situation, precaution to avoid a hazard, the result of not avoiding the hazard, or any combination of these messages. American National Standards Institute (ANSI) Z535.3, Criteria for Safety Symbols, provides guidance in the design and arrangement of safety symbols (Figure S.4). See also **Safety Sign**.

Safety System

A system important to safety, i.e., provided to ensure the safe shutdown of the equipment or otherwise control an operation or to limit the consequences of anticipated operational occurrences and design basis incidents. Safety systems consist of the protection system, the safety actuation systems, and the safety system support features. Components of safety systems may be provided solely to perform safety functions, or may perform safety functions in some plant operational states and non-safety functions in other operational states. Safety system support features are considered the collection of equipment that provides services such as cooling, lubrication, and energy supply required by the protection system and the safety actuation systems.

FIGURE S.4   Safety symbol sign.

Safety System Settings

The levels at which protective devices are automatically actuated in the event of anticipated operational occurrences or incident conditions, to prevent safety limits from being exceeded.

Safety Talk

A communication tool used to inform workers about health and safety requirements for the tools, equipment, materials, and procedures they use every day, for a particular job, or how to deal with specific problems on site. They do not replace formal training. A safety talk is usually five minutes in duration. See also **Safety Meeting**; **Tailgate Safety Meeting**; **Toolbox Safety Meetings**.

Safety Tapes

See **Barrier Tape**.

Safety Task

The sensing of one or more variables indicative of a specific postulated initiating event, the signal processing, the initiation and completion of the safety actions required to prevent the limits specified in the design basis from being exceeded, and the initiation and completion of certain services from the safety system support features.

Safety through Design

The integration of hazard analysis and risk assessment in the early design phases of a project and the implementation of the necessary actions to achieve an acceptable level of risk for injury or damages. Typically this entails facilities, hardware, equipment, tools, materials, layout, configurations, energy controls, environmental aspects, and products.

Safety Tongs

See **Tongs, Safety**.

Safety Tour

See **Safety Inspection**.

Safety Training

A site-specific training program developed on the basis of a needs assessment concerning the hazards that may be encountered at the specific location and emergency actions that may be required. Safety training requirements are identified

Safety Vest   252   Scaffold Safety

in Occupational Safety and Health Administration (OSHA) regulations 29 CFR 1910.120 (e); 29 CFR 1910.120 (p) (7), (p) (8) (iii); and 29 CFR 1910.120 (q) (6), (q) (7), and (q) (8).

Safety Vest

A high-visibility vest worn by individuals who are working in an environment where they are at risk to be injured by passing motorists. These vests are fluorescent in color, drawing attention to the otherwise unnoticeable person. Safety vests fall into one of four categories: Class 2, Class 3, Public Safety, and Economy vests. Class 2 safety vests are worn by individuals who work in areas of high traffic, dangerous weather conditions, and complex backgrounds. These people usually work in emergency response, law enforcement, construction, and utility fields. Class 2 safety vests are fluorescent lime, yellow, or orange, and are worn over the top of shirts or uniforms. There are several styles of Class 2 safety vests, including adjustable, mesh, illuminated, and flame retardant. Usually the safety vest is imprinted with the department or company name for identification purposes. Class 3 safety vests are used by individuals who are working in the dark and in areas of higher traffic or extreme weather conditions. These vests are full, half, or short sleeves, and fit like a shirt. This draws more attention to the wearer, providing an extra level of safety. They are available in fluorescent orange, yellow, or lime, and can be imprinted with the department or company name. This variety of vest is available in mesh, rubber, non-mesh, jacket, overcoat, and sweatshirt styles. Public safety vests are worn by police, paramedics, Department of Transportation (DOT) officials, and firefighters when high visibility is a necessity. Individuals are normally assigned a color depending upon their jobs. Police wear blue, paramedics wear green, DOT officials are assigned orange, and firefighters wear red, making it easier to distinguish the public safety personnel and their duties. Economy vests are the safety vests that are provided during special events. These vests are worn by people who are directing traffic or providing security services. These vests are fluorescent orange, yellow, or lime and are worn over the top of the clothing. Most of these vests are left unprinted or they may be printed with generic terms such as "Staff" or "Security," so they may be used interchangeably with other events. These vests may also be worn by motorcyclists who are riding at night, or during bad weather conditions, to increase their visibility. See also **ISEA (ANSI) 107, High Visibility Garment Standard**.

Safety Zone

An area free of the hazard of concern.

SARA Title III

The section of the Superfund Amendment and Reauthorization Act (SARA) that requires industry to develop comprehensive emergency response plans and mandates public disclosure of hazards of materials handled or stored in certain quantities. It requires facilities that store hazardous materials to provide officials and citizens with data on the types (flammables, corrosives, etc.), amounts on hand (daily, yearly), and their specific locations. Facilities are to prepare and submit inventory lists, Material Safety Data Sheets (MSDSs), and tier 1 and 2 inventory forms. It is also known as the Emergency Planning and Community Right-To-Know Act (EPCRA).

Scaffold Safety

In 2008, the Bureau of Labor Statistics' Census of Fatal Occupational Injuries (CFOI) reported that 88 fatalities occurred in the year 2007 from scaffolds and staging. In a Bureau of Labor and Statistics study, 72 percent of workers injured in scaffold accidents attributed the accident either to the planking or support giving way, or to the employee slipping or being struck by a falling object. Other concerns contributing to falls from scaffolds include improper maintenance or erection/dismantling procedures, incorrect methods for mounting or dismounting, overloading, absence of guardrails, scaffold component failures, defective personal protective equipment (PPE), or absence or improper use of PPE. The Occupational Safety and Health Administration (OSHA) general industry safety requirements for scaffolds are contained in 29 CFR 1910.28. There are also specific scaffold safety standards for shipyard employment, marine terminals, and longshoring.

Scald

A burn injury that may be caused by the direct exposure to hot liquid, steam, or a hot gas. It is estimated that hot water scalds account for 20 percent of all burns. Approximately 2000 children in the United States are treated for scald burns each year. A contributing factor is that children have not fully developed the thickness of their skin; therefore, they burn faster than adults. Scalding incidents occur most frequently in bathrooms and kitchens. The International Association of Plumbing and Mechanical Officials have amended the Uniform Plumbing Code to require anti-scald protection. Some manufactures offer an anti-scald thermostatic mixing valve to be installed in the hot water supply line to prevent inadvertent scalding water from reaching end-point devices in domestic water installations.

Scenario

A postulated or assumed set of conditions or events. Most commonly used in an analysis or assessment to represent possible future conditions and/or events to be modeled, such as possible incidents and their effects on the surrounding environment. A scenario may represent the conditions at a single point in time or a single event, or a time history of conditions and/or events (including processes). See also **Credible Scenario**.

Schedule Rating

A type of rating assigned under the Industrial Compensation Rating Schedule, as approved by the insurance commissioner, by which the Basic Manual Rate is modified to fit the physical conditions related to guarding of machines. It is also affected by the compliance of a safety organization to prescribe insurance standards. It has not been used in most states recently.

Scheduled Charge

The specific charge (in days) assigned to a permanent partial, permanent total, or fatal injury.

School Bus Safety

School buses are considered the safest form of motor vehicle transport per the National Highway Traffic Safety Administration (NHTSA). The greatest risk to school bus passengers is when they enter and leave the bus. Other concerns are driver training, illegal passing of school buses by motorists, and safety restraints for pre-school aged children traveling on school buses (Figure S.5).

School Crossing Safety

A location on a street designed for school children to cross safely, which is assisted by specific traffic regulations and typically school crossing guards. See also **Crossing Guard Safety**.

Screening

A type of analysis aimed at eliminating from further consideration factors that are less significant for protection or safety in order to concentrate on the more significant factors. This is typically achieved by consideration of very pessimistic hypothetical scenarios. Screening is usually conducted at an early stage in order to narrow the range of factors needing detailed consideration in an analysis or assessment.

S

FIGURE S.5   School bus safety.

Security Vulnerability Analysis (SVA)

In April 2007, the U.S. Department of Homeland Security (DHS) issued the Chemical Facility Anti-Terrorism Standard (CFATS). The DHS is to identify, assess, and ensure effective security at high-risk chemical facilities. Included in this standard is the requirement for facilities handling chemicals above a threshold amount, to submit an SVA for DHS review and approval along with a Site Security Plan (SSP). An SVA evaluates risk from deliberate acts that could result in major incidents. It is performed in a systematic and methodical manner to analyze potential threats and evaluates these threats against plant vulnerabilities. From this analysis, it determines possible consequences and whether safeguards to prevent or mitigate their occurrence are recommended. See also **Terrorism**.

Segmental Vibration (Hand-Arm Vibration)

Vibration applied to the hand/arms through a tool or piece of equipment. This can cause a reduction in blood flow to the hands/fingers (Raynaud's disease or vibration white finger). Also, it can interfere with sensory receptor feedback leading to increased handgrip force to hold the tool. Further, a strong association has been reported between carpal tunnel syndrome and segmental vibration. See also **Ergonomics; Whole Body Vibration.**

Self-Contained Breathing Apparatus (SCBA)

A respiratory protection device that consists of a supply of respirable air, oxygen, or oxygen-generating material worn by the worker. It is designed for entry into and escape from atmospheres Immediately Dangerous to Life or Health (IDLH) or oxygen deficient. It is one of the highest levels of respiratory protection. Wearers carry the air supply on their back.

Self-Heating Material

A material that when in contact with air and without an energy source, is liable to self-heat. Tested in accordance with paragraph 3.b (1) of Appendix E, 49 CFR 173.124, i.e., if during a 24-hour test period the material spontaneously ignites or exhibits a temperature of 200°C. See also **Pyrophoric Material; Spontaneous Combustible Material.**

Self-Insurance

A term used to describe the assumption of one's own financial risk.

Sensitivity Analysis

An analysis in which the relative importance of one or more of the variables thought to have an influence on the phenomenon under consideration is determined.

Sensitizer

A chemical that causes humans and animals to develop an allergic reaction in normal tissue after repeated exposure or contact to the chemical.

Separation Distance (Safety Light Screen)

The minimum distance from the midpoint of the defined area to the nearest hazard point that is required to allow the hazardous motion to come to a complete stop before a hand (or other object) can reach the nearest hazard point. Factors that influence the minimum separation distance include the machine stop time, the light screen system response time, and the light screen minimum object detection size.

Serious Injury

A classification for a work injury, which includes (1) all disabling work injuries; (2) nondisabling injuries in the following categories: (a) eye injuries from work-produced objects, corrosive materials, radiation, burns, etc., requiring treatment by a physician, (b) fractures, (c) any work injury that requires hospitalization for observation, (d) loss of consciousness (work related), and (e) any other work injury (such as abrasion, physical or chemical burn, contusion, laceration, or puncture wound) that requires (i) treatment by a medical doctor, or (ii) restriction of work or motion or assignment to another regularly established job.

Serious Injury Frequency Rate

The number of serious injuries, as defined in American National Standards Institute (ANSI) Standard Z16.2 per 1,000,000 employee-hours of exposure. When serious injury frequency rate is used, it should be clearly identified as serious injury frequency rate, to avoid confusion with other frequency rates. This rate relates serious injuries, as defined, to the employee-hours worked during the period and expresses the number of such injuries in terms of million-hour units by use of the following formula: SIFR = Number of Serious Injuries × 1,000,000/Employee-Hours of Exposure.

Serious Violation

Any Occupational Safety and Health Administration (OSHA) violation in which there is a substantial possibility that death or serious physical harm could result from either an exposure that exceeds permissible limits or from practices, methods, operations, or processes used in the workplace. See also **OSHA Violation(s)**; **Willful Violation**.

Severity

The magnitude of physical or intangible loss consequences resulting from an actual incident, combination of deviations, or estimated from a qualitative or quantifiable safety review. See also **Consequence**.

Severity Rate

The total days charged for work injuries as defined in American National Standards Institute (ANSI) Z16.2 per 1,000,000 employee-hours exposure. Days charged include actual calendar days of disability resulting from temporary total injuries and scheduled charges for deaths and permanent disabilities. These latter charges are based on 6000 days for a death or permanent total disability, with proportionately fewer days for permanent partial disabilities for varying degrees of seriousness. See standard disabling injury severity rate. SR = Total Days Charged for Work Injuries × 1,000,000/Employee-Hours Exposure.

Shade #

The comparative darkness or obscurity owing to interception of the rays of light. Used in reference to goggles, safety glasses, welder's lens, etc.

Shall

Usually meant to indicate a positive and definitive requirement of a code or standard that must be performed. Action is mandatory.

Shelter-in-Place

A method of protection, in which instead of escaping from a fire risk, toxic vapors, radiation, etc. (because avenues of escape are unavailable or time-consuming to reach) individuals protect themselves in the immediate vicinity to avoid injury such as in a structure or building. Shelter-in-place cannot be utilized where available oxygen supplies are insufficient or cannot be isolated from contaminants.

Sheltering

See **Shelter-in-Place**.

Shield System

A permanent or portable structure designed to withstand a cave-in in excavations or trenches. These structures can be pre-manufactured or job-built in accordance with the Occupational Safety and Health Administration (OSHA) regulation 29 CFR 1926.652 (c) (3) or (c) (4). It may also be called trench boxes or trench shields. See also **Trench Shield**.

Shock

A critical medical condition brought on by a sudden drop in blood flow through the body. There is usually a failure of the circulatory system to maintain adequate blood flow. This sharply curtails the delivery of oxygen and nutrients to vital organs. It also compromises the kidney and so curtails the removal of wastes from the body. Shock can be due to a number of different mechanisms including not enough blood volume (hypovolemic shock) and not enough output of blood by the heart (cardiogenic shock). The signs and symptoms of shock include low blood pressure (hypotension), over-breathing (hyperventilation), a weak rapid pulse, cold clammy grayish-bluish (cyanotic) skin, decreased urine flow (oliguria), and mental changes (a sense of great anxiety and foreboding, confusion, and, sometimes, combativeness). See also **Electric Shock**.

Shock Hazard

The potential release of energy caused by contact or approach to energized electrical conductors or circuit parts. See also **Electrical Shock**.

Shop Rules (Working Rules)
Either regulations established by an employer dealing with day-to-day conduct in the plant operations, safety, hygiene, records, etc., or working rules set forth in collective bargaining agreements and in some union constitutions.

Shoring
A system is designed to prevent excavation failure (cave-ins) by supporting trench walls with a system of vertical uprights or sheeting and cross braces (shores). Shores are structures that cross the trench and put pressure on the vertical uprights and sheeting. Hydraulic, timber, or mechanical systems are used that support the sides of an excavation, unstable buildings, or vessels out of water (to prevent collapse), to protect nearby individuals from injury and avoid property damage.

Shoring, Hydraulic
A pre-engineered shoring system of aluminum or steel hydraulic cylinders (cross-braces) used with vertical or horizontal rails. They are typically designed specifically to support side walls of an excavation to prevent cave-in.

Short-Term Disability Insurance
Provides short-term (typically 26 weeks) income protection to employees who are unable to work due to a non-work-related accident or illness.

Short-Term Exposure Limit (STEL)
A threshold limit value (TLV), recommended by the American Conference of Governmental Industrial Hygienists (ACGIH), that describes the maximum concentration of a contaminant to which workers can be exposed for a period of up to 15 minutes continuously without suffering from irritation, tissue change, or debilitating narcosis of sufficient degree to increase the likelihood of accidental injury, impair self-rescue, or materially reduce work efficiency. No more than four exposures are permitted per day, with 60-minute intervals required, and the TLV-TWA cannot be exceeded. "TWA" indicates a time-weighted average concentration for up to a 10-hour workday during a 40-hour workweek.

Signal Word
As defined in the American National Standards Institute (ANSI) 535.4 standard, a specific word that calls attention to a safety label or sign and designates a degree or level of hazard seriousness. The signal words designed in the standard include Danger, Warning, Caution, and Notice. Danger indicates a hazardous situation that, if not avoided, will result in death or serious injury. It is intended to be used for the most extreme situations. Warning indicates a hazardous situation that, if not avoided, could result in death or serious injury. Caution indicates a hazardous situation that, if not avoided, could result in minor or moderate injury. Additionally, Notice is the signal word used to address practices not related to personal injury. See also **ANSI Z535.4, Product Safety Signs and Labels; Safety Alert Symbol**.

Signaling Device
An alarm system component such as a bell, buzzer, horn, speaker, light, or text display that provides an audible, visible, or tactile output to announce a condition of concern.

Significant Threshold Shift (STS)
A change of hearing threshold level of 15 dB or greater, in either ear, at any frequency (1000 to 4000 Hz) between the reference audiogram and any subsequent audiogram.

SIL Selection
The process of defining tolerable risk, confirming existing risk (both likelihood and consequence), and assigning a Safety Integrity Level (SIL)-rated safety function as needed to achieve a tolerable level of risk. See also **Safety Integrity Level (SIL)**.

S

Silicosis

A disease of the lungs caused by the chronic inhalation of dust containing silicon dioxide.

Simulated Workplace Protection Factor (SWPF) Study

A respirator study, conducted in a controlled laboratory setting and in which Co and Ci sampling is performed while the respirator user performs a series of set exercises. The laboratory setting is used to control many of the variables found in workplace studies, while the exercises simulate the work activities of respirator users. This type of study is designed to determine the optimum performance of respirators by reducing the impact of sources of variability through maintenance of tightly controlled study conditions.

Single Point Failure (SPF)

A location in a system, that if a failure occurs, it will cause the entire system to fail, because backup or alternative measures to accomplish the task are not available.

Siren

A high-pitched wailing sound readily distinguishable from whistles, horns, or other monotone audible devices and easily discernable from the confusion of other sounds. Sirens are commonly employed to announce an emergency condition.

Site Evaluation

Analysis of those factors at a site that could affect the safety of a facility or activity on that site. This includes site characterization, consideration of factors that could affect safety features of the facility or activity so as to result in a release of hazardous materials and/or could affect the dispersion of such material in the environment, as well as population and access issues relevant to safety (e.g., feasibility of evacuation, location of people and resources).

Site Survey

A physical review of a facility to determine if safety policies, regulations, and practices are being applied. See also **Audit, Safety**.

Slip-on Cuffs

Slip-on cuffs (or band top) designs allow easy donning and doffing and are continuous with the rest of the glove (no seam is used).

Slips, Trips, and Falls

It is estimated that in the United States, more than one million people are injured from a slip, trip, or fall each year. In 2005, 17,700 fatalities occurred as a result of falls. They are considered the number one preventable cause of loss in the workplace and the leading cause of injury in public places. According the National Safety Council (NSC), they are the number one cause of death for people over the age of 75 and have replaced automobile accidents as the leading reason people receive emergency room care. Walking surfaces account for 55 percent of all such accidents and are the predominant cause of slips, trips, and falls. Prevention includes improved facility designs, effective management control programs, use or warning indicators, housekeeping and maintenance, standards relating to pedestrian safety, and use of slip resistance materials. The American National Standards Institute (ANSI) B101.0, Walkway Surface Auditing Guideline for the Measurement of Walkway Slip Resistance, has certified the National Floor Safety Institute (NFSI) standard of slip-meter testing. The tests gauge the slipperiness, or Coefficient of Friction (COF), of a walking surface. It rates three ranges—low, moderate, and high—of the COF. A high COF means the surface is not very slippery and there is little concern of an incident occurring. However, a low COF means the floor surface should be examined, as well as the cleaning methods used and the cleaning applications in order to improve the COF. See also **ANSI B101.0, Walkway Surface Auditing Guideline for** the **Measurement of Walkway Slip Resistance**; National Floor Safety Institute (NFSI).

Sloping

A method of protecting personnel from sidewall cave-in by forming sides of an excavation that are inclined away from the excavation. The safe angle of slope required varies with different types of soil, exposure to the elements, and superimposed loads.

Slow-Moving Vehicle Emblem

An identification placard for vehicles that by design move slowly, i.e., 25 mph or less, on public roads. It consists of a fluorescent yellow-orange triangle with a dark red reflective border. The yellow-orange fluorescent colors are highly visible for daylight exposure. The reflective border defines the shape of the fluorescent colors in daylight and creates a hollow red triangle in the path of motor vehicle headlights at night. The material, location, mounting, etc., of the emblem are to be in accordance with the American Society of Agricultural Engineers (ASAE) Emblem for Identifying Slow-Moving Vehicles, ASAE R276, 1967, or ASAE S276.2 (ANSI B114.1-1971) and is required by the Occupational Safety and Health Administration (OSHA) regulation 29 CFR 1910.145 (d) (10).

Smog

Irritating haze that results from the sun's effect on certain pollutants in the air, particularly industrial and automotive exhaust sources.

Smoke

An air suspension (aerosol) of particles, often originating from combustion or sublimation. Carbon or soot particles less than 0.1 μm in size result from the incomplete combustion of carbonaceous material such as coal or oil. Smoke generally contains droplets as well as dry particles. Tobacco, for instance, produces a wet smoke composed of minute tarry droplets.

Smoking Safety

The National Fire Protection Association (NFPA) estimates that there were 142,990 smoking material fires in the United States in 2006. These fires resulted in 780 fatalities and 1600 injuries. Smoking materials such as cigarettes, cigars, pipes, etc., are considered the leading cause of fire deaths in the United States. It is estimated that one of every four fire deaths in the United States is attributed to smoking materials. Older adults are at the highest risk of fire death or injury from smoking materials. Preventive measures include smoking outside; use of deep, wide ashtrays on a sturdy support; ensuring tobacco materials are completely extinguished after using; ensuring smoking materials do not fall into furniture padding materials; avoiding smoking where oxygen is being used; use of fire-safe cigarettes; avoiding smoking if you are sleepy, have been drinking, or taking medicine; and keeping matches and lighters from children.

SNAFU

An acronym typically meaning Situation Normal: All Fouled Up or something similar. Its interpretation refers to an intolerable situation, mistake, or cause of difficulty. It may have originated in the U.S. Army during World War II. The Army and Warner Bros. Cartoons produced training cartoons during WWII featuring a character called Private Snafu who always did the wrong thing.

Snuffing Steam

Pressurized steam used to smother and inhibit fire conditions.

Social Security Disability Benefits

Long-term financial assistance for totally disabled persons, granted by the U.S. Social Security Administration. These benefits may be reduced by workers' compensation payments in some states.

TABLE S.3
Common Sound Levels

dB Level	Common Sound Description
180	Rocket at 30 meters
150	Jet engine at 30 meters
140	Gunshot
130	Rock concert
120	Pain begins
110	Power tools
100	Subway
90	Lawn mower
80	Busy traffic
70	Vacuum cleaner
60	Office noise
50	Normal conversation speaking
40	Mosquito buzzing
20	Faint whisper
0	Threshold of hearing

Sonometer

A device for testing acuteness of hearing.

Sorbent

A material that removes gases and vapors from air passed through a canister or cartridge.

Sound

An oscillation in pressure, stress, particle displacement, particle velocity, etc., that is propagated in an elastic material, in a medium with internal forces (e.g., elastic, viscous), or the super-position of such propagated oscillations. Sound is also the sensation produced through the organs of hearing, usually by vibrations transmitted in a material medium, commonly air (Table S.3). See also **Noise**.

Sound Absorption

The change of sound energy into some other form, usually heat, in passing through a medium or on striking a surface. In addition, sound absorption is the property possessed by materials and objects, including air, of absorbing sound energy.

Sound Analyzer

A device for measuring the band-pressure level or pressure-spectrum level of a sound as a function of frequency.

Sound Attenuation

See **Attenuation (Sound)**.

Sound Intensity

The average rate at which sound energy is transmitted through a unit area perpendicular to a specified point.

Sound Level

A weighted sound-pressure level obtained by use of metering characteristics and the weighting A, B, or C as specified in American National Standards Institute (ANSI) Standard S1.4, Specifications for Sound Level Meters. If the frequency weighting employed is not indicated, A-weighting is usually implied.

Sound Level Meter

A device to measure instant noise levels. It is comprised of a microphone, amplifier, output meter, and frequency-weighting networks, which are used for the measurement of noise and sound levels. Sound-level meters are often made with various filtering networks that measure the sound directly on A, B, C, etc., scales. Sound-level meters may also incorporate octave-band filters for measuring sound directly in octave bands. Since sound levels are specific to the areas being measured, the sound levels are called area sampling. Sound level meters used for measuring noise for compliance requirements must meet American National Standards Institute (ANSI) Standard S1.4, Specifications for Sound Level Meters. See also **Noise Dosimeter**; **Weighted Measurements**.

Sour Gas

Term used for natural gas or a gasoline contaminated with odor-causing sulfur compounds. In natural gas, the contaminant is usually hydrogen sulfide (H_2S) and can be fatal in high concentrations; in gasoline, mercaptans are usually the source.

Spark

A small, incandescent particle from electrical failures or a moving ember from a fire source. During electrical fire investigations, the term spark is reserved for particles thrown out by arcs, whereas an arc is a luminous electrical discharge. If copper and steel are involved in arcing, the released spatters of melted metal begin to cool as they sail through the air. When aluminum materials are involved in the faulting, the particles may burn as they sail in the air and continue to be extremely hot until they burn out or are extinguished after they fall on a material. Therefore, aluminum sparks have a greater ability to ignite fine fuels than do sparks from copper or steel. Sparks from arcs are generally an inefficient ignition source. Besides adequate temperature, the size of the particles is important for the total heat content of the particles and the ability to ignite a fuel. See also **Arc**.

Specific Gravity

The ratio of the density of a material to the density of some standard material, such as water at a specified temperature, or (for gases) air at standard pressure and temperature. Abbreviated sp. gr. Also known as Relative Density.

Splash-Proof Goggles

Eye protection constructed of noncorrosive material that fits snugly against the face and has indirect ventilation ports to protect against liquids that may inadvertently be directed to the eyes.

Spontaneous Combustible Material

Materials that are either pyrophoric or self-heating. See also **Pyrophoric Material**; **Self-Heating Material**.

Spontaneous Combustion

The outbreak of fire without the application of heat from an external source. Spontaneous combustion may occur when combustible matter, such as hay or coal, is stored in bulk. It begins with a slow oxidation process (e.g., bacterial fermentation or atmospheric oxidation) under conditions not permitting ready dissipation of heat, e.g., in the center of a haystack or a pile of oily rags. Oxidation gradually raises the temperature inside the mass to the point at which a fire starts. Crops are commonly dried before storage or, during storage, by forced circulation of air to prevent spontaneous combustion by inhibiting fermentation. For the same reason, small pieces of coal are wetted to suppress aerial oxidation.

Spontaneous Ignition

Ignition resulting from a chemical reaction in which there is a slow generation of heat from oxidation of organic compounds until the combustion or ignition temperature of the material (fuel) is reached. This condition is reached only where there is sufficient air for oxidation but not enough ventilation to carry away the heat as fast as it is generated.

Sprain
A joint injury in which some of the fibers of a supporting ligament are ruptured, but the continuity of the ligament remains intact.

Sprinkler System
A combination of water discharge devices (sprinklers), distribution piping to supply water to the discharge devices or more sources of water under pressure, water flow controlling devices (valves), and actuating devices (temperature, rate of rise, smoke, or other type device). The system automatically delivers and discharges water in the fire area.

Spurious Trip
See **Safe Failure**.

Stair Railing
A vertical barrier erected along exposed sides of a stairway to prevent the fall of an individual. See also **Handrail**.

Standard Disabling Injury Frequency Rate
See **Disabling Injury Frequency Rate (DIFR)**.

Standard Railing
A vertical barrier erected along exposed edges of a floor opening, wall opening, ramp, platform, or runway to prevent the fall of an individual. See also **Handrail**.

Standard Threshold Shift (STS)
As defined in the Occupational Safety and Health Administration (OSHA) standard 29 CFR 1910.95 (g) (10) (i), a change in an individual's hearing threshold, relative to the baseline audiogram for that individual, of an average of 10 decibels (dB) or more at 2000, 3000, and 4000 hertz (Hz) in one or both ears.

Standby Person
An individual trained in emergency rescue procedures and outside a confined space who remains in communication with those inside for the purpose of rendering assistance or effecting rescue should the need arise from an incident. May also be called a Standby Man.

Static Electricity
An electrical charge that may arise as a result of friction or may be induced by certain processes.

Static Ignition Hazard
An electrical charge buildup of sufficient energy to be considered an ignition source. For an electrostatic charge to be considered as an ignition source four conditions must be present: (1) a means of generating an electrostatic charge, (2) a means of accumulating an electrostatic charge of sufficient energy to be capable of producing an incendiary spark, (3) a spark gap, and (4) an ignitable mixture in the spark gap. Removal of one or more of these features will eliminate a static ignition hazard. Static charges can accumulate on personnel and metallic equipment. If the static accumulation is separated by materials that are electrically nonconducting a dangerous potential difference may occur. These nonconducting materials or insulators act as barriers to inhibit the free movement of electrostatic charges, preventing the equalization of potential differences. A spark discharge can occur only when there is no other available path of greater conductivity by which this equalization can be effected (i.e., bonding or grounding) (Figure S.6).

FIGURE S.6   Static ignition hazard sign.

Static Sensitive Device
   Many electronic components can be static sensitive and can be damaged from common static charges, which may occur on individuals, tools, and other non-conductors or semiconductors. To avoid discharging the static onto the sensitive components personnel should use a ground mat, ground tool, anti-static garments, or an anti-static wrist strap. May also be called Electrostatic Sensitive Device (ESD).

Strain
   Overstretching or overexertion of some part of the body musculature.

Strategic Partnership Program, OSHA
   A cooperative program of the Occupational Safety and Health Administration (OSHA) that helps encourage, assist, and recognize the efforts of partners to eliminate serious workplace hazards and achieve a high level of worker safety and health. The strategic partnerships seek to have a broader impact by building cooperative relationships with groups of employers and employees. These partnerships are voluntary, cooperative relationships between OSHA, employers, employee representatives, and others (e.g., trade unions, trade and professional associations, universities, and other government agencies). See also **Strategic Partnership Program, OSHA**; **Voluntary Protection Program (VPP), OSHA**.

Stretcher
   Portable device to carry an injured or incapacitated individual by carrying or rolling on support frame, while keeping the patient immobile.

Subrogation
   The legal process by which a company endeavors to recover from a third party the amount paid to an insured under an insurance policy when such third party may have been responsible for the occurrence causing the loss.

Substitution
   A loss prevention strategy where a less hazardous substance, process, or work activity is utilized instead of a more dangerous one.

Supplemental Guarding
   Additional electro-sensitive safety devices and hard guarding measures used for the purpose of preventing a person from reaching over, under, or around the defined area of an installed safety light screen system and into the point of operation of the guarded machine.

Supplied Air
   Breathable air supplied to a worker's mask/hood from a source outside the contaminated area.

Supplied-Air Respirator
   Both continuous-flow and pressure demand respirators are of this type.

Supplied Air Suit
   A one- or two-piece suit that is impermeable to most particulate and gaseous contaminants and is provided with an adequate supply of respirable air.

S

Supported Gloves

Protective gloves that are constructed of a coated fabric to provide some rigidity.

Survivor Benefits

A series of payments to the dependents of deceased employees. Survivor benefits come in two types: First, the "transition" type pays the named beneficiary a monthly amount for a short period (usually 24 months). Transition benefits may then be followed by "bridge benefits," which are a series of payments that last until a specific date, usually the surviving spouse's 62nd birthday.

Suspect Carcinogen

A material believed to be capable of causing cancer, based on scientific evidence.

Suspension System

A suspension configuration that permits workers to sit and work safely while elevated. Unlike the fall arrest configuration, the suspension configuration distributes the worker's weight on areas of the body capable of bearing that weight for extended periods. A suspension system is designed to raise or lower and support a worker at an elevated work station. The connecting points of the system, such as shoulder or seat-strap D-rings, are not designed to properly distribute the impact forces that result in arresting a free fall. A suspension system alone cannot be relied upon to provide proper fall arrest protection; the worker must be properly attached to an independent fall arrest system if a free fall is a possibility.

Sword of Damocles

A cliché meaning impending danger. The original sword, according to legend, was suspended by a thread over the chair where Damocles sat at a banquet given by King Dionysius of ancient Syracuse. Damocles had carried on about the ease of the life of a king, and Dionysius chose the dangling sword as a way of demonstrating to him that life (including the life of kings) also entails perils and responsibilities.

System Safety

The application of operating, technical, and management techniques and principles to the safety aspects of a system throughout its life to reduce hazards to the lowest level possible through the most effective use of available resources.

System Safety Analysis

The hazard identification and evaluation of a complex process by means of a diagram or model that provides a comprehensive, overall view of the process, including its principal elements and the ways in which they are interrelated. There are four principal methods of analysis: failure mode and effect, fault tree, THERP, and cost-benefit analysis. Each has a number of variations, and more than one may be combined in a single analysis. See also **Cost-Benefit Analysis**; **Failure Mode and Effects Analysis (FMEA/FMECA)**; **Fault Tree Analysis (FTA)**; **THERP (Technique for Human Error Rate Probability)**.

System Safety Engineering

The application of scientific and engineering principles during the design, development, manufacture, and operation of a system to meet or exceed established safety goals.

System Safety Society

A nonprofit organization organized in 1962 that is dedicated to supporting the safety professional in the application of Systems Engineering and Systems Management to the process of hazard, safety, and risk analysis to identify, assess, and control associated hazards while designing or modifying systems, products, or services.

Systematic Failure

A failure that happens in a deterministic (non-random) predictable fashion from a certain cause, which can only be eliminated by a modification of the design or of the manufacturing process, operational procedures, documentation, or other relevant factors. Since these are not mathematically predictable, the safety lifecycle includes a large number of procedures to prevent them from occurring. The procedures are more rigorous for higher safety-integrity-level systems and components. Such failures cannot be prevented with simple redundancy.

Systemic Risk

The risk of an entire system, as opposed to risk associated with any one individual entity, group, or component of a system in contrast to a "specific risk" or "unique risk." Systemic risk is a risk of security that cannot be reduced through diversification. Systemic risk evaluates the likelihood and degree of negative consequences to the larger body.

S

T

Tabletop Drill

An emergency response exercise where the participants (emergency response groups, company personnel, etc.) meet in a room and discuss a particular emergency scenario. It is used so that the assigned personnel can act through their roles in a presented scenario in a low stress environment to allow team building and collaboration. Coaching and mentoring of team members is also enhanced. It may also be referred to as a Tabletop Exercise. See also **Emergency Drill**; **Fire Drill**.

Tag Line

Control ropes that are attached to items being hoisted and are used by individuals to guide the load to keep it from striking or being caught in a structure, as the items are being hoisted or lowered.

Tagging

See **Tag-Out**.

Tag-Out

The placement of a tag-out device on an energy isolation device (e.g., circuits or equipment) in accordance with an established procedure to identify and alert individuals that the energy isolation device and the equipment being controlled may not be operated or have their positions changed until the tag-out device is removed by an authorized individual. See also **Lockout/Tagout (LOTO)**.

Tag-Out Device

A prominent warning device such as a tag and means of attachment, which can be securely fastened to an energy-isolating device in accordance with an established procedure to indicate that the energy-isolating device and the equipment being controlled may not be operated until the tag-out device is removed by an authorized individual.

Tailgate Safety Meeting

Short, 10- to 15-minute, on-the-job meetings in construction and heavy industry held to keep employees apprised of work-related hazards. See also **Safety Meeting**; **Safety Talk**; **Toolbox Safety Meetings**.

Taproot® Investigation

A methodology of root cause analysis for incident investigations and to develop corrective actions. This is undertaken by developing a flowchart of events that have occurred and identifying conditions that impact these events, which enables the identification of causal factors. Taproot® closely focuses on the performance of systems that are in place. This includes systems that were in place but did not operate. The Taproot® flowchart is normally described as a snapshot of the circumstances that occurred during the incident. The flowchart is developed by using symbols representing events, conditions, incidents, and barriers to reconstruct the incident as it took place. Taproot® (in common with most other investigative tools) relies very heavily on the complete and accurate collection of factual information. It has been in use since 1991 by the chemical and petrochemical industries to investigate process safety incidents. It typically combines both inductive and deductive techniques for systematic investigation to identify the root causes of problems. See also **Root Cause Analysis (RCA)**.

T

Target Zero

A safety slogan or goal sometimes used in industry to highlight the measurement objective of no incidents for a particular reporting period. Its rationale is that you should never be planning to have an incident, and therefore your safety goal should always be zero.

Task Analysis

See **Job Safety Analysis (JSA)**.

Task Lighting

Lighting directed to a specific surface or area to provide illumination for visual or manual tasks. Task lighting is typically provided when the specific amount or quality of lighting cannot readily be obtained by general area lighting.

Telltale

A device used in sawmills to serve as a warning for overhead objects, as stated in Occupational Safety and Health Administration regulation 29 CFR 1910.265 (b) (43).

Tempered Glass

A type of glass that has been treated to improve its stability and resistance to heat, impact, and distortion. Glass sheets are tempered at about 650°C (1200°F) followed by a sudden chilling. This treatment increases the strength of the glass sheets approximately six times. When such glass does break (due to impact, expansion, etc.), it shatters into blunt granules to prevent injury instead of sharp-edged pieces. See also **Laminated Glass**; **Safety Glass**; **Tempered Glass, Window Film, Safety and Security**.

Temporary Disability Benefits

Paid when an employee loses wages because of an injury that prevents the employee from doing his or her usual job while recovering.

Temporary Emergency Exposure Limit (TEEL) Value

Acute exposure limits developed by the U.S. Department of Energy, Subcommittee on Consequence Assessment and Protective Actions (SCAPA), to describe the risk to humans resulting from once-in-a-lifetime, or rare, exposure to airborne chemicals. Each material is assigned a value for each chemical (i.e., PAC-0, -1, -2, and -3). Starting with "0," each successive benchmark is associated with an increasingly severe effect that involves a higher level of exposure. The four benchmarks present threshold levels for:

0: No adverse health effects
1: Mild, transient health effects
2: Irreversible or other serious health effects that could impair the ability to take protective action
3: Life-threatening health effects

See also **Acute Exposure Guideline Level (AEGL) Value**; **Emergency Response Planning Guideline (ERPG) Value**.

Temporary Threshold Shift (TTS)

The hearing loss suffered as the result of noise exposure, all or part of which is recovered during an arbitrary period of time when one is removed from the noise. It accounts for the necessity of checking hearing acuity at least 16 hours after a noise exposure.

Temporary Total Disability

The classification for any injury that does not result in death or permanent total or permanent partial disability, but which renders the injured person unable to perform a full day's work. This means that the injured employee cannot perform all the duties of a regularly established job that is open and available, or is unable to perform such duties during the entire time interval corresponding to the hours of the regular shift on any one or more days including weekends, holidays, and other days off, or plant shutdown, subsequent to the date of the injury. See also **Permanent Partial Disability, Permanent Total Disability**.

Terrorism

The use of force or violence against individuals, organizations, or property for the purposes of intimidation, coercion, or ransom by misguided extremist individuals or groups. The individuals or organizations in which the force or violence is inflected upon are usually not directly involved in the concerns and are typically affected by chance. Threats are also used to create fear in the public and to convince citizens that their government is powerless to prevent their actions, and to get immediate publicity for their causes. Terrorism is a violation of national and international laws. See also **Security Vulnerability Analysis (SVA)**.

Thermal Expansion

The change in size of an object as the temperature changes. Normally, as the temperature increases, the size of an object also increases. Conversely, the object will shrink as the temperature drops. As an object expands or contracts with a temperature change, its change in length depends on three quantities: the original length, the temperature change, and the thermal properties of the material composing the object. When designing bridges, power lines, or similar items that might be subjected to wide temperature variations, engineers must take into account the effects of thermal expansion. Depending on the temperature range in a particular area, they may choose building materials that expand with heat or not, different types of joint structures, and other kinds of reinforcements that might be required.

THERP (Technique for Human Error Rate Probability)

A technique for predicting the potential for human error in an activity. It evaluates quantitatively the contribution of the human error component in the development of an untoward system. Special emphasis is placed on the human component in product degradation. THERP involves the concept of basic error rate that is relatively consistent between tasks requiring similar human performance elements in different situations. Basic error rates are assessed in terms of contributions to specific system failures.

Basic THERP methodology steps:

1. Selecting the system failure
2. Identifying all behavior elements
3. Estimating the probability of human error
4. Computing the probabilities as to which specific human error will produce the system failure
5. Identifying specific corrective actions to reduce the likelihood of an error

Threat

A potential for loss (injury, damage, or other hostile action) from a deliberate act or condition.

Threshold

The level of stimulus energy that must be exceeded before a response occurs.

Threshold Limit Values (TLVs®)

The concentration of an airborne substance to which an average person can be repeatedly exposed without adverse effects. TLVs® may be expressed in three ways: (1) TLV-TWA: time-weighted average, based on an allowable exposure averaged over a normal 8-hour workday or 40-hour workweek; (2) TLV-STEL: short-term exposure limit or maximum concentration for a brief specified period of time, depending on a specific chemical (TWA must still be met); and (3) TLV-C: Ceiling Exposure Limit or maximum exposure concentration not to be exceeded under any circumstances (TWA must still be met). TLV® is a registered term of the American Conference of Governmental Industrial Hygienists (ACGIH®). See also **Time-Weighted Average (TWA)**.

Threshold of Hearing

The minimum sound level of a pure tone that an average ear with normal hearing can hear with no other sound present. The absolute threshold is not a discrete point, and is therefore classified as the point at which a response is elicited a specified percentage of the time. The threshold of hearing also varies with frequency. A normal child typically has a full hearing frequency range up to 20,000 Hz. By the age of 20, the upper limit for individuals may have dropped to 16,000 Hz. From the age of 20, hearing capability reduces gradually to approximately 8000 Hz by retirement age due to the normal aging process. This reduction in the upper frequency limit of the hearing range is accompanied by a decline in hearing sensitivity at all frequencies with age, the decline being less for low frequencies than for high. Hearing losses may also occur due to loud noise exposure to an individual.

Threshold of Pain

A subjective indication of when pain is beginning to be felt by an individual. The intensity at which a stimulus (e.g., heat, pressure) begins to evoke pain is termed threshold intensity. If a heat source exposed to your skin begins to hurt at 40°C, then that is considered the pain threshold temperature for that portion of your skin at that time. Forty-degrees Celsius is not the pain threshold, but the temperature at which the pain threshold was crossed. The intensity at which a stimulus begins to evoke pain varies from individual to individual and for a given individual over time.

Threshold Planning Quantity (TPQ)

The designated quantity of a Superfund Amendment and Reauthorization Act (SARA) Extremely Hazardous Substance as listed in 40 CFR 355, Appendix A and B, that if equaled or exceeded at a facility activates emergency response planning provisions required under SARA Title III. See also **SARA Title III**.

Tier Diagramming

An incident investigative technique used to identify both the root causes of an incident and the level of line management responsibility associated with contributing and root causes. An investigating team uses the tier diagram to hierarchically categorize the causal factors derived from an events and causal factors analysis. The diagram is typically divided into six tiers, numbered 0 through 5, which represent organizational responsibility ranging from the worker level to upper management. In a series of steps, causal factors are evaluated as potential root causes and linked to a level of responsibility in the line organization. See also **Causal Factor (CF)**; **Root Cause Analysis (RCA)**.

Time Delay Interlock

A type of machine interlock system that has a mechanical device that requires a specific time period for a guard to be released after the power has be shut off.

Time for Notification of Loss

Period allowed to an insured to inform an insurer of a loss. Typically insurance policies require immediate written notice or notice as soon as practical. Different types of policies may have their own time periods. Health insurance policies generally require notice within 20 days, windstorm insurance policies within 10 days, and hail insurance policies within 48 hours. The notification time period is to allow the insurer to investigate the loss and protect the property from further damage.

Time-Weighted Average (TWA)

The concentration of a substance to which a person is exposed in ambient air divided by the total time of observation. For occupational exposure a working shift of 8 hours is commonly used as the averaging time.

Tinnitus

Persistent ringing in the ears. A common aftermath from damage to receptor cells in the inner ear due to overexposure to noise. As a result of the damage, the brain generates its own sounds that are routed back to the ear, which is perceived as ringing, or possibly hissing or buzzing. Tinnitus is the early symptom or warning indication of overexposure to noise.

Tip of the Iceberg

A cliché meaning a small part of a large problem or phenomenon, as an analogy to icebergs, in which the major mass of the iceberg exists underwater and cannot be readily seen, and only the tip, which sticks out, can be seen. Often utilized in the safety profession to highlight the fact that initial small incidents portray or indicate a larger underlying root cause or systematic issue that needs to be addressed by management. See also **Iceberg Principle**.

Title 29

The section of the Code of Federal Regulations dealing with the regulations of the Occupational Safety and Health Administration (OSHA).

Toeboard

A vertical barrier at floor level erected along exposed edges of a floor opening, wall opening, platform, runway, or ramp (sides and ends) to prevent falls of materials. The installation of toeboards should conform to the American National Standards Institute (ANSI) standard A-1264.1, Safety Requirements for Workplace Floor and Wall Openings, Stairs, and Railings. Toeboards should be at least 4 inches high and should be made of wood, metal, or metal grille not exceeding 1-inch mesh. Toeboards at flywheels should be placed as close to the edge of the pit as possible. Wood toeboards for permanent installations should be of 1" × 4" stock or heavier.

Tolerable Risk

See **Acceptable Level of Risk**; **As Low as Reasonably Practical (ALARP)**.

Tongs, Safety

A device for feeding small objects to and removing them from a danger area. It is typically a metal U-shaped tool that is used to grasp items with one hand.

Toolbox Safety Meetings

Short on-the-job meetings in heavy industry and construction to keep employees apprised of work-related hazards. Their objective is to reinforce safety training and information on a particular topic. Employees are kept abreast of changes in regulations, safety procedures, equipment, personal protective equipment (PPE), and job assignments and responsibilities. They help employees to remember requirements, avoid risks, and prevent incidents. See also **Safety Meeting**; **Safety Talk**; **Tailgate Safety Meeting**.

Total Loss

The loss of all the insured property when it is damaged or destroyed to such an extent that it cannot be rebuilt or repaired to an equivalent condition prior to the loss. Also a loss involving the maximum amount for which a policy is liable.

Total Reaction Distance, Motor Vehicle

The distance traveled between the point at which the driver perceives that braking evasive action is required and the point at which the contact is made with the braking controls.

T

Total Reaction Time, Motor Vehicle
The time required for a vehicle to move the total reaction distance.

Total Recordable Case Rate (TRC)
The Occupational Safety and Health Administration (OSHA) Total Recordable Rate (sometimes referred to as Total Recordable Incident Rate [TRIR]) using the number of OSHA recordable injury and illness cases. The TRC is the total of all recordable injury and illness cases. This includes cases that involve days away from work, job transfer or restriction, and other recordable cases from the OSHA's Form 300 multiplied by 200,000 and then divided by the number of employee hours worked.

Total Stopping Distance, Motor Vehicle
The distance in which the vehicle comes to rest after the driver discovers a hazard that requires stopping. Includes driver reaction time, brake reaction time, and braking time.

Toxemia
Poisoning by way of the blood stream.

Toxic
Of, pertaining to, or caused by poison. Something that is poisonous or harmful.

Toxic Substance
A substance that demonstrates the potential to induce cancer, to produce short- and long-term disease or bodily injury, to affect health adversely, to produce acute discomfort, or to endanger life of humans or animals resulting from exposure via the respiratory tract, skin, eye, mouth, or other routes in quantities that are reasonable for experimental animals or that have been reported to have produced toxic effects in humans.

Toxic Substance Control Act (TSCA) of 1976
A regulation of the Environmental Protection Agency (EPA) that requires all chemicals manufactured in the United States, except other compounds such as those covered by the Federal Insecticide, Fungicide, and Rodenticide Act (FIFRA), to have developed data on their health and environmental effects. It also allows the EPA to require tests to restrict or prohibit the manufacture, use or distribution, export, and disposal of chemical substances and mixtures. See also **Federal Insecticide, Fungicide, and Rodenticide Act (FIFRA)**.

Toxicant
A substance that causes a degenerative alteration in any anatomic, physiologic, or biochemical system of a formed organism.

Toxicity
A relative property of a chemical agent with reference to a harmful effect on some biologic mechanism and the condition under which this effect occurs.

Toxicology
That branch of medical science that deals with the nature and effects of poisons.

Toxicology, Industrial
Study of the nature and action of toxic agents that may cause health impairment to workers.

Traffic Accident
Any accident involving one or more motor vehicles in motion on a roadway. A traffic accident may involve more than one unit if each unit comes in contact with some other unit involved while part or either is in contact with the road or sidewalk.

Traffic Safety
A generic term that mainly encompasses vehicle safety features, driver and pedestrian education and behavior, motor vehicle laws, and factors related to the design of roads and highways (Figure T.1). See also **Bicycle Safety**; **Motor Vehicle Safety**; **Motorcycle Safety**; **National Highway Traffic Safety Administration (NHTSA)**.

FIGURE T.1 Highway traffic safety.

Traffic Safety Cone
See **Cone, Safety**.

Transmission Loss, Sound
The ratio, expressed in decibels, of the sound energy incident on a structure to the sound energy that is transmitted.

Transportation Safety Institute (TSI)
An agency of the Department of Transport that offers safety, security, and environmental training, products, and services for both public and private sectors, nationally and internationally. The TSI was established in 1971. The institute offers transit, aviation, motor carrier, highway safety, hazardous materials, risk management, and other related training courses.

Trauma
Any injury, wound, or shock whether physically or emotionally inflicted. Trauma has both a medical and a psychiatric definition. Medically, "trauma" refers to a serious or critical bodily injury, wound, or shock. This definition is often associated with trauma medicine practiced in emergency rooms and represents a popular view of the term. In psychiatry, "trauma" has assumed a different meaning and refers to an experience that is emotionally painful, distressful, or shocking, which often results in lasting mental and physical effects.

Trench Shield
An engineered metal box that is placed in an excavation for the protection of workers. It does not provide structural strength to the excavation, but provides workers a safe worksite that protects them from collapsing material of the trench. A registered professional engineer must design the trench shield or trench box system, which can be pre-manufactured or built on site as necessary.

Trenching Hazards
Construction of a narrow excavation (in relation to its length) made below the surface of the ground. Where the width exceeds 15 feet, it is generally not considered a

T

trench. The primary hazard associated with excavation and trenching activities is a cave-in. A cave-in may result in entrapment and eventual suffocation or workers in an unprotected excavation. Associated hazards include falls, falling loads, mobile equipment, water accumulation, hazardous atmospheres, and access and egress obstructions. It is covered by the Occupational Safety and Health Administration (OSHA) regulation 29 CFR 1926.650, Subpart P. See also **Shield System**.

Triage

A methodology for the rapid assessment of emergency care priorities. Triage involves the sorting and allocation of treatment to patients, especially disaster victims, according to a system of priorities designed to maximize the number of survivors. It is usually based on the advanced determination of seriousness of injuries, kinds of treatment that can be afforded at the immediate location, and the availability and capability of emergency transportation to transfer victims to selected medical care facilities. As conditions at the location of a disaster fluctuate the sorting of victims may result in the re-determination of priorities that have been assigned.

Trigger Finger

A tendon disorder that occurs when there is a groove in the flexing tendon of the finger. If the tendon becomes locked in the sheath, attempts to move the finger cause snapping and jerking movements. It is usually associated with individuals using tools that have handles with hard or sharp edges.

Trip Device

A means by which an approach by an individual beyond the safe limit of working machinery causes the device to activate and stop the machine or reverse its motion.

Triple Modular Redundant (TMR) ESD System

Fault tolerant systems using three separate processors with triplicated input/output and bus structure. Each processor executes its individual application program, simultaneously verifying data; executing logic instructions, control calculations, clock, and voter or synchronization signals; and performing comprehensive system diagnostics. Process outputs are sent via triplicated paths to output modules where they are voted two out of three (2oo3), to ensure logic and output integrity. See also **Dual Modular Redundant (1oo2) ESD System**; **Emergency Shutdown System**.

Tunnel Guard

A form of distance guard whereby access to the area of hazard of the machine is prevented since materials are passed into the machine via a metal enclosure or tunnel. It is commonly used on metal cutting machines and usually incorporates an interlock, so if the tunnel guard is raised the machine stops operating.

Two-Hand Controls

Tripping devices for a machine that require simultaneous application of both hands to operate the control, so that the hands of the operator are kept out of the point-of-operation area while the machine is operating. A two-hand-control device protects only the hands of the machine operator (and not other individuals) when used as a safeguarding device. See also **Hostage Control Device**.

Type of Accident

The classification of an accident according to the manner of contact of the injured person with the agency. It involves the movement of an object, material, or person and association with an agency, such as fall, struck by, struck against, etc.

U

Ultimate Load

The minimum applied force necessary to cause failure of a material.

Ultimate Stress

The intensity of stress at the point of failure for a material.

Ultrasonic Testing

A type of nondestructive testing of materials using high frequency sound, which is used to check for defects and flaws.

Ultraviolet (UV) Radiation

Radiation in the electromagnetic spectrum from wavelengths of 100 to 3900 A. It is so named because the spectrum consists of electromagnetic waves with frequencies higher than those that individuals can identify as the color violet. UV light is found in sunlight and is emitted by electric arcs and specialized lights such as black lights. It is considered a danger to eyes and overexposure to the skin causes severe skin burns. UVA includes the wavelengths from 313 to 400 A, and UVB covers the wavelengths from 280 to 315 A. Both UVA and UVB destroy vitamin A in skin, which may cause further damage. UVA is also known to contribute to skin cancer via indirect DNA damage. UVC rays are the highest energy, most dangerous type of ultraviolet light. UVC rays are filtered out by the atmosphere.

UN Number

UN number stands for United Nations number. The UN number is a four-digit number assigned to a potentially hazardous material (such as gasoline, UN 1203) or class of materials (such as corrosive liquids, UN 1760). These numbers are used by firefighters and other emergency response personnel for identification of materials during transportation emergencies. UN numbers are internationally recognized. NA (North American) numbers are used only for shipments within Canada and the United States. PINs (Product Identification Numbers) are used in Canada. UN, NA, and PIN numbers have the same uses. Information concerning the UN number aspects can be seen in the DOT Emergency Response Guidebook. See also **CAS Registry Number**; **DOT Hazard Class**; **Emergency Response Guidebook (ERG), DOT**.

Unconfined Vapor Cloud Explosion (UVCE)

An explosion in which the cloud of vapor (gases or mist) ignites in air, resulting in a detonation accompanied by a blast wave and intense heat. UVCE explosions generally occur in the process industries (i.e., chemical and petroleum) and may cause vast devastation to the facility due to the blast effects of the incident. Unconfined vapor cloud explosions occur in the "open air" but have some degree of confinement, which allows the flame front to accelerate to achieve explosion parameters. The term "Vapor Cloud Explosion (VCE)" is therefore sometimes used to describe the same phenomenon and may be considered more appropriate. The vapor cloud explosion is the result of a flame front propagating through a premixed volume of air and flammable gas or vapor. The flame front must propagate with sufficient velocity to create a pressure wave. For a VCE to occur, certain conditions must be met. These include (1) release of flammable material, (2) sufficient mixing of the flammable material in air to allow rapid flame front propagation, (3) ignition, and (4) confinement of the flame path, which tends to accelerate the flame front. The blast effects of a VCE can vary greatly and are determined by the flame speed. The flame speed of the explosion is affected by the turbulence created within the vapor cloud as it passes through congested or confined areas where the vapor is released. Normal process industry plant design and equipment arrangements have been shown to create enough congestion and confinement to produce turbulent vapor cloud conditions and allow rapid flame propagation to occur. The timing of the vapor cloud ignition, whether immediate or delayed, will determine the amount of flammable material released and the magnitude of the vapor cloud explosion.

Underwriter

An insurance company employee who accepts or rejects risks on behalf of the insurance company. More broadly it is defined as anyone who makes insurance contracts.

Underwriters Laboratories (UL)

A worldwide recognized independent fire and safety testing laboratory. Underwriters Laboratories developed as a result of concerns of the Western Underwriters' Association for the installation and exhibits of electrical lighting at the 1893 World's Columbian Exhibition held in Chicago, Illinois. They employed an electrical engineer to evaluate the installations. A laboratory was provided in a room over a local fire station, which formally evolved into Underwriters Laboratories. Today UL evaluates more than 19,000 types of products, components, materials, and systems annually with 20 billion UL marks appearing on 72,000 manufacturers' products each year. UL's worldwide family of companies and network of service providers includes 64 laboratory, testing, and certification facilities serving customers in 98 countries. See also **Approved**; **Classified**; **Labeled**; **Listed**.

Underwriting

The process an insurance company uses to decide whether to accept or reject an application for a policy.

Unintentional Injury

The leading cause of death for individuals between the ages of 1 and 44, and the fifth leading cause of death in the United States. Unintentional injuries are among the top 10 causes of death in every age group.

Unsafe

Involving or causing danger or risk; liable to hurt or harm. In the workplace, unsafe is attributed to an unsafe act, behavior, or condition.

Unsafe Act or Behavior

A departure from an accepted, normal, or correct procedure or practice that has in the past actually produced injury or property damage or has the potential for producing such loss in the future; an unnecessary exposure to a hazard; or conduct reducing the degree of safety normally present. Not every unsafe act produces an injury or loss but, by definition, all unsafe acts have the potential for producing future injuries or losses. An unsafe act may be an act of commission (doing something that is unsafe) or an act of omission (failing to do something that should have been done). The most common unsafe behavior includes taking shortcuts, being overconfident, beginning a task with incomplete instructions, poor housekeeping, ignoring safety procedures, removing guards or not using personal protective equipment, drug and alcohol abuse, mental distractions, and failure to plan. The insurance industry estimates that 80 out of 100 incidents are directly attributable to the individual(s) involved in the incident. Stated another way, unsafe work behaviors cause four times as many incidents as unsafe work conditions. See also **Behavior-Based Safety**.

U

Unsafe Condition

Any physical state that deviates from that which is acceptable, normal, or correct in terms of its past production or potential future production of personal injury or damage to property; any physical state that results in a reduction in the degree of safety normally present. It should be noted that incidents are invariably preceded by unsafe acts and/or unsafe conditions. Thus, unsafe acts or unsafe conditions are essential to the existence or occurrence of an incident. Some examples of unsafe conditions include defective hand tools, uneven or slippery walking surfaces, faulty electrical wiring; improper storage of flammable or other hazardous materials, and unguarded machinery.

Unstable

A chemical that will vigorously polymerize, decompose, condense, or will become self-reactive under conditions of shocks, pressure, or temperature.

Unsupported Gloves

A type of protective, unlined glove without any type of fabric lining to provide internal support to the glove.

Upper Confidence Limit (UCL)

In sampling analysis, a statistical procedure used to estimate the likelihood that a particular value is above the obtained value.

Upper Explosive Limit (UEL)

The maximum proportion of vapor or gas in air above which propagation flame does not occur. It is the upper limit of the flammable or explosive range in which UELs are determined in accordance with American Society for Testing Materials (ASTM) E-681, Standard Test Method for Concentration Limits of Flammability of Chemicals. Limits of flammability may be used to determine guidelines for the safe handling of volatile chemicals. This test method covers the determination of the lower and upper concentration limits of flammability of chemicals having sufficient vapor pressure to form flammable mixtures in air at atmospheric pressure at the test temperature. They are used particularly in assessing ventilation requirements for the handling of gases and vapors. National Fire Protection Association (NFPA) 69 provides guidance for the practical use of flammability limit data, including the appropriate safety margins to use. See also **Explosive Limits**; **Lower Explosive Limit (LEL)**.

Upper Flammable Limit (UFL)

Synonymous with upper explosive limit. See also **Upper Explosive Limit (UEL)**.

Urticaria

An allergic reaction that results in raised bumps on the skin and mucous membranes. It is usually accompanied by itching. It is an allergen to food, plant inhalants (e.g., pollen), or chemicals.

U.S. Fire Administration (USFA)

An entity of the Department of Homeland Security's Federal Emergency Management Agency, whose mission is to provide national leadership to foster a solid foundation for our fire and emergency services stakeholders in prevention, preparedness, and response.

U

V

Vapor

The gaseous form of substances that are normally in the solid or liquid state (at room temperature and pressure). The vapor can be changed back to the solid or liquid states either by increasing the pressure or decreasing the temperature alone.

Ventilation

One of the principal methods used for the engineering control of occupational health hazards. The process causes fresh air to circulate to replace contaminated air that is simultaneously removed. There are several different kinds of ventilation. See also **Local Exhaust Ventilation**; **Natural Ventilation**.

Violation, OSHA

See **OSHA Violation**(s).

Violence, Workplace

Violence in the workplace is a serious safety and health issue. Its most extreme form, homicide, is the fourth-leading cause of fatal occupational injury in the United States. According to the Bureau of Labor Statistics Census of Fatal Occupational Injuries (CFOI), there were 564 workplace homicides in 2005 in the United States, out of a total of 5702 fatal work injuries. See also **Going Postal**.

Voluntary Protection Program (VPP), OSHA

A cooperative voluntary program of the Occupational Safety and Health Administration (OSHA) to promote health and safety at work sites, which began in 1982. VPP sets performance-based criteria for a managed safety and health system, invites sites to apply, and then assesses applicants against these criteria. OSHA's verification includes an application review and a rigorous on-site evaluation by a team of OSHA safety and health experts. OSHA approves qualified sites to one of three programs: Star, Merit, and Star Demonstration—recognition for worksites that address unique safety and health issues. Sites that make the grade must submit annual self-evaluations and undergo periodic on-site re-evaluations to remain in the programs. Acceptance and approval into the VPP is OSHA's official recognition of the outstanding efforts of employers and employees who have achieved exemplary occupational safety and health. VPP worksites generally have lost-workday case rates that range from one-fifth to one-third the rates experienced by average worksites. See also **Alliance Program, OSHA; Safety and Health Achievement Recognition Program (SHARP); Strategic Partnership Program, OSHA.**

V

W

Walk-Through Survey

A physical safety inspection of the workplace to identify and evaluate potential hazards.

Warehouse Safety

The most common hazard group for warehouse safety is slips, trips, and falls. The movement of materials on different levels and on different types of floor surfaces contributes to the possibility for individuals to lose their balance or stumble over an out-of-place item. There is also the possibility of falling objects in a warehouse. Items that aren't carefully stacked on floors, shelves, and other surfaces can fall on an individual. Warehouse equipment can also pose hazards. Conveyors, forklift trucks, and hand trucks can all cause incidents or injuries. Individuals also place themselves at risk if they lift and carry materials improperly, risking back injuries. They also have to be cautious with the equipment used to load, pack, and unpack—skids, pallets, strapping, and cutting tools. Materials stored in a warehouse can also pose dangers. Both physical and health hazards exist from the storage of hazardous substances or flammable or combustible materials. The following list shows the top 10 areas for which OSHA issues citations for warehouses: forklifts; hazard communication; electrical, wiring methods; electrical, system design; guarding floor and wall openings and holes; exits; mechanical power transmission; respiratory protection; lockout/tagout; portable fire extinguishers.

Warning

Communication and acknowledgment of dangers, such as operating procedures, practices, or conditions that may result in injury or damage if not carefully observed or followed.

Warning Label or Sign

As defined by American National Standards Institute (ANSI) 535.4, a sign or label to indicate a hazardous situation that if not avoided, could result in death or serious injury. Warning is to be indicated in black letters on an orange background. See also **Caution Label or Sign**; **Safety Alert Symbol**; **Safety Sign**.

Warning Signal

Typically means an auditory alarm (e.g., siren) or visual indication (e.g., flashing lights) to warn of a hazard aspect occurring over a wide area (e.g., toxic gas release in a chemical plant; tsunami approach, etc.) and the need for immediate actions to prevent injury or damage (e.g., evacuation, shelter-in-place, etc.). The effectiveness of the signal is dependent on its capability to be observed, the population's awareness of the signal characteristics, and the actions to be taken upon its activation.

Warning Tag

As defined by Occupational Safety and Health Administration (OSHA) regulation 29 CFR 1910. 145 (f) (7), warning tags are used to represent a hazard level between caution and danger. See also **Caution Tag**; **Danger Tag**.

Water Hammer

A shock wave produced in water systems due to the sudden change in flowing conditions. It occurs where there is a sudden or abrupt change in flow velocity or pressure.

The shock wave may be several times the normal pressure and will travel to the end of the line where it will be reflected back. The cycle will be repeated until all the energy is dissipated or relieved. The rapid closing of valves and piping layouts that are conductive to water hammer should be avoid. Surge tanks, pneumatic chambers, relief valves, or shock absorbers and arrestors are sometimes used to control water hammer concerns.

Water Reactive Chemical

A chemical that reacts with water to release a gas that is either combustible or presents a health hazard.

Weighted Measurements

Two weighting curves are commonly applied to measures of sound levels to account for the way the ear perceives the "loudness" of sounds. A-weighting: A measurement scale that approximates the "loudness" of tones relative to a 40-dB sound pressure level (SPL) 1000-Hz reference tone. A-weighting has the added advantage of being correlated with annoyance measures and is most responsive to the mid-frequencies, 500 to 4000 Hz. C-weighting: A measurement scale that approximates the "loudness" of tones relative to a 90-dB SPL 1000-Hz reference tone. C-weighting has the added advantage of providing a relatively "flat" measurement scale that includes very low frequencies.

Welder's Flashburn

An eye injury that may occur to welders who have failed to use adequate eye protection. The intense ultraviolet light from a welder's electric arc absorbed by the eye causes a superficial and painful keratitis, a condition in which the eye's cornea, the front part of the eye, becomes inflamed. It may be also known as Arc Eye, Welder's Flash, Bake Eyes, Corneal Flash Burns, or Flash Burns.

Welding Shields

Used to protect a welder's eyes and face from sparks, burns, and debris. They are fitted with filtered lenses for protection from infrared and radiant light. They are required to meet American National Standards Institute (ANSI) Z87.1, Occupational and Educational Eye and Face Protection Devices, and be clearly marked with the manufacturer's name (see Figure W.1).

What-If Analysis (WIA)

A safety review method by which "What-If" investigative questions (i.e., brainstorming or checklist approach) are asked by an experienced and knowledgeable team of the system or component under review where there are concerns about possible undesired events. Recommendations for the mitigation of identified hazards are provided. See also **Brainstorming**; **Hazard and Operability Study (HAZOP)**; **Safety Flowchart**.

Whole Body Vibration

Exposure of the whole body to vibration (usually through the feet/buttocks when riding in a vehicle). Whole body vibration may increase the risk for injury, including low back pain and internal organ disruption. See also **Ergonomics**; **Segmental Vibration (Hand-Arm Vibration)**.

5 Whys Technique

See **Five Whys Technique**.

Will, Shall, Must

To be understood as a mandatory condition, as opposed to Could, Would, and Should, which are directive type of words that can be taken as a suggestion only and to be used when caution is implied.

Willful Misconduct

Deliberate failure to comply with statutory regulations.

FIGURE W.1   Welder's helmet.

Willful Violation

An intentional and knowing violation of an Occupational Safety and Health Administration (OSHA) standard(s). Occurs when the employer commits an intentional and knowing violation of safety law or when the employer did not consciously violate a safety law but was aware that an unsafe or hazardous condition existed and made no reasonable effort to eliminate the condition. See also **OSHA Violation(s)**.

Window Film, Safety and Security

Polyester films that are applied to glass in order to hold them together if the glass is shattered, similar to laminated glass. The main difference between film and laminated glass is that the film can be applied to the glass after manufacture or installation, i.e., it is a retrofit product. These films are used on trains, buses, cars, and buildings. They can be applied to toughened, annealed, or laminated glass. The films are available in various thicknesses, from 100 μm (or 4 mils) through to 375+ μm (15 mils). The film thickness is selected for level of protection desired and the dimensions of the glass pane. Manufacturers recommend 100-μm film for glass up to 3 mm (1/8 inch) and 175-μm film for glass over 6 mm (1/4 inch). Standards for manufacture include American National Standards Institute (ANSI) Z97, Safety Requirements for Architectural Glazing Material, or Consumer Product Safety Commission (CPSC) 16 CFR 1201, Cat II (400 ft lbs). See also **Laminated Glass; Safety Glass; Tempered Glass**.

Work Injury

Any injury suffered by an individual that arises out of and in the course of employment.

Work Permit

A written record by which an individual in charge of a unit, equipment, or building area authorizes a worker or work crew to do a specific job at that work site. It identifies what precautions (safe work practices) are to be taken to ensure that the working conditions are safe for the type of work to be performed in a specific job location during a specific time interval. It outlines the safety equipment required and to be used for that specific job location. See also **Hot Work**.

Work Recovery Cycles

The job pattern that defines how work is organized with respect to lighter tasks or rest. High work-to-recovery ratios, measured as continuous time on each type of activity, have higher potential for fatigue. See also **Rest Allowances**.

Work-Related Musculoskeletal Disorder Hazard

Workplace conditions or physical work activities that cause or are reasonably likely to cause or contribute to a work-related musculoskeletal disorder.

Work-Related Musculoskeletal Disorders (WMSD, WRMSD)

Injuries and disorders of the muscles, nerves, tendons, ligaments, joints, cartilage, and spinal disc due to physical work activities or workplace conditions in the job. Examples include carpal tunnel syndrome related to long-term computer data entry, rotator cuff tendonitis from repeated overhead reaching, and tension neck syndrome associated with long-term cervical spine flexion.

Work Restrictions

Typically a doctor's description of the work that an employee can and cannot do. Specifically, per 28 CFR 1910.900, they are limitations, during the recovery period, on an injured employee's exposure to Musculoskeletal Disorder (MSD) hazards. Work restrictions may involve limitations on the work activities of the employee's current job (light duty), transfer to temporary alternative duty jobs, or temporary removal from the workplace to recover. Generally, temporarily reducing an employee's work requirements in a new job in order to reduce muscle soreness resulting from the use of muscles in an unfamiliar way is not a work restriction. The day an employee first reports an MSD is not considered a day away from work, or a day of work restriction, even if the employee is removed from his or her regular duties for part of the day. See also **Musculoskeletal Disorder (MSD)**.

Work Strain

The natural physiological response of the body to the application of work stress. The locus of reaction may be remote from the point of application of work stress. Work strain is not necessarily traumatic, but may appear as trauma when excessive, either directly or cumulatively, and must be considered in the design of equipment and task design.

Work Stress

Biomechanically, any external force acting on the body during the performance of a task. It always produces work strain.

Workers' Compensation

A quasi-judicial system where employers have a statutory obligation to pay medical expenses, disability, and other benefits for work-related injuries or illness. Administrative aspects and extent of coverage of workers' compensation laws vary from state to state.

Workers' Compensation Insurance

A form of insurance imposed on employers by legislative statute, which insures medical disability and other coverage for work-related injuries or illness. Depending on the state, an employer may be self-insured or pay premiums to a carrier or the state to administer the program.

Working Surface

Any surface or plane on which an employee walks or works. It is a factor used in classifying incidents.

Workplace Hazardous Materials Information System (WHMIS)
A legislated hazard communication system including labels, Material Safety
Data Sheets, and a worker training program. See also **Hazardous Materials
Identification System (HMIS®)**.

Workplace Illness
See **Occupational Illness**.

Workplace Injury
See **Occupational Injury**.

Workplace Protection Factor (WPF) Study
A study, conducted under actual conditions of use in the workplace, that measures
the protection provided by a properly selected, fit-tested, and functioning respirator, when the respirator is worn correctly and used as part of a comprehensive
respirator program that is in compliance with the Occupational Safety and Health
Administration (OSHA) Respiratory Protection Standard 29 CFR 1910.134.
Measurements of Co and Ci are obtained only while the respirator is being worn
during performance of normal work tasks (i.e., samples are not collected when the
respirator is not being worn). As the degree of protection afforded by the respirator
increases, the WPF increases.

Workplace Safety Program
A program that aims to develop a long-term plan that is successful in protecting
people from injury and death, that complies with regulations, and that controls the
associated financial costs of loss.

Workplace Stress
The harmful physical and emotional response that occurs when there is a poor match
between job demands and the capabilities, resources, or needs of the worker. The
combination of high demands in a job and a low amount of control over the situation can lead to stress. The American Institute of Stress quotes two recent studies
that state an average of 20 workers are murdered each week in the United States,
making homicide the second highest cause of workplace deaths and the leading one
for females. It is estimated at least 18,000 non-fatal violent crimes such as sexual and
other assaults also occur each week while the victim is working, or about a million
a year. Particularly dangerous occupations of police officers and cab drivers have
higher rates of homicide and non-fatal assaults than other occupations. However,
postal workers who work in a safe environment have experienced so many fatalities
due to job stress that "going postal" has crept into our language. Desk rage and phone
rage have also become increasingly common terms.

Workplace Wellness Program
An employee health improvement program offered by some employers as a combination of educational, organizational, and environmental activities designed to support
behavior conducive to the health of employees in a business and their families. It
consists of education, medical screenings, and behavior modification programs
designed to achieve better health and reduce the associated health risks for employees. The programs also help employers reduce health insurance premiums, since it is
estimated that they reduce the likelihood of employee health impacts particularly in
preventing cardiovascular disease.

X, Y, Z

Xenobiotic

A chemical or a chemical mix that is not a normal component of the organism that is exposed to it. Xenbiotics, therefore, include most drugs (other than those compounds which naturally occur in the organism), as well as other foreign substances.

Z List

The Occupational Safety and Health Administration (OSHA) Toxic and Hazardous Substances (Tables Z.1, Z.2, and Z.3) of air contaminants, found in 29 CFR 1910.1000. These tables record permissible exposure limits (PELs), time-weighted averages (TWAs), and ceiling concentrations for the materials listed. Any material found in these tables is considered to be hazardous. The tables can be viewed on OSHA's Web site:

Table Z.1: http://www.osha.gov/pls/oshaweb/
owadisp.show_document?p_table=standards&p_id=9992
Table Z.2: http://www.osha.gov/pls/oshaweb/
owadisp.show_document?p_table=STANDARDS&p_id=9993
Table Z.3: http://www.osha.gov/pls/oshaweb/
owadisp.show_document?p_table=STANDARDS&p_id=9994

Zero Energy

The state of a piece of equipment when all sources of energy (i.e., electrical, mechanical, hydraulic, etc.) is isolated from the particular piece of equipment or is effectively blocked and all sources of stored energy are depleted.

Zero Mechanical State

The mechanical potential energy in all elements of a machine or piece of equipment is dissipated, so that opening or activation of any device will not produce a movement that could cause injury.

Safety-Related Organizations

American Board of Industrial Hygiene (ABIH)
6015 West St. Joseph, Suite 102
Lansing, MI 48917-3980
www.abih.org
E-mail: abih@abih.org

American Conference of Governmental Industrial Hygienists (ACGIH®)
1330 Kemper Meadow Drive
Cincinnati, OH 45240
www.acgih.org
E-mail: amil@acgih.org

American Industrial Hygiene Association (AIHA)
2700 Prosperity Avenue, Suite 250
Fairfax, VA 22031
www.aiha.org
E-mail: infonet@aiha.org

American Institute of Stress (AIS)
124 Park Avenue
Yonkers, NY 10703
www.stress.org
E-mail: Stress@optonline.net

American Insurance Association (AIA)
2101 L Street, NW, Suite 400
Washington, DC 20037
www.aiadc.org
E-mail: info@aiadc.org

American Ladder Institute
401 North Michigan Avenue
Chicago, IL 60611
www.laddersafety.org
E-mail: jrapp@smithbucklin.com

American Society of Safety Engineers (ASSE)
1800 East Oakton Street
Des Plaines, IL 60016
www.asse.org
E-mail: customerservice@asse.org

Board of Certified Safety Professionals (BCSP)
208 Burwash Avenue
Savoy, IL 61874
www.bcsp.org
E-mail: Use Web page contact form.

Chemical Safety and Hazard Investigation Board (CSB)
2175 K Street NW, Suite 400
Washington, DC 20037
www.csb.gov
E-mail: info@csb.gov

Construction Safety Council (CSC)
4100 Madison Street
Hillside, IL 60162
www.buildsafe.org
E-mail: Not available at time of publication.

Consumer Product Safety Commission (CPSC)
4330 East West Highway
Bethesda, MD 20814
www.cpsc.gov
E-mail: Use Web page.

Department of Transportation (DOT)
1200 New Jersey Avenue, SE
Washington, DC 20590
www.dot.gov
E-mail: Use Web page.

Electrical Safety Foundation International (ESFI)
1300 North 17th Street, Suite 1752
Rosslyn, VA 22209
www.esfi.org
E-mail: info@esfi.org

Factory Mutual
270 Central Avenue
PO Box 7500
Johnston, RI 02919
www.fmglobal.com
E-mail: Use Web page.

Federal Emergency Management Association (FEMA)
500 C Street, SW
Washington, DC 20472
www.fema.gov
E-mail: See Web page for specific e-mail contact address.

Fire Safety Institute
PO Box 674
Middlebury, VT 05753
www.firesafetyinstitute.org
E-mail: info@firesafetyinstitute.org

Home Safety Council
1725 Eye Street NW, Suite 300
Washington, DC 20006
www.homeafetycouncil.org
E-mail: info@homesafetycouncil.org

Human Factors and Ergonomics Society (HFES)
PO Box 1369
Santa Monica, CA 90406-1369
www.hfes.org
E-mail: info@hfes.org

Institute for Safety and Health Management (IHSM)
4841 East County 14 1/4 Street
Yuma, AZ 85365
www.ishm.org
E-mail: Use Web page.

Insurance Institute of America (IIA)
The Institutes
720 Providence Road, Suite 100
Malvern PA 19355-3433
www.igie.org/InsuranceInstituteOfAmerica.htm
E-mail: customerervice@cpcuiia.org

International Code Council (ICC)
500 New Jersey Avenue, NW, 6th Floor
Washington, DC 20001
www.iccsafe.org
E-mail: webmaster@iccsafe.org

International Safety Equipment Association (ISEA)
1901 North Moore Street
Arlington, VA 22209-1762
www.safetyequipment.org
E-mail: Use Web page contact form.

International Society for Fall Protection (ISFP)
9453 Coppertop Loop NE
Bainbridge Island, WA 98110
www.isfp.org
E-mail: info@isfp.org

The Loss Prevention Foundation
8037 Corporate Center Drive, Suite 400
Charlotte, NC 28226
www.losspreventionfoundation.org
E-mail: Use Web page contacts.

Mine Safety Health Administration (MSHA)
1100 Wilson Boulevard, 21st Floor
Arlington, VA 22209-3939
www.msha.gov
E-mail: Use Web page e-mail contact list.

National Fire Protection Association (NFPA)
1 Batterymarch Park
Quincy, MA 02169-7471
www.nfpa.org
E-mail: Use Web page inquiry form.

National Floor Safety Institute (NFSI)
PO Box 92607
Southlake, TX 76092
www.nfsi.org
E-mail: info@nfsi.org

National Highway Traffic Safety Administration
1200 New Jersey Avenue, SE
West Building
Washington, DC 20590
www.nhtsa.dot.gov
E-mail: Use Web page inquiry form.

National Institute for Occupational Safety and Health (NIOSH)
395 E Street, SW, Suite 9200
Patriots Plaza Building
Washington, DC 20201
www.cdc.gov/niosh
E-mail: cdcinfo@cdc.gov

National Safety Council (NSC)
1121 Spring Lake Drive
Itasca, IL 60142-3201
www.nsc.org
E-mail: info@nsc.org

National Transportation Safety Board (NTSB)
490 L'Enfant Plaza, SW
Washington, DC 20594
www.ntsb.gov
E-mail: See Web page for contacts.

Occupational Safety and Health Administration (OSHA)
200 Constitution Avenue, NW
Washington, DC 20210
www.osha.gov
E-mail: Use Web page.

Office of Pipeline Safety
Pipeline and Hazardous Materials Safety Administration (PHMSA)
U.S. Department of Transportation
East Building, 2nd Floor
1200 New Jersey Avenue, SE
Washington, DC 20590
www.phmsa.dot.gov
E-mail: Use Web page inquiry form.

Office of Railroad Safety
Federal Railroad Administration
1200 New Jersey Avenue, SE
Mail Stop 25
Washington, DC 20590
www.fra.dot.gov
E-mail: rrs.correspondence@fra.dot.gov

Risk and Insurance Management Society, Inc.® (RIMS)
1065 Avenue of the Americas, 13th Floor
New York, NY 10018
www.rims.org
E-mail: Use Web page inquiry form.

System Safety Society
PO Box 70
Unionville, VA 22567-0070
www.system-safety.org
E-mail: systemsafety@system-safety.org

Underwriters Laboratories (UL)
333 Pfingsten Road
Northbrook, IL 60062
www.ul.com
E-mail: Use Web page inquiry form.

U.S. Fire Administration (USFA)
16825 South Seton Avenue
Emmitsburg, MD 21727
www.usfa.fema.gov
E-mail: Use Web page inquiry form.

Standards Cited

AMERICAN NATIONAL STANDARDS INSTITUTE (ANSI) STANDARDS

ANSI A10.14, Construction and Demolition Operations—Requirements for Safety Belts, Harnesses, Lanyards, and Lifelines for Construction and Demolition Use

ANSI A10.32, Fall Protection Systems for Construction and Demolition Operations

ANSI A14.1, Safety Requirements for Portable Wooden Ladders (ALI)

ANSI A14.2, Safety Requirements for Portable Metal Ladders (ALI)

ANSI A14.3, Safety Requirements for Fixed Ladders (ALI)

ANSI 14.4, Safety Requirements for Job-Made Ladders (ALI)

ANSI A14.5, American National Standard, Safety Requirements for Portable Reinforced Plastic (ALI)

ANSI A14.7, Safety Requirements for Mobile Ladders (ALI)

ANSI A1264.1, Safety Requirements for Workplace Walking/Working Surfaces and Their Access

ANSI B101.0, Walkway Surface Auditing Guideline for the Measurement of Walkway Slip Resistance

ANSI ISA 84.00.01, Functional Safety, Safety Instrumented Systems for the Process Industry (ISA)

ANSI Z1, Quality Assurance (ASQ)

ANSI Z9, Safety Standard for Exhaust Systems (AIHA)

ANSI Z9.2, Fundamentals Governing the Design and Operation of Local Exhaust Ventilation Systems

ANSI Z16, Accident Statistics (NSC)

ANSI Z16.2, Standard for Information Management for Occupational Safety and Health

ANSI Z34, Principles Underlying Valid Certification and Labeling of Products and Services (ACIL)

ANSI Z49.1, Safety in Welding, Cutting, and Allied Processes (AWS)

ANSI Z83, Underwater Safety (DEMA)

ANSI Z87.1, Standard for Occupational and Educational Eye and Face Protection Devices (ASSE)

ANSI Z88, Safety Standards for Respiratory Protection (LLNL)

ANSI Z88.2, Respiratory Protection

ANSI Z88.6, Respirator Use—Physical Qualifications for Personnel

ANSI Z89, Safety Standard for Industrial Head Protection (DU)

ANIS Z89.1, Protective Headware for Industrial Workers

ANSI Z90, Vehicular Head Protection

ANSI Z97 Safety Requirements for Architectural Glazing Material (SGCC)

ANSI Z117, Confined Space (ASSE)
ANSI Z124, Safety in Tree Trimming Operations (ISAr)
ANSI Z133, Safe Use of Lasers (LIA)
ANSI Z223, Lockout Protection (NSC)
ANSI Z245, Fall Protection (ASSE)
ANSI Z359, Committee for the Control of Cumulative Trauma Disorders
ANSI Z359.1, Safety Requirements for Personal Fall Arrest Systems, Subsystems, and Components (ASSE)
ANSI Z365, Control of Cumulative Trauma Disorders (NSC)
ANSI Z375, Diving Instructional Standards and Safety (RSTC)
ANSI Z358.1, Emergency Eye Wash and Shower Equipment
ANSI Z400.1, Hazardous Industrial Chemicals—Material Safety Data Sheets: Preparation
ANSI 535.1, Safety Colors
ANSI Z535.2, Environmental and Facility Safety Signs
ANSI Z535.3, Criteria for Safety Symbols and Labels
ANSI Z535.4, Product Safety Signs and Labels
ANSI Z535.5 Safety Tags and Barricade Tapes
ANSI Z535.6 Product Safety Information in Product Manuals and Instructions and Other Collateral Material

AMERICAN PETROLEUM INSTITUTE (API)

RP 500, Recommended Practice for Classification of Locations for Electrical Installations at Petroleum Facilities Classified as Class I, Division 1, and Division 2
RP 521, Guide for Pressure Relieving and Depressurizing Systems

AMERICAN SOCIETY FOR TESTING MATERIALS (ASTM) STANDARDS

ASTM E-681, Standard Test Method for Concentration Limits of Flammability of Chemicals
ASTM F 2412, Test Methods for Foot Protection
ASTM F 2413, Specification for Performance Requirements for Protective Footwear

CODE OF FEDERAL REGULATIONS (CFR), UNITED STATES

16 CFR, Parts 1101 to 1406, Consumer Products Safety Act of 1972
16 CFR 1203, Consumer Products Safety Commission (CPSC), Bicycle Helmet Standard
16 CFR 1512, Consumer Products Safety Commission (CPSC), Bicycle Standard
29 CFR 1910, General Industry Safety Standards (OSHA)
29 CFR 1926, Construction Safety Standards (OSHA)
16 CFR, 2051, Consumer Products Safety Improvement Act of 2008
49 CFR 172, Department of Transportation, Hazardous Materials

NATIONAL FIRE PROTECTION ASSOCIATION (NFPA) STANDARDS

NFPA 51B, Standard for Fire Prevention in Use of Cutting and Welding Processes

NFPA 68, Explosion Protection by Deflagration Venting

NFPA 69, Standard on Explosion Prevention Systems

NFPA 70, National Electrical Code

NFPA 70E, Standard for Electrical Safety in the Workplace

NFPA 101, Life Safety Code

NFPA 1982, Standard on Personal Alert Safety Systems (PASS) for Fire Fighters

NFPA 704, Standard System for the Identification of the Hazards of Materials for Emergency Response

NFPA 1500, Fire Department Occupational Safety and Health Program

Acronyms

1oo2	1 out of 2
2oo3	2 out of 3
3Es	Engineering, Education, and Enforcement
ABIH	American Board of Industrial Hygiene
ACGIH®	American Conference of Governmental Industrial Hygienists
AEA	Action Error Analysis
AED	Automatic External Defibrillator
AEGL	Acute Exposure Guideline Level (value)
AFCI	Arc Fault Circuit Interrupter
AHA	American Heart Association
AHJ	Authority Having Jurisdiction
AIA	American Insurance Association
AIHA	American Industrial Hygiene Association
AIS	American Institute of Stress
ALARA	As Low as Reasonably Achievable
ALARP	As Low as Reasonably Practical
ALI	American Ladder Institute
ALS	Advanced Life Support
ANSI	American National Standards Institute
AOE	Arising Out of Employment
APF	Assigned Protection Factor
ARM	Associate in Risk Management
ASAE	American Society of Agriculture Engineers
ASOII	Annual Survey of Occupational Injuries and Illnesses
ASP	Associate Safety Professional
ASSE	American Society of Safety Engineers
ASTM	American Society for Testing Materials
ATHEANA	A Technique for Human Error Analysis
BBS	Behavior-Based Safety
BCBSA	Blue Cross and Blue Shield Association
BCSP	Board of Certified Safety Professionals
BEI®	Biological Exposure Index
BI	Bodily Injury or Business Interruption
BLEVE	Boiling Liquid Expanding Vapor Explosion
BLS	Bureau of Labor Statistics
BMS	Burner Management System
BOP	Blowout Preventer
CAS	Chemical Abstracts Service
CCA	Cause-Consequence Analysis
CCP	Critical Control Point
CCPS	Center for Chemical Process Safety

CDC	Centers for Disease Control and Prevention
CERCLA	Comprehensive Environmental Response, Compensation, and Liability Act
CF	Casual Factors
CFATS	Chemical Facility Anti-Terrorism Standard
CFC	Causal Factors Chart
CFOI	Census of Fatal Occupational Injuries
CFR	Code of Federal Regulations
CFSE	Certified Function Safety Expert
CFSP	Certified Function Safety Professional
CHEMTREC	Chemical Transportation Emergency Center
CHP	Chemical Hygiene Plan
CIH	Certified Industrial Hygienist
COE	Course of Employment
CPCU	Chartered Property Casualty Underwriter
CPR	Cardiopulmonary Resuscitation
CPSC	Consumer Products Safety Commission
CSB	Chemical Safety and Hazard Investigation Board
CSC	Construction Safety Council
CSHM	Certified Safety and Health Manager
CSO	Construction Safety Orders
CSP®	Certified Safety Professional®
CTD	Cumulative Trauma Disorder
DART	Days Away from Work, Restricted or Transfered
DAW	Days Away from Work
DAWC	Days Away from Work Cases
dB	Decibel
dBA	Decibels on the A Scale
DETAM	Dynamic Event Tree Analysis Method
DHS	Department of Homeland Security
DIFR	Disabling Injury Frequency Rate
DII	Disabling Injury Index
DISR	Disabling Injury Severity Rate
DOL	Department of Labor
DOT	Department of Transport
DRII	Disaster Recovery Institute International
DYLAM	Dynamic Event Logic Analytical Methodology
EAP	Employee Assistance Program
EAS	Emergency Alert System
EEL	Emergency Exposure Limit
EIV	Emergency Isolation Valve
ELF	Extremely Low Frequency
ELSI	End of Service Life Indicator
EOC	Emergency Operations Center
EPA	Environmental Protection Agency
EPCRA	Emergency Planning and Community Right-to-Know Act

EPF	Effective Protection Factor
ERG	Emergency Response Guidebook
ERP	Emergency Response Plan
ERPG	Emergency Response Planning Guideline (value)
ERT	Emergency Response Team
ESD	Electrostatic Sensitive Device or Emergency Shutdown (System)
ESFI	Electrical Safety Foundation International
ESFR	Early Suppression, Fast Response
ETA	Event Tree Analysis
FACP	Fire Alarm Control Panel
fc or ft-c	Footcandle
FC	Fail Close
FEMA	Federal Emergency Management Association
FIFRA	Federal Insecticide, Fungicide, and Rodenticide Act
FM	Factory Mutual
FMEA	Failure Mode and Effects Analysis
FMECA	Failure Mode and Effects Criticality Analysis
FMEDA	Failure Modes Effects and Diagnostics Analysis
FO	Fail Open
FRSA	Federal Railway Safety Act
FS	Factor of Safety or Fail Safe or Fail Steady
FSI	Fire Safety Institute
FTA	Fault Tree Analysis
GFCI	Ground Fault Circuit Interrupter
GISO	General Industry Safety Orders
GSP	Graduate Safety Practitioner
HACCP	Hazard Analysis and Critical Control Point
HAZCOM	Hazard Communication
HAZID	Hazard Identification
HAZMAT	Hazardous Material
HAZOP	Hazard and Operability Analysis
HAZWOPER	Hazardous Waste Operations and Emergency Response
HCA	High Consequence Area
HCP	Hearing Conservation Program
HCS	Hazard Communication Standard
HEPA	High Efficiency Particulate Air
HFE	Human Factors Engineering
HFES	Human Factors and Ergonomics Society
HHS	Health and Human Services (Department of)
HIB	Hazard Information Bulletin
HIP	Hazard Identification Plan
HIPS	High Integrity Protective System
HMIS®	Hazardous Materials Identification System
HPD	Hearing Protection Device
HPR	Highly Protected Risk
HSC	Home Safety Council

IBC	International Building Code
IC	Incident Commander
ICC	International Code Council
ICP	Incident Command Post
ICS	Incident Command System
IDI	Industrial Disabling Injury
IDLH	Immediately Dangerous to Life or Health
IFC	International Fire Code®
IH	Industrial Hygiene
IIA	Insurance Institute of America
IIPP	Injury and Illness Prevention Program
IMS	Incident Management System
IMT	Incident Management Team
IPL	Independent Protection Layer
IR	Infrared
ISEA	International Safety Equipment Association
ISHM	Institute for Safety and Health Management
ISFP	International Society for Fall Protection
ISO	Insurance Services Office or International Organization for Standardization
ISRS	International Safety Rating System
JHA	Job Hazard Analysis
JSA	Job Safety Analysis
KPI	Key Performance Indicator
LC	Lethal Concentration
LCL	Lower Confidence Limit
LD	Lethal Dose
LEL	Lower Explosive Limit
LEPC	Local Emergency Planning Committee
LFL	Lower Flammable Limit
LOPA	Layers of Protection Analysis
LOTO	Lockout/Tagout
LPE	Loss Prevention Engineer
LSC	Life Safety Code
LTA	Less Than Adequate
LTI	Lost Time Injury
MFL	Maximum Foreseeable Loss
MMEA	Misuse Mode and Effects Analysis
MOC	Management of Change
MORT	Management Oversight Risk Tree
MOU	Memorandum of Understanding
MPC	Maximum Permissible Concentration
MPD	Maximum Permissible Dose
MPS	Manual Pull Station
MSD	Musculoskeletal Disorder
MSDS	Material Safety Data Sheet

MSHA	Mine Safety and Health Administration
MSI	Musculoskeletal Injury
MTBF	Mean Time between Failure
MTTF	Mean Time to Failure
MUC	Maximum Use Concentration
MUTCD	Manual on Uniform Traffic Control Devices
MVA	Motor Vehicle Accident
MW	Microwave
NEC	National Electrical Code
NEMA	National Electrical Manufacturers Association
NFPA	National Fire Protection Association
NFSI	National Floor Safety Institute
NHTSA	National Highway Traffic Safety Administration
NIHL	Noise-Induced Hearing Loss
NIMS	National Incident Management System
NIOSH	National Institute for Occupational Safety and Health
NMR	Near Miss Report
NOAEL	No Observable Adverse Effect Level
NRR	Noise Reduction Rating
NRTL	Nationally Recognized Testing Laboratory
NSC	National Safety Council
NTSB	National Transportation Safety Board
ODI	Occupational Disabling Injury
OEL	Occupational Exposure Limit
OIICS	Occupational Injury and Illness Classification System
OPS	Office of Pipeline Safety
OSH Act	Occupational Safety and Health Act of 1970
OSHA	Occupational Safety and Health Administration
OSHRC	Occupational Safety and Health Review Commission
PASS	Personal Alert Safety Systems
PD	Property Damage
PEL	Permissible Exposure Limit
PFD	Personal Flotation Device or Probability of Failure on Demand
PHA	Preliminary Hazard Analysis or Process Hazard Analysis
PHL	Preliminary Hazards List
PHMSA	Pipeline and Hazardous Materials Safety Administration
PM	Particulate Matter
PML	Probable Maximum Loss
ppb	Parts per Billion
PPE	Personal Protective Equipment
PPF	Program Protection Factor
ppm	Parts per Million
PSA	Professional Safety Academy
PSM	Process Safety Management
PSSR	Pre-startup Safety Review
PtD	Prevention through Design

QLFT	Qualitative Fitting Test
QNFT	Quantitative Fitting Test
QRA	Quantitative Risk Analysis
RBI	Risk-Based Inspection
RCA	Root Cause Analysis
RCM	Root Cause Map
RCRA	Resource Conservation and Recovery Act
RCS	Risk Control System
RDI	Restricted Duty Injury
REL	Recommended Exposure Limit
RFI	Radio Frequency Interference
RIMS	Risk and Insurance Management Society, Inc.®
RMP	Risk Management Plan
ROPS	Rollover Protection System
RSI	Repetitive Strain Injury
RTECS	Registry of Toxic Effects of Chemical Substances
SACHE	Safety in Chemical Engineering Education
SAE	Society of Automotive Engineers
SARA	Superfund Amendment and Reauthorization Act
SCBA	Self-Contained Breathing Apparatus
SFAIRP	So Far As Is Reasonably Practical
SFF	Safe Failure Fraction
SHARP	Safety and Health Achievement Recognition Program
SHE	Safety, Health, and Environment
SHIB	Safety and Health Information Bulletin
SIF	Safety Instrumented Function
SIL	Safety Integrity Level
SIS	Safety Instrumented System
SMORT	Safety Management Organization Review Technique
SMS	Safety Management System
SOLAS	Safety of Life at Sea
SPF	Single Point Failure
SSP	Site Security Plan
STEL	Short-Term Exposure Limit
STS	Standard Threshold Shift or Significant Threshold Shift
SVA	Security Vulnerability Assessment
SWPF	Simulated Workplace Protection Factor
SWL	Safe Working Load
TEEL	Temporary Emergency Exposure Limit
THERP	Technique for Human Error Rate Probability
TLV®	Threshold Limit Value
TMR	Triple Modular Redundant
TPQ	Threshold Planning Quantity
TRC	Total Recordable Case (rate)
TRIR	Total Recordable Incident Rate
TSCA	Toxic Substance Control Act

TSI	Transportation Safety Institute
TTS	Temporary Threshold Shift
TWA	Time-Weighted Average
UCL	Upper Confidence Limit
UEL	Upper Explosive Limit
UFL	Upper Flammable Limit
UL	Underwriters Laboratories
UN	United Nations
USFA	U.S. Fire Administration
UV	Ultraviolet
UVCE	Unconfined Vapor Cloud Explosion
VPP	Voluntary Protection Program
WIA	What-If Analysis
WHMIS	Workplace Hazardous Materials Information System
WP	Work Permit
WPF	Workplace Protection Factor
WRMSD	Work-Related Musculoskeletal Disorders

Bibliography

American Society of Safety Engineers (ASSE), *The Dictionary of Terms Used in the Safety Profession*, 4th Edition, ASSE, Park Ridge, IL, 2000.

American Society of Testing Materials, (ASTM), ASTM E1445–08, *Standard Terminology Relating to Hazard Potential of Chemicals*, American Society of Testing Materials International, West Conshohocken, PA, 2008.

American Society of Testing Materials, (ASTM), ASTM F1646–05, *Standard Terminology Relating to Safety and Traction for Footwear*, American Society of Testing Materials International, West Conshohocken, PA, 2008.

Brauer, Roger L., *Safety and Health for Engineers*, 2nd Edition, Wiley-Interscience, Hoboken, NJ, 2005.

Finucane, Edward W., *Definitions, Conversions, and Calculations for Occupational Safety and Health Professionals*, 3rd Edition, CRC Press/Taylor & Francis Group, Boca Raton, FL, 2006.

Genium Publishing Corp., *The Workplace Safety and Health Dictionary*, Genium Publishing Corp., Schenectady, NY, 1996.

Institute of Industrial Engineers, (IIE), *American National Standard Institute (ANSI) Z94.13, Occupational Health and Safety Terminology*, ANSI, New York, NY, 2000.

International Atomic Energy Agency (IAEA), *IAEA Safety Glossary; Terminology Used in Nuclear, Radiation, Radioactive Waste and Transport Safety*, IAEA, Vienna, Austria, 2006.

International Code Council (ICC), *International Fire Code*, International Code Council, Inc., Cenage Delmar Learning, Albany, NY, 2009.

International Fire Service Training Association (ISFTA), *Fire Service Orientation and Terminology*, 4th Edition, Fire Protection Publications, Norman, OK, 2004.

Karwowski, Waldemer (editor), *International Encyclopedia of Ergonomics and Human Factors*, Taylor & Francis, London, UK, 2001.

Luigi, Parmeggiani (editor), *Encyclopedia of Occupational Health and Safety*, 3rd Edition, Volumes I and II, International Labour Office, Geneva, Switzerland, 1989.

Mannan, S. (editor), *Lees' Loss Prevention in the Process Industries*, 3rd Edition, Hazard Identification, Assessment and Control, Butterworth-Heinemann, Oxford, UK, 2005.

National Fire Protection Association (NFPA), *NFPA's Illustrated Dictionary of Fire Service Terms*, 6th Edition, Jones & Bartlett Publishers, Boston, MA, 2006.

National Safety Council, *Occupational Safety & Health*, 3rd Edition, National Safety Council, Itasca, IL, 2000.

National Safety Council, *Supervisors' Safety Manual*, 10th Edition, National Safety Council, Itasca, IL, 2009.

Nolan, Dennis P., *Encyclopedia of Fire Protection*, 2nd Edition, Thomson-Delmar Learning, Albany, NY, 2006.

Occupational Safety and Health Administration (OSHA), 29 CFR 1910, *Occupational Safety and Health Standards*, OSHA, U.S. Department of Labor, Washington, DC.

Plog, Barbara A. (editor), *Fundamentals of Industrial Hygiene*, 5th Edition, National Safety Council, Itasca, IL, 2002.

Rubin, Dr. Harvey W., *Dictionary of Insurance Terms*, Barron's Educational Series, Inc., 5th Edition, Hauppauge, NY, 2008.

Stramler, James H., *The Dictionary for Human Factors/Ergonomics*, CRC Press, Boca Raton, FL, 1993.

Stranks, Jeremy, *Health and Safety at Work, Key Terms*, Butterworth-Heinemann, Oxford, UK, 2002.

U.S. Department of Labor, Bureau of Labor Statistics, *Occupational Injury and Illness Classification System Manual*, U.S. Government Printing Office, 2007.

U.S. Department of Labor, Occupational Safety and Health Administration, *Job Hazard Analysis, OSHA 3071*, U.S. Government Printing Office 2002.

U.S. Department of Labor, Occupational Safety and Health Administration (2003), *Code of Federal Regulations 29 Parts 1900 to 1910, and 1926.S.* Government Printing Office Washington, DC, 2004.

For Product Safety Concerns and Information please contact our EU
representative GPSR@taylorandfrancis.com
Taylor & Francis Verlag GmbH, Kaufingerstraße 24, 80331 München, Germany

www.ingramcontent.com/pod-product-compliance
Ingram Content Group UK Ltd.
Pitfield, Milton Keynes, MK11 3LW, UK
UKHW021017180425
457613UK00020B/961